The Concept of Probability

Fundamental Theories of Physics

An International Book Series on The Fundamental Theories of Physics: Their Clarification, Development and Application

Editor: ALWYN VAN DER MERWE
              University of Denver, U.S.A.

Editorial Advisory Board:

ASIM BARUT, *University of Colorado, U.S.A.*
HERMANN BONDI, *Natural Environment Research Council, U.K.*
BRIAN D. JOSEPHSON, *University of Cambridge, U.K.*
CLIVE KILMISTER, *University of London, U.K.*
GÜNTER LUDWIG, *Philipps-Universität, Marburg, F.R.G.*
NATHAN ROSEN, *Israel Institute of Technology, Israel*
MENDEL SACHS, *State University of New York at Buffalo, U.S.A.*
ABDUS SALAM, *International Centre for Theoretical Physics, Trieste, Italy*
HANS-JÜRGEN TREDER, *Zentralinstitut für Astrophysik der Akademie der Wissenschaften, G.D.R.*

The Concept of Probability

*Proceedings of the Delphi Conference,
October 1987, Delphi, Greece*

edited by

E. I. Bitsakis
University of Ioannina, Greece

and

C. A. Nicolaides
*National Hellenic Research Foundation and
National Technical University, Athens, Greece*

KLUWER ACADEMIC PUBLISHERS
DORDRECHT / BOSTON / LONDON

Library of Congress Cataloging in Publication Data

The Concept of probability : proceedings of the Delphi Conference,
   October 1987, Delphi, Greece / edited by Eftichios Bitsakis and
   Cleanthes Nicolaides.
        p.   cm. -- (Fundamental theories of physics)
   Includes index.
   ISBN 9027726795
   1. Probabilities--Congresses. 2. Quantum theory--Congresses.
I. Bitsakes, Eutychēs I., 1927-   . II. Nicolaides, Cleanthes A.
III. Series.
QC174.17.P68C66   1989
530.1'2--dc19   88-37516

ISBN 90-277-2679-5

Published by Kluwer Academic Publishers,
P.O. Box 17, 3300 AA Dordrecht, The Netherlands.

Kluwer Academic Publishers incorporates
the publishing programmes of
D. Reidel, Martinus Nijhoff, Dr W. Junk and MTP Press.

Sold and distributed in the U.S.A. and Canada
by Kluwer Academic Publishers,
101 Philip Drive, Norwell, MA 02061, U.S.A.

In all other countries, sold and distributed
by Kluwer Academic Publishers Group,
P.O. Box 322, 3300 AH Dordrecht, The Netherlands.

All Rights Reserved
© 1989 by Kluwer Academic Publishers
and copyrightholders as specified on appropriate pages within.
No part of the material protected by this copyright notice may be reproduced or
utilized in any form or by any means, electronic or mechanical
including photocopying, recording or by any information storage and
retrieval system, without written permission from the copyright owner.

Printed in The Netherlands

PREFACE

This volume contains articles from invited speakers at a meeting which took place in Delphi, during the week of October 12-16, 1987. The theme of the meeting was "The concept of probability" and was organized by the "Group of Interdisciplinary Research" (Physics Department, University of Athens) and the Theoretical and Physical Chemistry Institute of the National Hellenic Research Foundation, Athens. (The Group of Interdisciplinary Research organized two previous Meetings, 1) on the Concept of physical reality (1982) and 2) on the question of determinism in Physics (1984)).

This small gathering, which was attended by scientists, mathematicians and philosophers from more than 22 countries, took place on the occasion of the 100th year from the birthday of E.Schrödinger.

As the father of wave-mechanics, Schrödinger thrushed us into an era of physics where knowledge of the Ψ-function is considered, for most situations, as the ultimate aim and the ultimate truth. Yet, he, as well as another towering figure of 20th century physics, A.Einstein, never really felt confortable with the interpretation of the meaning of Ψ and of the information that it contains. With Einstein playing the leading role a debate about concepts and interpretation started as soon as quantum mechanics was born. Central theme to this debate is the concept of probability, a concept which permeates- explicitly or implicitly- all science and even our decision making in everyday life.

The articles cover a broad spectrum of thought and results - mathematical, physical, epistemological, experimental, specific, general,- many of them outside the accepted norm. Regardless of their degree of validity, we hope that the mosaic of ideas, information, arguments and proposals which they contain, will prove to be a useful and timely addition to the existing literature on probability and its physical implications.

It is our pleasure to thank the speakers and all the participants for their contribution to a succesful and pleasant meeting. We are grateful for the financial support from the Greek Ministries of Culture and of Science and Technology, as well as from the National Hellenic Research Foundation and the International Center at Delphi.

Finally, we are thankful to Professor Thomas Brody for his invaluable help in the preparation of this volume as well as to Dr.G.Roussopoulos for his help in the organization of the conference.

Athens, September 1988
E.I.Bitsakis
C.A.Nicolaides

CONTENTS

Preface   v

PART 1   FOLLOWING SCHRÖDINGER'S THOUGHTS

Letter from Schrödinger to Einstein, 18 November 1950   3

Schrödinger's thoughts on perfect knowledge
  W. Duch   5

Concerning Schrödinger's question: is Democritus of Planck the founder of quantum
theory?
  H-J. Treder   15

Schrödinger's reception of Greek atomism
  R. Wahsner   21

PART 2   PROBABILITY AND QUANTUM MECHANICS

Probability theory in quantum mechanics
  L. E. Ballentine   31

Measurement and amplitudes
  S. Gudder   43

Representations of quantum logics and transition probability spaces
  S. Pulmannova   51

Probability in quantum mechanics
  G. T. Rüttimann   61

PART 3   ASPECTS OF THE ARGUMENTS ON NONLOCALITY, BELL'S THEOREM
          AND EPR CORRELATIONS

Bell's theorem: a counterexample that agrees with the quantum formalism
  T. D. Angelidis   71

Quantum probability and quantum potential approach to quantum mechanics
  A. Kyprianidis   91

Time and enhancement: two possible local explanations for the EPR puzzle
  S. Pascazio   105

On Bell-type inequalities in quantum logics
  J. Pykacz   115

Physical meaning of the performed experiments concerning the EPR paradox
  F. Selleri   121

Comments on the "uncontrollable" character of non-locality
  J. P. Vigier   133

PART 4   REAL OR GEDANKEN EXPERIMENTS AND THEIR INTERPRETATION

Some basic differences between the Copenhagen and de Broglie interpretation of quantum mechanics leading to practical experiments
  J. R. Croca   143

Search for new tests of the EPR paradox from elementary particle physics
  D. Home   159

On the role of consciousness in random physical processes
  R. G. Jahn and B. J. Dunne   167

Recent experiments in the foundations of quantum mechanics
  A. G. Zajonc   179

Numerical simulations of reduction of wavepackets
  Y. Murayama and M. Namiki   189

Description of experiments in physics: a dynamical approach
  J. von Plato   199

The physical quantities in the radndom data of neutron interferometry
  J. Summhammer   207

PART 5   QUESTIONS ABOUT IRREVERSIBILITY AND STOCHASTICITY

Intrinsic irreversibility in classical and quantum mechanics
  I. E. Antoniou and I. Prigogine   223

A new challenge for statistical mechanics
  W. T. Grandy, Jr.   235

Quantum stochastic calculus as a unifying force in physics and probability
  R. L. Hudson   243

On the search of the time operator since Schrödinger
  Z. Marić   255

Stochastic optics: a wave theory of light based on classical probabilities
    T. W. Marshall and E. Santos   271

Quasiprobability distributions in quantum optics
    G. J. Milburn and D. F. Walls   289

Stochastic-dynamical approach to quantum mechanics
    M. Namiki   301

PART 6   EPISTEMOLOGY, INTERPRETATION AND CONJECTURE

Relativity and probability, classical and quantal
    O. Costa de Beauregard   315

Classical and quantum probabilities
    E. Bitsakis   335

The ensemble interpretation of probability
    T. A. Brody   353

QM axiom representations with imaginary & transfinite numbers and exponentials
    W. M. Honig   371

A change in paradigm: a realistic Copenhagen interpretation (realism without hidden variables)
    J. Horváth and M. Zágoni   389

Pythia and Tyche: an eternal golden braid
    L. M. Kirousis and P. Spirakis   395

Quanta of action and probability
    L. Kostro   405

Comments on Popper's interpretations of probability
    M. Rédei and P. Szegedi   417

Van Fraassen's constructive empiricism and the concept of probability
    G. Roussopoulos   427

Unification of the concepts of quantum ensembles and potential possibilities
    A. A. Tyapkin   437

Subject index   443

Part 1

Following Schrödinger's Thoughts

From the book "LETTERS ON WAVE MECHANICS" ed.K.Przibram
for the Austrian Academy of Sciences(1963)
Translated by M.J.Klein, Phil.Library, N.Y. (1967)

                                    Innsbruck, Innrain 55
                                       18 November 1950

Dear Einstein,

    It seems to me that the concept of probability is terribly mishandled these days. Probability surely has as its substance a statement as to whether something is or is not the case- an uncertain statement, to be sure. Bur nevertheless it has meaning only if one is indeed convinced that the something in question quite definitely either is or is not the case. A probabilistic assertion presupposes the full reality of its subject. No reasonable person would express a conjecture as to whether Caesar rolled a five with his dice at the Rubicon. But the quantum mechanics people sometimes act as if probabilistic statements were to be applied just to events whose reality is vague.
    The conception of a world that really exists is based on there being a far-reaching common experience of many individuals, in fact of all individuals who come into the same or a similar situation with respect to the object concerned. Perhaps instead of "common experience" one should say "experiences that can be transformed into each other in a simple way". This proper basis of reality is set aside as trivial by the positivists when they always want to speak only in the form: if "I" make a measurement then "I" "find" this or that. (And that is to be the only reality).
    It seems to me that what I call the construction of an external world that really exists is identical with what you call the describability of the individual situation that occurs only once-different as the phrasing may be. For it is just because they prohibit our asking what really "is", that is, which state of affairs really occurs in the indi-

vidual case, that the positivistis succeed in making us settle for a king of collective description. They accuse us of metaphysical heresy if we want to adhere to this "reality". That should be countered by saying that the metaphysical significance of this reality does not matter to us at all. It comes about for us as, so to speak, the intersection pattern of the determinations of many-indeed of all conceivable-individual observers. It is a condensation of their findings for economy of thought, which would fall appart without any connections if we wanted to give up this mode of thought before we have found an equivalent that at least yields the same thing. The present quantum mechanics supplies no equivalent. It is not conscious of the problem at al; it passes it by with blithe disinterest.

It is probably justified in requiring a transformation of the image of the real world as it has been constructed in the last 300 years, since the re-awakening of physics, based on the discovery of Galileo and Newton that bodies determine each other's accelerations. That was taken into account in that we interpreted the velocity as well as the position as instantaneous properties of anything real. That worked for a while. And now it seems to work no longer. one must therefore go back 300 years and reflect on how one could have proceeded differently at that time, and how the whole subsequent development would then be modified. No wonder that puts us into boundless confusion!
Warmest regards!

Yours

E. Schrödinger

SCHRÖDINGER'S THOUGHTS ON PERFECT KNOWLEDGE

Włodzisław Duch

Instytut Fizyki UMK
ul. Grudziądzka 5
87-100 Toruń
Poland

> What is one is one
> What is not one is also one
>
> Chuang Tsu

ABSTRACT. Perfect knowledge of the many-body system is contained in the wavefunction but, as Schrödinger has already emphasized, best possible knowledge of a whole does not include the best possible knowledge of its parts. Separability from the point of view of quantum mechanics is discussed. General "entangled" systems are analysed in terms of knowledge. If the state vector is defined for a whole system its parts are described only by mixed density operators. Correlations violating Bell's inequality are necessary to avoid superluminal signaling and result from the lack of independent reality of subsystems. Model calculations on two separated atoms and on non-interacting gas show that the perfect knowledge of the whole system or the total wavefunction is not sufficient to calculate local properties without actually solving the local problem.

1. INTRODUCTION

The history of modern civilization spans roughly about 5000 years. Civilization is clearly in its infancy because a single lifetime may still span as much as 2% of its history. Only a few men have been privileged to contribute so much to the development of human thought as Erwin Schrödinger, whose centenary of birth we are now celebrating. He has introduced the concept of wavefunction to physics and never stopped bothering himself with its meaning. Quantum Mechanics (QM) is a strange theory already at the level of single-particle phenomena but, accepting the wave-particle dualism in one form or another [1], one can still form a reasonable picture of reality. It is the holistic nature of QM at the multi-particle level or, technically speaking, the existence of multi-particle nonfactorizable states, that is more bothering. A few months after the famous Einstein, Podolsky and Rosen (EPR) paper [2] appeared Schrödinger made a profound analysis of the problem [3]. He summarized the conclusion in one sentence: "Best possible knowledge of a whole does not include best possible knowledge of its parts - and that is what keeps coming back to haunt us".

It is my feeling that we have just started to explore the consequences of this statement. The present paper is an attempt to look at it from modern perspective. In the second section separability in QM is discussed. Next, two sections deal with the spatially extended quantum states, EPR paradox and Bell inequality. In section 5 general "entangled systems" are considered. Finally Schrödinger's statement about the knowledge of a whole is exemplified in section 6 on a model of non-interacting gas.

2. SEPARABILITY IN QUANTUM MECHANICS

The question "When can one consider two physical systems to be separate?" is rather subtle. Naively, if I could isolate one system and perform experiments on this system without influence from the second system I would call it "a separate system". From the QM point of view such a definition would be naive indeed: how can one be sure that the results of experiments are really not determined by what happens to the other system? Maybe the wavefunction of my system was changed by someone experimenting with the second system? The only way to know is by computing correlations between the results of measurements on the two systems and checking if these correlations are trivial, i.e. if they can be explained assuming that the systems are independent. I will come to this point in the next section.

Leaving aside the subtleties concerned with isolation of physical systems let us consider the question of separation. The common knowledge is that "when two systems interact, their ψ - functions... consist, to mention this briefly, at first simply of the product of the two individual functions...", as Schrödinger [3] has put it. Quite recently Rosen [4], discussing separability, made a similar statement. Is it really true?

There is no smooth connection between distinguishability and indistinguishability, and thus between the simple product of two functions and a total function with a definite permutational symmetry. If there was, is, or will be a possibility of an interaction the product function is a false start. No matter how small, the interaction makes the total Hamiltonian symmetric and forces a definite permutational symmetry on the total wave function. By setting the interaction to zero one switches off the possibility of any interaction in the future. But, to rephrase the famous remark of Pauli [5], "Was Gott vereint hat, soll der Mensch nicht trennen" (what God has united men should not separate). We cannot start with the product function without "playing God's role".

<u>Formal proof</u>: consider two systems, S_A and S_B, with N_A and N_B particles, respectively. Each system is described by its own function, χ_A antisymmetric in N_A particles and χ_B in N_B. It is easy to show that the product function $\chi = \chi_A \chi_B$ is always "far" from the antisymmetric function $\psi = \hat{A}\chi$ looking at the norm $\|\chi - \psi\|$. If χ and ψ are normalized then the antisymmetrizer $\hat{A}$ is

(1) $\quad \hat{A} = (N_A! N_B!/N!)^{1/2} \sum_P \epsilon(P) P$

where $N = N_A + N_B$ and P is either identity or permutes particles of S_A with those of S_B. Therefore

(2) $\quad ||\chi-\psi||^2 = \langle \chi-\hat{A}\chi | \chi-\hat{A}\chi \rangle = 2-2\text{Re}\langle \chi | \hat{A}\chi \rangle \geq 2-2^{1/2}$

because the integral $\langle \chi | \hat{A}\chi \rangle = (N_A! N_B!/N!)^{1/2}$ is not smaller than $2^{-1/2}$.

Quantum mechanics is thus unable to describe two separated systems. The minority attitude towards this fact is to beat the drum: "quantum mechanics is dead", as Piron does [6]. His criticism is based primarily on his own quantum logic approach and on the axiomatic approach of Aerts [7], who emphasized many times that "it is impossible in quantum mechanics to describe two separated physical systems". Aerts presented a theory of the quantum-logic type that allows for such a description. The majority adopts a "so what?" attitude. As long as QM calculations agree with experimental results everything is O.K. Axiomatic approaches are not important because they do not give numbers. I would like to adopt here the Middle Way: with the "so what" attitude we may easily overlook things that are important but we never dreamed of. On the other hand it is too soon to mourn over the death of quantum mechanics before we will find predictions in direct conflict with experimental results. After all, separation may be just one of these illusions acquired in the childhood. It is simply impossible to measure correlations among more than a few bodies but concentrating on just two particles correlations proving their nonseparability should be seen. Let us look closer at the consequences of quantum mechanics.

3. SPATIALLY EXTENDED QUANTUM STATES

Suppose that two spatially separated particles are in a pure quantum state. There are basically two ways in which such a state could be prepared. First, by breaking one system into fragments, for example in the two-photon cascade emission or photodissociation of H_2 into $H + H$ in a triplet state, spatially extended systems in pure states are obtained. A second possibility, peculiar to QM, is to prepare the two systems in such a way that in future they will fit together into a single system in a definite way. As an example one may think of a particle or an odd-electron atom with total spin $S = 1/2$ crossing a magnetic field that separates $M_S = \pm 1/2$ states. If one has two atoms in $|S=1/2, M_S=+1/2\rangle$ states coming from the opposite directions even before they start to interact they must be in $|S=1, M_S=1\rangle$ combined state. It is not a question of interaction at an earlier time but of the wave functions that we are allowed to write. The first possibility of forming a spatially separated quantum state is due to the common past and the second due to the common future [8].

Are there observable consequences of the existence of such quantum states? Obvious consequences were noted by Einstein, Podolsky and Rosen [2] and by Schrödinger [3] already in 1935. There are also less obvious consequences noted by Bell [9] in 1965. To see the obvious consequences let us write the state of the two systems as

(3) $\quad |\psi\rangle = 2^{-\frac{1}{2}}(|a_1\rangle|b_1\rangle + |a_2\rangle|b_2\rangle)$

where $|a_k\rangle$ ($|b_k\rangle$) are states of the system S_A (S_B). If a measurement on S_A finds the system in the state $|a_k\rangle$ then S_B should be in $|b_k\rangle$, so a measurement on one system selects the state of the other system! This is the essence of the EPR paradox, a feature of quantum mechanics entirely unacceptable to Einstein [10]. This is also the reason why the definition of separated systems given previously is naive. It is tempting to assume that the state $|\psi\rangle$ describes an ensemble of systems, half of them in the $|a_1\rangle|b_1\rangle$ and half in $|a_2\rangle|b_2\rangle$ state. Einstein himself has advocated such interpretation. Unfortunately this would remove all interference and exchange effects. Superposition of states is not just the question of our knowledge but of potential reality.

Formal argument against ensemble interpretation of $|\psi\rangle$ is the following: density operator ρ_e corresponding to the two ensembles $|a_1\rangle|b_1\rangle$ and $|a_2\rangle|b_2\rangle$ is:

(4) $\quad \rho_e = \frac{1}{2}\{|a_1\rangle|b_1\rangle\langle a_1|\langle b_1| + |a_2\rangle|b_2\rangle\langle a_2|\langle b_2|\}$

This operator does not correspond to the pure state (3) nor any other pure state because it is not idempotent ($\rho_e^2 \neq \rho_e$). It means that ρ_e does not represent the maximal knowledge about the total system. It misses precisely the correlation between the two systems, included in the true density operator $|\psi\rangle\langle\psi|$

(5) $\quad \rho = \rho_e + \frac{1}{2}\{|a_1\rangle|b_1\rangle\langle a_2|\langle b_2| + |a_2\rangle|b_2\rangle\langle a_1|\langle b_1|\}$

One may still think that there is some physical mechanism that reduces the pure state ρ to the mixture ρ_e as the distance between the two systems S_A and S_B increases, thus localizing the states around S_A and S_B. A proposition in this spirit has been most recently advocated by Piccioni [11], but it has a longer history. Einstein himself suggested that "the current formulation of the many-body problem in quantum mechanics may break down when particles are far enough apart" (in a private communication to Bohm [12]). The possibility of such a localization process has been also considered by other authors [1]. In particular some physicist believe that when the wave packets of the two systems are localized and do not overlap there should be no correlations between measurements on the two systems, i.e. the reduced density operator ρ_e should be used. In this context Bohm and Hiley [13] analyzed the anisotropy measurements of the gamma rays, showing that pure states extend over macroscopic distances, more than an order of magnitude exceeding the width of the wave packet associated with each photon. The situation is qualitatively different from that which one finds in the self-interference experiments of Jánossy-Náray type or in the neutron interferometry (see the review on "empty waves" by Selleri [14]) because the experiments involving single-particle states are easily explained using the wave picture. The wavefunction in the two-particle case exists in 6-dimensional space and cannot be pictured as a real wave.

Less obvious consequences of the existence of the extended pure states were found by J.S. Bell. I will discuss them briefly now.

4. EPR AND BELL INEQUALITY REVISITED

Let E_{ab} be the usual correlation coefficient between simultaneous measurements [cf. 15]. Under a very general assumptions of locality and realism one may then prove the Bell inequality: combination of four correlation coefficients with the parameters (in particular angles) (ab), (ab'), (a'b) and (a'b') must not exceed 2:

(6) $\quad C(a,b,a',b') = E_{ab} - E_{ab'} + E_{a'b} + E_{a'b'}.$
$\quad |C(a,b,a',b')| \leq 2, \quad |E_{ab}| \leq 1$

Correlations computed with the extended quantum states (3) violate this inequality. In singlet and triplet spin states correlation coefficients are

(7) $\quad \begin{array}{l}|0,0\rangle \text{ state: } E_{ab} = -\cos(a-b) \\ |1,0\rangle \text{ state: } E_{ab} = -\cos(a+b) \\ |1,\pm 1\rangle \text{ state: } E_{ab} = \cos a \cos b\end{array}$

In the singlet state inequality (6) is strongly violated because $C(0,45,90,135) = 2 \cdot 2^{\frac{1}{2}}$. The same violation is observed in the triplet $|10\rangle$ state. Thus the existence of pure, spatially extended, nonfactorizable states is manifested by non-trivial correlations between the results of measurements on the two systems, correlations greater than allowed by Bell's inequality. Correlations that do not violate the Bell inequality are trivial if the two systems interacted in the past because one can explain them easily by local realistic models.

Is it possible to find non-trivial correlations between the systems that never interacted? In triplet $|1,\pm 1\rangle$ states correlations are just as large as allowed by the Bell inequality, with $C(a,b,a',b')$ reaching at most 2. This is not surprising because $|1,+1\rangle = |\alpha\alpha\rangle$ and $|1,-1\rangle = |\beta\beta\rangle$, both are factorizable and therefore there are no interference terms in the density matrix [cf. 16]. But these are the only pure states that we know how to prepare externally! Mixture of all 4 states does not give any correlation, as one may expect. Going to higher spins, like s = 3/2, does not help either. Thus unless there is a way to prepare a pure state or a non-trivial mixture there will be no way to observe EPR correlations without previous interaction.

5. ENTANGLED SYSTEMS

Let us look from a more general point of view at the problem of "entanglement", as Schrödinger [3] calls it, manifested in EPR correlations violating local realism. If such correlations exist, as all experiments performed so far seem to indicate, then the ensemble interpretation of quantum mechanics cannot be maintained and local modifications of quantum mechanics are not admissible.

Two particles with spin ½ are in one of four spin states $|S,M\rangle$ but in experimental situations we usually have an ensemble described by a mixed state ($|\alpha\rangle$ and $|\beta\rangle$ are the single-particle spin functions):

(8) $\quad \rho(1,2) = \frac{1}{4}\big(|0,0\rangle\langle 0,0| + \sum_M |1,M\rangle\langle 1,M|\big) =$

$\quad\quad \frac{1}{2}(|\alpha\rangle\langle\alpha| + |\beta\rangle\langle\beta|)\;\frac{1}{2}(|\alpha\rangle\langle\alpha| + |\beta\rangle\langle\beta|) = \rho_1(1)\rho_1(2)$

In general for N particles with spin ½ the number of states degenerate at large distance is 2^N and the mixed state is factorized

(9) $\quad \rho(1,2..N) = 2^{-N}\sum_S\sum_M |S,M\rangle\langle S,M| = \prod_i \rho_1(i)$

It is very hard to avoid mixed states in real experimental situations and mixed spin-density operators are nicely separated into products of single-particle operators which represent mixtures, i.e. half of the particles with spin α and half with spin β in an ensemble of particles. This is an answer to the question why EPR correlations are not seen more often.

Factorizability of the density matrix is a stronger requirement than separability which should require no more than the lack of correlations between measurements on the two spatially separated systems. Indeed if one takes two spatial states, $|a\rangle$ localized around the system S_A and $|b\rangle$ around S_B, antisymmetric two-particle singlet and triplet functions give

(10) $\quad |\psi_{00}\rangle = 2^{-\frac{1}{2}}(|ab\rangle+|ba\rangle)2^{-\frac{1}{2}}(|\alpha\beta\rangle-|\beta\alpha\rangle)$
$\quad\quad |\psi_{1M}\rangle = 2^{-\frac{1}{2}}(|ab\rangle-|ba\rangle)|1,M\rangle$
$\quad\quad \rho = \frac{1}{4}\sum_{SM} |\psi_{SM}\rangle\langle\psi_{SM}| = \frac{1}{2}(|ab\rangle-|ba\rangle)(\langle ab|-\langle ba|)\rho_1(1)\rho_1(2)$

The spin part is factorized but the spatial part is not. Off-diagonal matrix elements $|ab\rangle\langle ba|$ correspond to the exchange of particles between systems S_A and S_B and may be neglected at large separations of the two systems. Of course the (negligibly small) possibility of such an exchange always exists and the cross terms in the density operator simply remind us of this, but all operators $\hat{F}$ corresponding to physical observables have vanishing matrix elements $\langle a|\hat{F}|b\rangle = 0$. One may call it practical separability. Claims that quantum mechanics is not able to describe separated systems [6,7] are true but irrelevant. It is rather axiomatic approach to quantum mechanics ("new quantum theory" as Piron calls it [6]) that is unable to describe the process of separation because it washes out all physics inherent in this process. It is completely artificial to base separation on pure states requiring factorizability at large separations. Real processes must involve other bodies exerting forces breaking the system or processes like photodissociation, with all its subtleties, that no axiomatic approach can describe.

Let us come back to Schrödinger's analysis of entanglement. For the two systems in a pure state $|\psi_{AB}\rangle$ with the density operator ρ_{AB} one has

(11) $\quad |\psi_{AB}\rangle = \sum_n \lambda_n |A_n\rangle|B_n\rangle; \quad \rho_{AB} = \sum_{nm} \lambda_n \lambda_m^* |A_n\rangle|B_n\rangle\langle A_m|\langle B_m|$

where ρ_{AB} is idempotent. If more than one λ_n value is non-zero there are non-trivial correlations between the two systems – they are entangled. A pure state represents the best possible or perfect knowledge about the whole system and that of course includes the correlations between the two subsystems. Maximal knowledge of S_A must include also the knowledge of its correlation with S_B, therefore it should not be possible to describe the S_A system separately by a wavefunction. Indeed, reducing ρ_{AB} to the subsystem,

(12) $\quad \rho_A = Tr_B(\rho_{AB}) = \sum |\lambda_n|^2 |A_n\rangle\langle A_n|$

and similarly for ρ_B. These density operators correspond to mixed states because they are not idempotent. Taking

(13) $\quad 0 \leq \mu(\rho) = Tr(\rho - \rho^2) \leq 1$

as the measure of loss of information one has $\mu(\rho) = 0$ for a pure state and $\mu(\rho) = 1$ or complete loss of knowledge for the worst mixed state possible. If the number of particles in both subsystems is large $\mu(\rho_A) \simeq 1$. Almost all information about the whole system concerns the correlations between the two subsystems. Of course density operators do not tell us which part of the knowledge is interesting – it may be just the tiny local one, contained in the reduced density matrix.

Perfect knowledge of the whole is not the best possible knowledge of all parts even when the parts are not interacting. What does it imply in practice? It implies that non-trivial EPR correlations are necessary to avoid non-locality! This seems to be rather paradoxical because one is tempted to think about EPR correlations as proving non-locality. In fact it is the existence of extended quantum states that saves Einstein's locality introducing instead non-local correlations that cannot be used to send information [17].

7. REDUCING THE PERFECT KNOWLEDGE.

As another illustration of Schrödinger's statement that the best possible knowledge of the whole does not give the best possible knowledge of the parts consider two hydrogen-like atoms separated by a large distance R. It is the starting point of the Heitler-London theory of hydrogen molecule [cf. 18]. In this case the wavefunction is

(14) $\quad |\psi\rangle = 2^{-\frac{1}{2}}(|a\rangle|b\rangle \pm |b\rangle|a\rangle)|SM\rangle$

where $|a\rangle$ is the 1s orbital centered on the atom A and $|SM\rangle$ is the corresponding spin function. The spinless Hamiltonian

(15) $\quad H(1,2) = H_A(1) + H_B(2) + V(1,2); \quad T_i = -\frac{1}{2}\Delta_i$
$\quad H_A(1) = T_1 - Z_A/r_{A1}; \quad H_B(2) = T_2 - Z_B/r_{B2};$
$\quad V(1,2) = -Z_A/r_{A2} - Z_B/r_{B1} + Z_AZ_B/R + 1/r_{12}$

where the meaning of various terms is obvious [18]. For a large separation R the overlap integral $\langle a|b\rangle$ is zero and the energy is

(16) $\langle\psi|H|\psi\rangle = E_A + E_B$

We know that $|\psi\rangle$ is never reduced to a simple product of $|a\rangle$ and $|b\rangle$. Still we do hope that a local energy operator exists. Such is not the case! Take for example H_A and calculate

(17) $\langle H_A \rangle = \langle\psi|H_A|\psi\rangle = \frac{1}{2}(\langle ab|H_A|ab\rangle + \langle ba|H_A|ab\rangle) =$
   $\frac{1}{2}(E_A + \langle b|H_A|b\rangle)$

Calculating the last element directly or via the virial theorem one obtains

(18) $\langle b|H_A|b\rangle = \langle b|T+V_A|b\rangle = \langle b|T|b\rangle = -\langle b|H_B|b\rangle = -E_B$

This result may seem surprising but the state $|b\rangle$ belongs to the continuum of the H_A Hamiltonian. As a result the average value is

(19) $\langle\psi|H_A|\psi\rangle = -\langle\psi|H_B|\psi\rangle = \frac{1}{2}(E_A - E_B)$

Thus the sum of the local energies for the two atoms is zero! Where is all the energy gone? It is easily verified that

(20) $\langle\psi|V|\psi\rangle = E_A + E_B$

Calculating local energies using local operators and non-local wave-function gives paradoxical results: all the energy is in the interaction, although the interaction terms are negligibly small. Extension of this simple model to a gas of N noninteracting hydrogen atoms in the 1s states with $-1/2$ atomic unit of energy gives

(21) $\langle\psi_{1s}(1..N)|H_A(1)|\psi_{1s}(1..N)\rangle = (N-2)/2N$
   $\langle\psi_{1s}(1..N)|V_A(1)|\psi_{1s}(1..N)\rangle = 1-N$

For the excited states $|2s\rangle$ with the energy $-1/8$ au the local energy comes out as $(N-2)/8N$, i.e. is lower than in the ground state!

These surprising results may be interpreted in different ways. Since the total wave function of the world is not reducible to a product function strictly speaking local energies are not well defined and a kind of "quantum Mach principle" is introduced: it is the interaction, even if negligibly small, that gives meaning to the total properties like energy. If $|\psi\rangle$ is the exact multi-particle state and one-particle approximation is not introduced from the beginning it is not easy to obtain local properties from the knowledge of the whole. Perfect knowledge of the whole should be complemented by the knowledge of local states.

From another point of view the decomposition (15) of the total Hamiltonian, although the most natural one, leaves the interaction $V(1,2)$ in a non-symmetric form. The problem is that there is no better decomposition and thus no way of reducing the description of the whole to the description of one of the non-interacting parts. To obtain a local density operator one has to know local states and project the reduced density operator on these local states. For the two atoms

(22) $\rho_1 = Tr_2|\psi\rangle\langle\psi| = \frac{1}{2}(|a\rangle\langle a|+|b\rangle\langle b|)$
  $\rho_A = (|a\rangle\langle a|)$ $2\rho_1 = |a\rangle\langle a|$

In any case although the perfect knowledge is at hand it does not solve the local problem.

8. CONCLUSIONS

Application of quantum mechanics to systems that are not large enough to be treated classically and not small enough to be treated as isolated demands a careful look at separability. Even for small molecules forming weakly interacting van der Waals complexes separability is not a trivial problem. Using perturbation theory for intermolecular interactions one is tempted to think that, if the exchange integrals are small, one may skip antisymmetrization and start from the product function. Computational experience and theoretical analysis [19] have proved that such approach (termed "polarization approximation") is completely inadequate. This fact creates great complications in applications of perturbation theory to many-body systems. Technically these problems are connected with the non-existence of local energy operators when the antisymmetric wavefunctions are used. One could expect problems in the direct determination of the differences between the properties of the whole and the parts (interaction energies) on the basis of the general considerations given in the last section.

So far quantum mechanics has never failed, therefore such strange consequences as non-separability or macroscopic superpositions of pure states [20] should be taken seriously. It is hard to accommodate such notions in the western philosophical tradition; despite experimental verifications still a large number of experts are not ready to accept it [21]. Whole branches of physics, like stochastic electrodynamics, appear motivated by desire to understand physical picture at the quantum level [22]. Such efforts, although so far not successful, should nevertheless be respected: no path should be left unexplored. But, one might ask, is quantum mechanics really "beyond human comprehension"? Non-separability and the complementarity of knowledge about the whole and its constituents would certainly not come as a surprise to the Taoist sages like Czuang-Tsu who wrote:

"The knowledge of the ancients was perfect. How perfect? At first, they did not know that there were things. This is the most perfect knowledge; nothing can be added. Next, they knew that there were things, but did not make distinctions between them. Next, they made distinctions between them but did not pass judgments upon then. When judgments were passed, Tao was destroyed."

One may say: when local predictions are made the Unity (correlations with the other systems) is destroyed.

ACKNOWLEDGEMENTS: I am grateful to C. Nicolaides for an invitation to Delphi and the discussions we had in Athens and to J-P. Vigier, A. Kyprianidis and many other participants of the Delphi conference for the fun we had together discussing physics. This paper was partially sponsored by the Institute for Low Temperature Research, project CPBP 01.12.

REFERENCES

1. The Wave-Particle Dualism, Eds: S. Diner, D. Fargue, G. Lochak, F. Selleri. D. Reidel, Dordrecht 1982
2. A. Einstein, B. Podolsky, N. Rosen, Phys. Rev. 47,777(1935)
3. E. Schrödinger, Naturwissenschaften 23,807; 823; 844(1935); Proc. Cambridge Phil. Soc. 31,555(1935); ibid. 32,446(1936)
4. N. Rosen, in: Symposium on the Foundations of Modern Physics, Eds. P. Lahti, P. Mittelstaedt, World Scientific, Singapore 1985, p. 17
5. H-J. Treder, Astron. Nachr. 304,145(1983)
6. C. Piron, in [4], p. 207
7. D. Aerts, Found. Phys. 12,1131(1982); D. Aerts, in [4], p. 305
8. O. Costa de Beauregard, Found. Phys. 15,871(1985)
9. J.S. Bell, Physics 1,195(1965)
10. A. Pais, 'Subtle is the Lord', Oxford Univ. Press 1982
11. O. Piccioni, W. Mehlhop, in [4], p. 197
12. D. Bohm, Y. Aharonov, Phys. Rev. 108,1070(1957)
13. D.J. Bohm, B.J. Hiley, Il Nuovo Cim. 35,137(1976)
14. F. Selleri, Found. Phys. 12,1087(1982)
15. F. Selleri, 'History of the EPR Paradox', in: Quantum Mechanics Versus Local Realism: the Einstein-Podolsky-Rosen Paradox, Plenum Publ. Co, London-New York 1987
16. V. Capasso, D. Fortunato, F. Selleri, Int. J. Theor. Phys. 7,319(1973)
17. W. Duch, 'Violation of Bell inequalities in interference experiments', in: Open Problems in Physics, Eds. L. Kostro et.al, World Scientific, Singapore (in print); Correlations violating Bell's inequality are necessary to save locality, Phys. Rev. Lett (submitted).
18. A.S. Davydov, Quantum Mechanics, Moscow 1963
19. P. Arrighini, Intermolecular Forces and Their Evaluation by Perturbation Theory, Lecture Notes in Chem. 25, Springer 1981
20. A.J. Legget, in: Quantum Concepts in Space and Time, Ed. R. Penrose, C.J. Isham, Clarendon Press, Oxford 1986
21. W. Duch, D. Aerts, Phys. Today 6,13(1986)
22. T.A. Brody, Rev. Mexicana de Fisica 29,461(1983)

CONCERNING SCHRÖDINGER'S QUESTION:
IS DEMOCRITUS OR PLANCK THE FOUNDER OF QUANTUM THEORY?

Hans-Jürgen Treder
Einstein-Laboratorium,
Deutsche Akademie der Wissenschaften,
Rosa-Luxemburg-Straße 17a,
Potsdam 1509, DDR

ABSTRACT. The question how far Democritean atomism provides the conceptual background for quantum physics is examined. First, Schrödinger's profound studies of this philosophy and the relation of its concepts to quantum atomism and to statistical physics is reviewed. Then the contradiction is brought out between a relativity theory in which Democritean atoms are inconceivable and a quantum theory which implies their existence; as a result, paradoxes still plague quantum field theory. Schrödinger's and Heisenberg's way out of the difficulty leads, via Plato's ideal structures, to taking universal constants as the fundamental building blocks of physics.

1. SCHRÖDINGER'S RECEPTION OF DEMOCRITEAN ATOMISM

"Only atoms and empty space possess reality"

In his work "Die Natur und die Griechen (Kosmos und Physik)", Erwin Schrödinger declared the Ionian natural philosophers, the members of the school of Miletus, to be the discoverers of the "comprehensibility of natural phenomena". According to him, Thales is at the same time the first scientist and the first mathematician. I would like to interpret Schrödinger's term "comprehensibility of nature" in the sense of its intrinsic comprehensibility, without any *deus ex machina*. Schrödinger emphasised at the same time that a simple collection of pure facts, such as was already at the disposal of Egyptian and Chaldean priests, does not constitute a science. It was, according to him, the synthesis of philosophic speculation and empirical accumulation of facts that constituted the Ionian philosophers' original contribution.

What the natural sciences methodically offer is, as Schrödinger comments in his book, the objectivisation of statements: anybody can verify their validity any time, anywhere. This is brought about by the quantisation of concepts—something that in the framework of Greek mathematics always was a digitisation, too; for it is possible to dispute about the quality of something, but the numbers 1 and 2 can be objectively distinguished. And it was precisely for this reason that Schrödinger saw in the atomism of Democritus the greatest step forward in scientific thinking after the Ionians.

Democritus' atomism in principle is built only on quantities, namely the number and size of the atoms and their velocities. Here Democritus was far ahead of his time in that he took, preceding Galileo in assuming something like a law of inertia, each atom's velocity to be constant, unless a collision with another atom prevents its free motion. For Democritus, the cosmos is a world of quantities uniquely given which continue their motion according to their own inertia until they are perturbed by other particles of the same nature.

Further on in his work Schrödinger entered into the difficulties of ancient atomism. He noted that this atomism began the trend towards the extreme sensualism which Epicurus and after him T. Lucretius introduced and which turned the showing up of the ineffectiveness of the gods into an acceptation that no theoretical understanding of our world is possible. (To have fought against this is to the credit of Plato's Academy.) Another problem for Democritean atomism springs from its admirable reduction of everything to quantitative form. At least as Epicurus and Lucretius saw it, and later on also Gassendi, Democritean atoms possess, indissolubly linked to their geometric structure, the property of absolute impenetrability and rigidity. Democritus takes as his starting point that two atoms can never occupy the same place at the same time, and that the structure of atoms (now in the sense of Max Planck) "is dependent on nothing". Here Kant had already observed that the principle of the excluded middle cannot move even the lightest feather, so that the mutual impenetrability of atoms implies, in Newtonian mechanics, a field of infinitely strong forces. As Fritz Hund put it: the problem of atomism is the stability and impenetrability of the atoms.

In his book "Der Teil und das Ganze" Werner Heisenberg maintains, as against Schrödinger's view, that Plato's treatment of Democritus by emphasising algebraic and topological structures as cosmological invariants overcomes the difficulties of Democritean atomism and so leads from Democritus in the direction of Planck (I return to this below). Yet although his epistemology was even more idealist that Heisenberg's, Schrödinger had no very high opinion of Platonic mathematics in comparison with that of Democritus. "Democritus was deeply interested in geometry, and not simply as an enthusiastic admirer like Plato; he was himself a notable geometrist." Schrödinger mentions that Democritus extended his atomistic conception to time and space, and could therefore carry out quadratures, cubatures, rectifications and complanations by using (before its time) the principle of Cavalieri. Admittedly his atomistic method of integrating never yields exact values; thus the relation between volume and surface depends on the relative size or smallness of the Democritean atoms (1).

Besides Democritus, Schrödinger always mentions the name of his somewhat legendary teacher Leucippus. Wherever we link the name "Democritus" to atomistic conceptions, we could in fact say "Leucippus and Democritus". But if Democritus' writings are conserved only in a few authentic fragments, or in the critical and mostly negative comments from the Academics and the peripatetics on the one hand, and the enthusiastic praise of the Epicureans on the other, of Leucippus' texts nothing whatever can be reconstructed.

2. ATOMISM, STOCHASTICITY AND PROBABILITY

In discussing the philosophy of physics Schrödinger wrote that cognition is a game with loaded dice. In his philosophy, "the Lord cheats". This saying of his sheds light on how he saw the relation between stochasticity and probability.

Schrödinger belonged to the great Viennese school of physics that helped to found atomistic physics and at the same time raised the problem of how statistics and stochasticity could be related. Its father figures were Josef Stefan and above all his pupil and successor Ludwig Boltzmann. It was Boltzmann who saw that, given the atomic structure of matter, thermodynamics must mean that the universal validity of Newtonian mechanics in the microcosm leads to statistical laws and so to stochastic behaviour at the macroscopic level (the central point here being Boltzmann's principle of microscopic reversibility). Schrödinger stressed this in his inaugural lecture at the Berlin Academy of Science, where he mentioned his teachers, F. Exner and F. Hasenöhrl, both pupils of Boltzmann (and in their turn teachers of many of the best-known older atomists, men such as M. von Smoluchowski and P. Natanson or E. von Schweidler, as well as his fellow students A. March and V.F. Hess); he noted that Exner and Hasenöhrl had learnt from Boltzmann that atomism is taken to mean that macrophysical quantities are statistical averages over microscopic events. This is fully valid for the thermodynamical functions such as entropy, pressure or temperature, but less evident for mechanical quantities. The conservation law of classical mechanics are independent of all internal properties, as follows from Newton's principle o the equality of action and reaction. And even Hasenöhrl's example (discussed by Schrödinger in th lecture cited above) of a body which does not fall but rises in the earth's gravitational field is not simply compatible with with the universal validity of Newtonian principles, it requires them (2).

Now according to Boltzmann and Planck, the principles of classical (or of relativistic) mechanics have absolute validity, while those of thermodynamics are statistical in nature. But Exner was unwilling to recognise this distinction. For him—and in the cited lecture Schrödinger likewise took up this position—there are no "absolutely random processes", and for this reason, that there are no exact laws of nature. Yet it turns out that according to atomic theory the first law of thermodynamics has absolute validity; no hypothesis that conservation of energy is merely statistical in nature, such as Bohr, Kramers and Slater (1924) proposed as an alternative to Einstein's photon theory, can hold water; indeed, such ideas were not only shown to be wrong by the Geiger-Bothe experiments on the Compton effect but disproved precisely by Schrödinger's own wave mechanics. From their results Bothe and Geiger (1924) drew the definitive conclusion that the atomic hypothesis is valid—that the Einsteinian photon exists—in independence of chemical argument but as a consequence of thermodynamics and Planck's cavity-radiation law.

As against this Schrödinger took up again Exner's remark that, just as Democritean atomism leads to the construction of approximations in place of the infinitesimal calculus developed by Newton and Leibniz, so modern atomism admits of no sharply defined initial conditions in dynamics. True, there is no uncertainty relation, but no measurement procedure can be indefinitely precise. Determinism is drowned by stochastic background noise. Schrödinger noted (*contra* Planck) that there is therefore no difference of principle between classical and quantum mechanics.

In his response Planck emphasised not only that there *are* exact physical data—conditioned by the universal constants of nature—but that just because of this "absolute coincidences" (coincidences whose random nature does not arise from our ignorance) exist. Schweidler had found statistical laws of this absolute character in the regularities of radioactive decay, and Einstein had derived others from his atomistic deduction of the Planck radiation law. And these "a priori probabilities" can be calculated precisely from the wave mechanics of Schrödinger (as transition probabilities).

Yet Schrödinger's wave mechanics yields the Bohr model of the atom and so, in the last analysis, the "infinitely sharp" ground states corresponding to the finiteness of the Planck constant h. The probabilistic uncertainty in quantum mechanics, expressed in Heisenberg's uncertainty relations

$$\Delta p \Delta q \geq h, \qquad \Delta E \Delta t \geq h$$

is not at all of stochastic character, but springs from the existence of the universal constant of nature h.

Exner's and Schrödinger's arguments, too, presuppose universal constants, namely those linked to the Democritean atoms. But according to quantum theory h "precedes any atomistic assumption" (Einstein). For this reason Schrödinger quite soon came to accept Planck's criticisms and took up Einstein's suggestion of a deterministic physics in the configuration space of the Schrödinger equation. He himself showed (following up a hint due to Planck) that in this space everything evolves deterministically. But this Schrödingerian image of the world is not sufficient.

Every measurement takes place in ordinary three-dimensional space, plus a one-dimensional time. To measure means (this is the joint idea of Bohr, Dirac, Heisenberg and Schrödinger, as well as their pupils) to project from the $3N$-dimensional configuration space of wave mechanics onto the classical three-dimensional space of Newtonian particle mechanics. Schrödinger observed that the possibility of formulating this projection mathematically is linked to Democritean atomism: not every solution of the Schrödinger equation of the N-body problem in a $3N$-dimensional configuration space makes physical sense; only those do that are symmetric (or, if spin—the Pauli principle—is taken into account, antisymmetric) under permutations of the particles. Bose-Einstein (or Fermi-Dirac) statistics demonstrate that in quantum mechanics Democritean atomism applies. According to Schrödinger, the "second quantisation", which is a consequence of the first one, involves the existence of Democritus' atoms as indistinguishable particles. That in the solution manifold of the Schrödinger equation for the N-body problem only a symmetric (or antisymmetric) ansatz should possess physical meaning, is—for

Schrödinger—straight out of Democritus. Second quantisation connects Democritus to Planck.

3. DEMOCRITUS, PLANCK AND PLATO

In his "Statistical Thermodynamics", Schrödinger emphasised that what quantum mechanics had contributed as fundamentally new to classical atomic theory was above all the principle of the indistinguishability of equal particles. This is intended in the sense of Leibniz' *principium individuationis*: all "equal atoms" (or particles or molecules) are to be "counted only once". In this way Schrödinger sees quantum physics as ideally fulfilling certain Democritean conceptions: there are indistinguishable atoms, while in classical atomistics every atom may be dintinguished from every other one by its space-time localisation. Indistinguishability, that is to say consistent atomism, has profound consequences at the macroscopic level. Schrödinger himself stresses the Gibbs paradox and mentions by way of further examples Bose-Einstein condensation, the Planckian cavity-radiation law, and the Fermi sphere. The effectiveness of the atomic view in chemistry shows up in the theoretical explanation of homeopolar valence, based on the indistinguishability of electrons according to the Pauli principle. The indistinguishability of atoms is connected with their stability, and both are consequences of quantum mechanics. We cannot in any way mark the Democritean atoms without overstepping the energy threshold given by the relation $E = h\nu$ due to Planck, Einstein and Bohr : one cannot draw a chalk mark on an atom. Any such marking deforms the atom and so contradicts its inherent stability. To deform an atom—in inelastic collisions—one must exceed this energy threshold the existence of which was demonstrated in the collision experiments of James Franck and Gustav Hertz (whose centenary is also being celebrated this year).

Quantum theory, then, makes—as made evident by the atom models of Bohr, Schrödinger and Dirac—the unchangeability of atoms under external influences depend on the energy thresholds experimentally demonstrated by Franck and Hertz. But the same quantum theory contains, corresponding to the wave aspect that de Broglie and Schrödinger described, the uncertainty relations of Heisenberg, of the form

$$\Delta\left(\frac{\partial H}{\partial q}\right) \cdot \Delta\left(\frac{\partial H}{\partial p}\right) \geq h$$

From these relations one derives, in the quantum-theoretical form stressed by Schrödinger, that this matter of the indistinguishability of atoms depends on the measurement precisions. More specifically, quantum mechanics shows that the indistinguishability of particles is linked to the velocity and position uncertainties which result in the attempts to determine the localisation and trajectories of the particles.

However, no real "atom" is effectively unchangeable, in the quantum sense, nor is it unmarkable. A chemical atom is "indivisible" in the energy range up to a few electron-volts. An atomic nucleus is stable up to energies of a few thousand or even million electron-volts. For the break-up of baryons gigavolyts and more are necessary.

The atom concept of Democritus and Epicurus contains open questions to which Schrödinger devoted much attention. In principle the concept would lead to a kind of "block" geometry, a zero-dimensional manifold, and so "impoverish mathematics" (on this point, Schrödinger agreed with Einstein and Weyl). Isaac Newton's position was very close to that of Democritean atomism, only that he wanted to understand it within the framework of classical mathematics. For Newton, particles were (as in the present-day quantum field theories of special relativity) pointlike singularities, represented by Dirac delta functions. An atom of finite extension and mass is, for Newton's and Faraday's dynamic theorising, a region containing infinitely large repulsive forces; therefore dynamics underlies geometry.

As a result, the theory of special relativity once more places doubt on the Democritean and Newtonian conception of "atoms". The Lorentz contraction has as a consequence that there is no "smallest length", and from the principle of special relativity—Einstein's definition of causality—it follows that neither are there any instantaneous interactions. But the atom of Democritus (or the dynamide of Newton) presupposes instantaneous interactions. Now the Lorentz contraction and the transmission of interactions with the one fundamental velocity c are two sides of the same coin for Einstein; in other words, in a relativistic field theory there cannot exist any Democritean atoms.

This is the paradox of quantum field theory, repeatedly underlined by P.A.M. Dirac: this theory implies on the one hand that there must be Democritean elementary particles, and on the other hand it makes their existence inconceivable (this is the problem behind renormalisation).

Within relativistic field theory there cannot be any atoms of finite diameter; all indivisible particles must be pointlike. Democritean atomism, however, requires a "block structure", which cannot be Lorentz invariant. Relativity theory and finite extension of particles are mutually exclusive. Once again, this difficulty could be resolved along Platonic lines, and the questions raised by Democritus' atomism could be answered in the sense of Plato. Beyond space and time, and prior to all dynamics, there exist structures that determine a priori—both in the sense of Plato and in that of Kant—the possible structures within the universe. Thus the contradiction between Democritean and Platonic atomism vanishes. Schrödinger's and Heisenberg's views lead to the same result: it is in fact justified to reduce Democritean atomism to that of Plato. The place of the atoms that in the sense of Democritus persist eternally, independent of all external influence, is taken by the elementary constants which, in Planck's words, depend on "nothing, nothing at all".

Quantum mechanics and the special relativity theory take as their starting point that the Planck constant h and the velocity of light c, respectively, are such universal constants. Yet, oddly, relativistic quantum theory provides the basis for stipulating a Democritean atomism (I refer to Schrödinger's argument), while from h and c we cannot derive either the rest masses of the elementary particles (this is dimensionally impossible) or Sommerfeld's fine-structure constant (3). Einstein called Planck's constant h the basis of every atomism. The quantisation of phase space thus precedes that of massive particles; yet special-relativistic quantum theory achieves nothing beyond what Dirac, Pauli and Schrödinger had seen. The atomism of Schrödinger requires that measurements of time and space be also quantised; for without it the finite transmission speed of relativity is not compatible with the existence of particles with finite extension. Heisenberg, referring to A. March and D.D. Ivanenko, introduced into relativistic field theory a new universal constant, the absolute rest length, a Compton wavelength $l = h/mc$; a knowledge of h, c, and l then implies also a unit m for the rest masses of the particles.

Thus in the end Schrödinger's question, "Democritus or Planck", leads to Heisenberg's answer: "Plato". For anything to be countable and thus computable, there must exist, as Democritus stipulated, invariant and invariable units of measurement. But originally these units are neither the atoms nor the elementary particles, but the universal constants proposed for this role by Planck in 1899: the quantum of action h, the velocity of light c, and Newton's gravitational constant f (Heisenberg considered that instead of f the Compton wavelength of a baryon should be used).

But it is just the "super-grand-unified theories" which, basing on the Planckian units, were intended to do what Democritean atomism demanded. Both Heisenberg's and Planck's conception of spatio-temporal structure hold the discrimination between extended particles (Democritean atoms) and pointlike singularities (the dynamides of Newton and Faraday) to be in principle unachievable. Thus Schrödinger's view, that quantum field theory should give rise to atomic structure, or that—as Einstein formulated it—Planck's constant h "precedes all atomism", acquires full validity. According to Planck, phase space is quantised, and its quantum is given by his constant as the elementary volume h^{3N}. Einstein demonstrated that the velocity of light c is independent of kinematics, and thus is one of the basic bricks of physics. Lastly, general relativity theory presupposes the universality of Newton's gravitational constant f (the universality of f is the Newton-Einstein version of the Galilean principle of equivalence). Seen in this way, the bricks out of which physics is built would be, not the Democritean atoms, but Plato's universal structure. But it would be better to say that the contradiction which Schrödinger observed between Democritus and Plato—along the lines of the discussions between Bohr, Dirac, Einstein, Heisenberg, Pauli and Schrödinger himself—is resolved in the conception of three universal constants that possess all thinkable properties of physical and mathematical invariance or invariability.

NOTES

(1) Today this problem has become that of the relation between Hausdorff dimensions and fractals

(2) Max Planck explained this point in his response to Schrödinger's lecture (in *Physiker über Physiker*, vol. II, ed. C. Kirsten).

(3) None of the ingenious attempts at such derivations, from Eddington to Heisenberg, nave achieved any success.

SCHRÖDINGER'S RECEPTION OF GREEK ATOMISM

Renate Wahsner
Einstein-Laboratorium,
Deutsche Akademie der Wissenschaften,
Rosa-Luxemburg-Strasse 17a,
Potsdam 1509, DDR

ABSTRACT. The Eleatics thought that the real world was knowable only because it was motionless, the Heracliteans that it was in constant motion and thus unknowable. The Greek atomists, chiefly Democritus and Epicurus, found the way out of this gnoseological dilemma and so provided a basis for conceiving motion which remains significant in present-day physics; this explains Schrödinger's intense interest in them. It is noted how Newton overcame the chief limitation of Greek atomism, which could not assign an active role to the interactions between atoms, and the parallelism to a guiding thought in Marx's work is pointed out.

1. INTRODUCTION

Nowadays, atomism is commonly regarded as an out-of-date sort of rationality, as a variety of mechanistic metaphysics. The usual objection is that the mechanistic conception, according to which the whole world may be explained as composed of motion of indivisible and hence smallest particles, has been refuted by modern physics, and by quantum mechanics in particular. Yet one of the co-founders of quantum mechanics, Schrödinger, writes: "Quantum theory dates twenty-four centuries further back, to Leucippus and Democritus. They invented the first discontinuity—isolated atoms embedded in empty space. Our notion of the elementary particle has historically descended from their notion of the atom: we have simply held onto it. And these particles have now turned out to be quanta of energy, because—as Einstein discovered in 1905—mass and energy are the same thing. So the idea of discontinuity is very old." [1]

According to Schrödinger, the work of Planck and Einstein has allowed present-day atomistic thought, under the name of quantum mechanics, to extend its conceptual domain from "ordinary matter" to all kinds of radiation [2] and thereby for the first time made it possible to apply the ancient Greek discovery of atomism consistently.

Schrödinger's labours may be summed up as an answer to the question of how continuity and discontinuity are related. As a pupil of Boltzmann's he was fully convinced of the profound significance in physics of the Boltzmannian atomism—both, that is, of Boltzmann's thermodynamics and of Boltzmann's gnoseological point of view. Under the impact of such ideas Schrödinger became the founder of wave mechanics. But when he became aware of the fact that it was an illusion to believe that wave mechanics could be interpreted as a pure continuum conception, he began to study the epistemological bases of science.

True, the founders of wave mechanics "lived for a time in the fond hope that they had succeeded in finding the way back to a classically continuous description of nature", as Schrödinger put it in a talk on 'Science and Humanism' [3]. But he was not content simply to accept that this hope had failed, he began to search for the reason for its failure. He found this reason in the difficulty of circumscribing adequately the idea of a continuum. In his view one could not simply reject the possibility "that the problems facing physics today spring from the conceptual enigma posed by the continuum" [2]. We mostly overlook this possibility, because we consider that we are familiar with the concept of a continuum and know how to handle it [3]. The ancient thinkers, on the other hand, knew that the continuum concept has its problems. Indeed, they invented atomism in order to overcome them. This invention was outstanding enough to carry the weight of science to the present day; and so Schrödinger sings the praise of atoms and quanta as the age-old counterspell against the magic of the continuum [1,3].

It may seem surprising to call what forms part of the foundation of modern science an "age-old spell". But does this surprise not have its basis in our lack of surprise that the world is comprehensible, that motion can be circumscribed by laws, *i.e.* by stable relations? The comprehensibility of our world is so self-evident for our modern scientist that it does not even occur to him to ask what assumptions underlie it. Yet just this is the question that in Schrödinger's view must be put if the difficulties of modern physics are to be understood and hence also solved.

The ancient atomistic method of conceiving the continuum by way of the discrete is rooted, according to Schrödinger, in two principles: on the one hand the assumption that what happens in the world may be comprehended, and on the other the fundamental rule of eliminating the knowing subject from the desired image of the world and letting it play only the role of an external observer (objectivisation) [2,3].

The first principle found its justification when Ionian natural philosophy overcame the mythological way of thinking. This ascendancy initiated the transition to a rational view of the world and so provided a basis both for philosophy and for science [4].

The second principle arose from atomism, which regarded the entire world as made up of atoms and the vacuum, nothing more. Thus the knowing subject was seen as a mere natural object and hence eliminated *qua* subject. But when ancient atomism attempted also to explain the soul as built up of atoms, it entered into contradiction with its own fundamental principle and gave rise to paradoxes [2]. Schrödinger considered that omitting the knowing subject is an artifice, yet one that only a fool would not make use of [3]. But he took Ionian natural philosophy to have done no more than create a precondition for scientific rationality, and only traced out the later development of this rationality; hence he simply failed to see the knowing subject in the modern picture of the world—the subject from which abstraction is made in fact only in the sciences, not in philosophy.

Schrödinger's analysis of what ancient atomism means for modern physics leads to this conclusion: However much the concept of an atom may have changed (the atoms of quantum mechanics are no longer individuals whose motion we may follow through time) we cannot do without this concept; taking atomism here not so much in the sense of the theory that matter is composed of minimal particles [1] but in the sense of an "artifice" by means of which the conceptual difficulties of the continuum may be mastered. Here Schrödinger calls "artifice" a law of thought that has gnoseological justification. He describes it as a necessary tool that deforms our picture of the world only insofar as we forget that we have used it and do not know what it consists in.

To sum up, we may say: Schrödinger considered not so much that the world possesses an atomistic structure, but rather that atomism is anchored as a law of thought in the fundamental principles of science.

2. THE LAW OF THOUGHT OF ANCIENT ATOMISM

As has been shown elsewhere [5], Greek atomism was the first to develop the principle of physical representation of nature; the central idea in this principle was to reduce the entire world, all that exists, to the atom and the vacuum, to body and space, to being and not-being. Their philosophical system was the atomists' reaction to the recognition, arrived at by their predecessors, the Eleatics and the

Heracliteans, that motion cannot be conceived without contradiction and therefore, as they concluded, cannot be conceived at all. This recognition took form in two opposing viewpoints. The Eleatics considered that in reality the world was motionless and comprehensible, conceivable, only because of that. In the eyes of the Heracliteans, on the other hand, the world is in constant motion, but precisely for that reason it cannot be known or conceived but can at most be adequately reflected in sensual perception. Motion seemed unthinkable both in the rationalist-Eleatic and in the empiricist-Heraclitean conception, because not-being was seen as not thinkable. For what—like not-being—was not an object, what thus has no specifiable features, cannot be conceived in thought. Consequently the idea that something is and is not, that something by being in motion is at one place and simultaneously is not there, is likewise taken to be an unthinkable contradiction [6]. The atomists solved the problem of motion, of conception and of reality which their predecessors had posed from opposing viewpoints—a problem which sprang from the insight that thought processes (always in some way discontinuous) and real existence (in the last analysis always continuous) cannot directly correspond. Thus the atomists solved a problem springing from a fundamental question in philosophy which here was explicitly posed for the first time.

The ancient atomists found a highly ingenious escape from this dilemma: by means of their reduction of everything that exists to atoms in a void they exhibited the possibility of conceiving not only what is but also what is not as object, so that being and not-being can be thought of as coexisting in the world. Atoms represent being, the void represents not-being, and one is as real as the other. The reality of what is not splits up what is, so that the pieces of what is may now have different and contrary descriptions. What is is continuous and discontinuous, indivisible and divisible, constant and variable, infinite and finite. (And it is no longer impossible to think that something, by being in motion, both is and not is.) The first term in each pair concerns the individual atom, the atom by itself, the second concerns conpositions of atoms, that is to say macroscopic bodies and more in general atoms in relation among each other. That it is the same being to which these different descriptions apply is guaranteed by the fact that all atoms are qualitatively equal: they all have only size and form, the differences among them are merely quantitative.

For the representation of physical nature there result two significant consequences from this splitting up of the world into void and fulness, into real non-being and real being: one, that all natural processes are universally comparable; and two, that the diversity perceived by our senses can be derived from a minimum of presuppositions and is the result of quantitatively determinable differentiations.

The invariability of the atoms guarantees that amid the ebb and flow of phenomena they, at least, remain constant. This constancy renders thinkable the existence of objective natural laws, objective laws that capture what is general and necessarily repeatable. Moreover, the atoms are predecessors of the concept of physical quantities, without which the laws of physics cannot be formulated as equations linking mathematical magnitudes. Just like the atoms, physical quantities are not immediately perceptible to the senses and require thought; they are thought objects, matters of reason. The atoms embody the conception, later realised in the measurable quantities of physics, of comparing what differs with respect to one quality, in other words of qualitatively setting equal what is being compared, in order quantitatively to capture the difference. There is however no way to associate a value with the primary atomic qualities, since they are not measurable. This distinguishes the atoms from physical quantities and hints at the limits within which the original atomism was able to provide a gnoseological foundation for physics.

By constructively synthesising rationalism and empiricism (Eleatic and Heraclitean thought), by extracting the rational nuclei of these two philosophical systems and joining them in a new unity, atomism created one basic presupposition for the development of individual sciences. It rendered possible a rational consideration not merely of the whole world but of parts of it (the hypothesis of comprehensibility emphasised by Schrödinger). It thus created the possibility of thinking the world, not just philosophically but also physically. For since the essence of all natural phenomena is given by the atoms and the void, it is not constituted by a universal interaction that establishes the global connections but already by a few atoms and the empty space between them. A basis was thus found for a science whose object is not the universe as a totality but single, discrete regions of it. It is a matter of principle that any region whatever can be the object of the science. This is an indispensable condition, expressed in the requirement of universal comparability.

We may resume the significance to physics of Greek atomism as follows: It provided a basis for physical thinking by separating the moments that effectively exist only conjointly, by *separating opposing conceptual specifications and then objectifying, reifying, them*. Opposing specifications become linked yet clearly delimited independent existences.

This separation of moments for the first in the history of human thought permitted discriminating conceptually between matter and space. It also made it possible to grasp motion conceptually, in such a fashion that the contradiction constituting motion [5] can be thought physically, *i.e.* by making motion measurable and computable. Without the concept of space the simultaneous reality of being and not-being could not have been maintained. But now space appeared as the condition for making motion possible and therefore also as the condition for the possibility of comprehending, of thinking about the world—the moving world—in ways that shift the logical contradictions away from the centre of the stage and make them innocuous, at least for the moment.

To accept that concept and reality are not identical and to accept that motion is a contradictory idea thus formed the condition for developing the principle that underlies physical thinking. The atomists' solution for making motion thinkable and bringing thought and reality into coincidence was one possible solution of this problem of motion, thought and reality. It was that particular solution that took nature as its undoubted presupposition. It conceived the entire universe as nature; conscience, the knowing subject, functions merely as an external observer (*cf.* Schrödinger's principle of objectifiability). This is just the point of view of science. The other possible (and indeed necessary) solution of the problem is the philosophical one, embodied in the birth of dialectics and likewise initiated by the contributions of the Eleatics and Heracliteans. The distinction between the two solutions as regards the atomic hypothesis may be simply stated: Philosophical atoms cannot exist, but physical ones must exist [5].

3. THE LIMITATIONS OF ANCIENT ATOMISM

The atomist conception made motion physically thinkable; but it could not establish a necessary connection between this law of thought and sensual perception. In other words, while ancient atomism developed the principle of physical thinking, it could not establish a principle of physical perception [5]. For this purpose atomism had to be modified.

In the original atomist view the atom is in essence postulated as a single, unrelated individual. But making the particle's individuality absolute in this way cannot provide a sufficient gnoseological basis for physics, at least not in a way that would link theory and measurement. The limit of the atomist view appears in its attempt to explain the diversity of the world, the processes of birth and decay that bring it about, by means of properties that may be ascribed to one body, to one individual atom.

The isolation of the physical body, conceived as an atom, was eliminated by the *physical* (not the mechanicist) concept of force, by the laws of physical interactions. The physical concept of force comprises the idea that natural bodies can act on each other, that in referring or better said relating to each other they exert a force, and therefore are not something strictly isolated and essentially singular. Basing himself on the notions—mostly related to the development of the experimental method—that had evolved in the intervening centuries, Newton developed this conception, which goes beyond the viewpoint of ancient atomism, by devising, in the shape of classical mechanics, the first dynamical theory of physics. In particular, when he discusses the concept of gravitation in the framework of his physical theory he arrives at the result that the force with which bodies act on each other cannot be viewed as an inherent or primary atomic property, not—in other words—as a property to be ascribed to an individual body. This follows from the insight that bodies are heavy only relative to each other and not by themselves. He drew this conclusion from his law of gravitation, both formulated mathematically and based on experiment, after he had made repeated but vain attempts to derive gravitation from primary, individual atomic properties by using the so-called aether models. Newton concluded that gravitation and other physical forces or dynamic interactions are *active principles*, and so recognised that while physics requires passive principles (primary atomic properties), they do not suffice; active principles of physical interactions are also needed [8]. Thus the notion of "physical explanation" had to change radically and only now became identified with the notion of a "scientific explanation" — within a particular science, that is to say—as opposed to the earlier meaning of "explanation in natural philosophy" [9].

It is interesting here to note that Karl Marx arrived at an analogous conclusion in his doctoral thesis "Difference between Democritean and Epicurean Natural Philosophy" [10], when he discussed the question what a philosophy has to be like in which the force concept bridges the gap between philosophy and reality and so lets theoretical principles become objective. At the end of his thesis he voiced his conviction that in the nature of things it is not singularity that dominates, and that therefore all genuine science goes out the window if one takes as basic the principle of the isolated individual or if one assumes human consciousness to be merely singular [11]. Thus Marx did not find the philosophical conception that he had been seeking in the two atomist systems he analysed: while the Epicurean version of atomism makes individual freedom (of man or of the atom) possible, it is entirely subjective since it is built on the principle of the insolated individual. Democritean atomism, on the other hand, is objective but has no room for any kind of freedom. However, as Marx recognised, a philosophy that wants its realisation cannot be content with criticising the world, it must attempt to change it. This requires objective laws, in which what is and what is general are objective. Like Democritus, Marx wished to trace out the *real* possibilities in this world, since only they give rise to real necessity. He concluded that what is needed is a philosophy which conceives the human being in such a way that his consciousness is compatible with an independent nature. And so he asked: how are we to picture man without seeing him as an isolated particular?

Finding the answer occupied Marx throughout his life. In the course of his work he came to see that human nature is not given once and for all, but that it is created through the process of human appropriation of nature, of man's work (of the effective interaction between man and nature). That therefore man cannot be conceived of as an isolated individual was first stated by Marx in the Paris manuscripts of 1844; its classic expression came a year later, in the well-known words of his sixth thesis on Feuerbach: "Human nature is not an abstraction inherent in the isolated individual. In its reality it is the network of social relations" [12].

Marx's meaning here runs parallel to the already mentioned insight, consequent upon Newton's gravitational theory, that bodies are not in themselves heavy but only relative to each other. The gravity of classical mechanics is a property that only comes to be in the mutual relations of bodies and cannot exist apart from these relations. (It is in this sense a property or an effect of the system.) The relational character of the concept is, as has been said, the necessary modification of ancient atomism, in which gravity appeared as a property of the isolated atom.

Nevertheless ancient atomic theory did contain a germ of this necessary modification, to be found in Epicurus, who added to Democritean atomism the notion of a systemic property. Epicurus saw the composition of atoms (to form a macroscopic body) as a whole that is more than its parts, the individual atoms. For him not only the essence but also the phenomenon really exists.

This is not the case for Democritus, for whom the composition coincides with the sum of the parts. It embodies, just as the individual atoms do, the essence that is expressed in the so-called primary qualities, which differ in this from secondary ones and from the phenomena. The latter, according to Democritus, are not proper to things in themselves but result from perception. Yet he does not explain why or how the atomic properties give rise for instance to the perception "red", he only maintains that it is so. Somehow or other, in his view, this process occurs within the subject. There is thus an unbridgeable break between sense perception and thought. A phenomenon is nothing but subjective appearance.

With Epicurus, on the other hand, the notion that the whole is more than its parts turns the secondary qualities into something objective, something that in principle at least can be studied by physics. Secondary qualities arise not in perception but through the composition of elementary bodies. But atoms as such are not perceptible, they are thought objects. What mediates between essence and phenomenon and between phenomenon and perception remains inexplicable for Epicurus. The transition from the thinkable to the perceptible occurs— "somehow".

Thus in the last analysis Epicurus solves the open gnoseological problem as little as Democritus. However, Democritus' system has the advantage of opening its peculiar way, so fruitful for physics, for conceptually capturing motion and continuum, a way not open to Epicurus [10]. For only in the Democritean system do atom and composition of atoms both belong to the "realm of essence", only there are the opposed conceptual moments ascribed to entities of the same kind; in the Epicurean system they are split into essence and phenomenon, so that the contradiction of their being, of what is, of their essence, is not accessible to thought.

Epicurus was the first to try to remedy the defects of the atomist system by means of the principle that the whole is greater than the sum of its parts. But only Newton succeeded in using this principle to modify atomism in a scientifically constructive way and so to synthesise Democritus and Epicurus.

The parallelism of Newton's modification of atomism in classical mechanics and the development of Marx' philosophy and theory of society suggests what surely is basic in every science: a basis for a science cannot be established if it conceives its subject matter as isolated individuals. Only a principle of social or collective individuals [1] can provide a basis. It is an indispensable condition for the foundations of any science, indeed a typical characteristic of what constitutes a science, that it should picture the objects of its researches not merely as given individual objects but also in their totalities, in their objective relations, as systems (although this can evidently not precede the theoretical development, the theory being its underpinning).

It is occasionally said that the so-called systems theory opens a new epoch of scientific development, since now it is no longer the individual that is the object under study but the system. Such statements have little justification; they only show that their authors have neither fully grasped their physics nor seriously read their Marx.

4. CONCLUDING REMARKS

With his atomism Democritus established an important principle of physics that has not needed revision to the present day. Insofar as quantum mechanics also has its basis in this principle, Democritus may be called a co-founder of quantum mechanics. But his discovery or invention is of equal significance for a field theory such as the theory of general relativity, in which the existence of gravitons as quanta of the field remains problematic. It takes the form here of the duality, as undeniable as ever, of space and matter [2]. The atomist principle of thought appears here not in the subdivision of space and time into discrete space-time "corpuscles" but in the splitting of the conceptual moments in the physical description of motion [14].

Atomism, as modified by Newton, is significant for physics because it offers a notable solution to a gnoseological problem. Yet it must not be forgotten that this is a solution only within the realm of physics and leads to mechanistic rigidities if the physical atom or other elementary concept of any physical theory is taken as an ultimate essence. It is up to philosophy to explain the status of the physical atom without itself supposing an atomic essence.

If Democritus' great merit has not yet received adequate recognition, this may be for the same reason that Schrödinger cites for the loss of the old atomistic manuscripts: "not so much because it all happened so long ago, rather perhaps because their views could be incorporated less easily than those of Plato or Aristotle, say, into the cultural atmosphere that had the ancient texts in its care" [1].

NOTES

(1) The terms "social" and "collective" are here intended to be taken, not solely in the meaning they have in the social sciences, but as a general opposition to "isolated" or "individual".

(2) This was explicitly stated by Einstein in a talk he gave at Leiden in the year 1920. Here for the first time he formulated his opinion that the theory of general relativity had not made space fully dynamic. And he added that "besides the observable objects something else that is not observable must be considered to be real" [13].

REFERENCES

[1] E. Schrödinger, "Was ist ein Naturgesetz?", in *Beiträge zum naturwissenschaftlichen Weltbild*, R. Oldenburg, Munich and Vienna 1962.
[2] E. Schrödinger, *Die Natur und die Griechen*, Paul Zsolnay, Vienna 1955.
[3] E. Schrödinger, "Science and Humanism", in *Physics in Our Time*, Cambridge University Press, Cambridge 1951.
[4] R. Wahsner, *Mensch und Kosmos – die copernicanische Wende*, Akademie-Verlag, Berlin 1978, pp. 12–31.
[5] R. Wahsner, *Das Aktive and das Passive. Zur erkenntnistheoretischen Begründung der Physik durch den Atomismus, dargestellt an Newton und Kant*, Akademie-Verlag, Berlin 1981.
[6] G.W.F. Hegel, *Wissenschaft der Logik*, part II, Verlag Felix Meiner, Leipzig 1951, p. 59.
[7] R. Wahsner and A. Griese, "Raum, Zeit und Gesetz", in A. Griese and H. Laitko (eds.), *Gesetz – Erkenntnis – Handeln*, Dietz-Verlag, Berlin 1972.
[8] I. Newton, *Opticks*, Query 31, Dover Publications Inc., New York, N.Y. 1952.
[9] H.-H. v. Borzeszkowski and R. Wahsner, *Newton und Voltaire. Zur Begründung und Interpretation der klassischen Mechanik*, Akademie-Verlag, Berlin 1980.
[10] R. Wahsner, "Nicht die Einzelheit herrscht in der Natur der Dinge. Zum Wissenschaftsprinzip des kollektiven Individuums", preprint PE-EL 03-87.
[11] K. Marx, "Differenz der demokritischen und epikureischen Naturphilosophie", in K. Marx and F. Engels, *Werke*, Dietz-Verlag, Berlin 1968, supplementary volume, part I, p. 304.
[12] K. Marx, "Thesen über Feuerbach", in K. Marx and F. Engels, *Werke*, Dietz-Verlag, Berlin 1958, vol. 3, p. 6.
[13] A. Einstein, *Äther und Relativitätstheorie*, Springer-Verlag, Berlin 1920, p. 11.
[14] H.-H. v. Borzeszkowski und R. Wahsner, "Physikalische Bewegung und dialektischer Widerspruch", *Dt. Zs. f. Philos.* **30** (1980) 634; *Physikalischer Dualismus und dialektischer Widerspruch. Studien zum physikalischen Bewegungsbegriff*, Wiss. Buchgesellschaft, Darmstadt (in press).

Part 2

Probability and Quantum Mechanics

PROBABILITY THEORY IN QUANTUM MECHANICS*

Leslie E. Ballentine
Department of Physics,
Simon Fraser University
Burnaby, B.C., V5A 1S6
Canada

ABSTRACT. The abstract theory of probability and its interpretation are briefly reviewed, and it is demonstrated that the formalism of quantum mechanics satisfies the axioms of probability theory. This refutes the suggestions which have occasionally been made that "classical" probability theory does not apply to quantum mechanics. Several erroneous applications of probability theory to quantum mechanics are examined, and the nature of the errors are exposed.

1. Introduction

It is generally agreed that probability must be employed at a fundamental level in the interpretation of quantum mechanics. Yet the concept and theory of probability are usually treated very loosely and superficially. I have not seen any textbook which demonstrates just how the axioms of probability theory are satisfied by the formalism of quantum mechanics. The first objective of this paper is to remedy that shortcoming, and in order to do so I first give a brief outline of the theory of probability, and those aspects of its interpretation that are relevant to this task.

It has occasionally been claimed that "classical" probability theory does not apply to quantum mechanics. Those claims are sometimes based on misinterpretations of quantum mechanics, but more often on misinterpretations of probability theory. Some of those erroneous claims are examined in the latter part of this paper.

2. Axiomatic Probability Theory

The mathematical content of the theory of probability concerns the properties of a function $P(A|B)$, which may be read as "the probability of A

---

* This is a slightly revised version of the paper that was originally published in the American Journal of Physics 54, 883-889 (1986). It is reprinted by permission of the copyright holder, the American Association of Physics Teachers.

conditional on B". A and B may be *events*, in which case P(A|B) is "the probability that event A will happen under the conditions specified by the occurrence of event B". Alternatively, A and B may be *propositions*, in which case P(A|B) is "the probability that A is true given that B is true". These alternative interpretations, as well as the interpretation of "probability" will be discussed in the next section. But all such interpretations can be based upon a common mathematical formalism, which derives from a set of axioms. For the sake of definiteness I will use the language of events in this section.

It is desirable to treat sets of events as well as elementary events. Therefore we introduce notations for certain composite events: ~A ("not A") denotes the non-occurrence of A; A&B ("A and B") denotes the occurrence of both A and B; AvB ("A or B") denotes the occurrence of at least one of the events A or B. For brevity these sets of events will also be referred to as "events". The three operators (~,&,v) are called *negation, conjunction,* and *disjunction.*. In evaluation of complex expressions the negation operator has the highest precedence. Thus ~A&B = (~A)&B, and ~AvB = (~A)vB.

Several different but mathematically equivalent forms of the axioms can be given[1]. The particular choice used here is influenced by the work of R.T.Cox[2].

$$0 \le P(A|B) \le 1 \tag{1}$$
$$P(A|A) = 1 \tag{2}$$
$$P(\sim A|B) = 1 - P(A|B) \tag{3}$$
$$P(A\&B|C) = P(B|A\&C) P(A|C) \tag{4}$$

Axiom (2) states the convention that the probability of a certainty (the occurrence of A given the occurrence of A) is one, and (1) says that no probabilities are greater than the probability of a certainty. Axiom (3) expresses the intuitive notion that the probability of non-occurrence of an event increases as the probability of its occurrence decreases. It also implies P(~A|A)=0, that is to say, an impossible event (the non-occurrence of A given that A occurs) has zero probability. Axiom (4) states that the probability that two events both occur (under some condition C) is equal to the probability of occurrence of one of the events multiplied by the probability of the second event given that the first event has occurred. (In the work of Cox[1] these quantitative axioms are derived from more fundamental qualitative postulates.)

The probabilities of negation (~A) and conjunction (A&B) of events each required an axiom. However, no further axioms are required to treat disjunction because AvB = ~(~A&~B); in word, "A or B" is equivalent to the negation of "neither A nor B". Thus from (3) we have
$$P(AvB|C) = 1 - P(\sim A\&\sim B|C) \tag{5}$$
which can be evaluated from the existing axioms to yield (see Appendix A)
$$P(AvB|C) = P(A|C) + P(B|C) - P(A\&B|C). \tag{6}$$

If P(A&B|C)=0 we say that A and B are mutually exclusive on condition C. Then (6) reduces to the rule of *addition of probabilities for mutually exclusive events*,

$$P(A \vee B | C) = P(A|C) + P(B|C). \tag{7}$$

This is often used as an axiom instead of (3). Indeed it is possible to derive (3) from (7) (see Appendix B). We thus have two equivalent sets of axioms: (1)-(4), or (1),(2),(4) and (7). The former set of axioms is more elegant because it applies to all events, whereas (7) applies only if A and B are mutually exclusive. Nevertheless the latter set is commonly used, and it has advantages in certain situations.

A very important concept in probability and its applications is that of *independence* of events. If $P(B|A\&C) = P(B|C)$, that is to say, if the occurrence of A has no influence on the probability of B, then we say that B is independent of A (under condition C). From (4) we then obtain

$$P(A\&B|C) = P(B|C) P(A|C). \tag{8}$$

The symmetry for this formula implies that *independence* is a mutual relationship; if B is independent of A then also A is independent of B. This notion is called *statistical* or *stochastic independence* in order to distinguish it from other notions such as causal independence.

3. Interpretation of Probability Concepts

The abstract probability theory, consisting of axioms, definitions, and theorems, must be supplemented by an *interpretation* of the term "probability". This provides the correspondence rule by means of which the abstract theory can be applied to practical problems. There are many different interpretations of probability because anything that satisfies the axioms may be regarded as a kind of probability.

One of the oldest interpretations is the *limit frequency* interpretation. If the conditioning event C can lead to either A or ~A, and if in n repetitions of such a situation the event A occurs m times, then it is asserted that $P(A|C) = \lim_{n \to \infty}(m/n)$. This provides not only an interpretation, but also a definition of probability in terms of a numerical frequency ratio. The axioms of the abstract theory can be derived as theorems of the frequency theory. In spite of its superficial appeal, the limit frequency interpretation has been widely discarded, primarily because there is no assurance that the above limit really exists for the actual sequences to which one wishes to apply probability theory.

The defects of the limit frequency interpretation are avoided without losing its attractive feature (close contact with observable data) in the *propensity* interpretation[3]. The probability $P(A|C)$ is interpreted as a measure of the tendency, or propensity, of the physical conditions described by C to produce the result A. It differs mathematically from the older limit frequence theory in that "probability" remains a fundamental undefined term, and is not redefined or derived from anything more fundamental. However its relationship to frequency emerges, suitably qualified, in a theorem (the *law of large numbers*). It differs conceptually from the frequency theory in viewing probability (propensity) as a characteristic of the physical situation C that may potentially give rise to a sequence of events, rather that a property (frequency) of an actual sequence of events. This fact is emphasized by always writing probability in the conditional form $P(A|C)$, and never merely as P(A). The propensity interpretation is particularly well suited for application to quantum mechanics.

In addition to the probability P, one must choose an interpretation of the arguments A and C of the function P(A|C). So far we have spoken of them as being *events*, but one can also treat them as *propositions*. In many cases the difference between the two interpretations is merely verbal. Corresponding to the event A is the proposition "event A has occurred". But one can usefully consider propositions that do not correspond to events. For example one can consider the probability (conditional on specified experimental evidence) that the electronic charge is within one part in a thousand of its conventional published value. This is the point of view that is used in inductive inference, and we shall have some occasion to apply it.

In spite of the seemingly greater generality of the interpretation of the arguments as propositions, it is often more convenient in quantum mechanics to interpret them as events. For example, let A be the proposition "the position of the particle lies between q_1 and q_2". Let B be the proposition "the momentum of the particle lies between p_1 and p_2". Quantum mechanics provides a means of computing the probability that A is true, and also for computing the probability that B is true. But it does not provide any formula for the probability that the compound proposition A&B, (A and B), is true. Whether or not it is a defect of the present formulation of quantum mechanics, this limitation can be reasonably accommodated within the event interpretation. In computing the probability P(A|C) of an event A, one must specify all the physical conditions C which are relevant. This may reasonably be held to include the configuration of any measuring apparatus, since it can influence the outcome of an event. Since different apparatuses are used to measure position than to measure momentum, one will be dealing with $P(A|C_q)$ and $P(B|C_p)$, where C_q includes the configuration of the position measuring device and C_p includes the configuration of the momentum measuring device. But one has no occasion to consider events A (detection of position within a certain range) and B (detection of momentum within a certain range) under a common condition C, and so one does not need to compute P(A&B|C) in this case.

4. Probability and Frequency

Although no direct connection between frequency and probability is postulated in the propensity interpretation, a close connection emerges through a theorem known as the *law of large numbers*. The derivation of the simplest form of this theorem is outlined below.

Let X be some quantity which, under some condition C, may take on a range of non-negative values, with $P(x<X<x+dx|C) = g(x)\,dx$. Then for any $\epsilon > 0$ the probabilistic average of X (denoted $\langle X \rangle$) satisfies

$$\langle X \rangle = \int_0^\infty g(x)\, x\, dx \geq \int_\epsilon^\infty g(x)\, x\, dx \geq \epsilon \int_\epsilon^\infty g(x)\, dx = \epsilon\, P(X \geq \epsilon | C).$$

Thus we have $P(X \geq \epsilon | C) \leq \langle X \rangle / \epsilon$, which is known as *Chebyshev's inequality*. We may apply this inequality to non-negative variable $|X - \langle X \rangle|^2$, instead of X,

obtaining
$$P(|X-\langle X\rangle|\geq \epsilon|C) \leq \langle |X-\langle X\rangle|^2\rangle/\epsilon^2 \tag{9}$$

Now let us consider an experiment E which may have outcome A, with probability $P(A|E)=p$. In a sequence of n identical repetitions of E (denoted E^n) the event A may occur m times, ($0\leq m\leq n$). We refer to $f=m/n$ as the frequency of outcome A in a realization of the experimental sequence E^n. One expects, intuitively, that f should be close to p as n becomes large. The following theorem justifies and makes more precise this intuitive expectation. Let K_i have the value 1 if the outcome of the i'th repetition of E is A, otherwise K_i is zero. The frequency of A is given by $f = \sum_{i=1}^{n} K_i/n$.

It is not difficult to show that $\langle f\rangle=p$. Substitution of f for X in (9) then yields

$$P(|f-p|\geq\epsilon|E^n) \leq \langle [\sum_{i=1}^{n}(K_i-\langle K_i\rangle)]^2\rangle/(n\epsilon)^2$$

$$= \sum_i\sum_j \langle(K_i-p)(K_j-p)\rangle/(n\epsilon)^2 \tag{10}$$

Since the various repetitions of E are independent, we have
$\langle(K_i-p)(K_j-p)\rangle = \langle(K_i-p)\rangle\langle(K_j-p)\rangle$ for $i\neq j$. Thus (10) becomes
$$P(|f-p|\geq\epsilon|E^n) \leq \langle(K_i-p)^2\rangle/n\epsilon^2, \tag{11}$$
the average on the righthand side being independent of i. This result, which is an instance of the *law of large numbers*, asserts that the probability of the frequency of A being more than ϵ away from p converges to zero as n becomes infinite. From the conceptual point of view, this is the most important theorem in probability theory, establishing the connection between abstract probabilities and frequencies in observable data.

It should be noted, in passing, that the full proof of (11) uses axiom (4) only to the extent that it is needed to derive the addition rule (7). So if, as is sometimes done, we were to choose (1),(2),(4) and (7) as axioms, instead of (1)-(4), we could say that axiom (4) was not needed to derive the law of large numbers.

5. Probability in Quantum Mechanics

In quantum mechanics a dynamical variable R is represented by a self-adjoint operator R, whose eigenvalues are the possible values of R.

$$R|r_n\rangle = r_n|r_n\rangle \tag{12}$$

According to a standard postulate of quantum mechanics, the probability of obtaining the particular value $R=r_n$ is given by

$$P(R=r_n|\Psi) = |\langle r_n|\Psi\rangle|^2, \tag{13}$$

in the simplest case of a discrete non-degenerate eigenvalue spectrum and a pure state represented by the vector Ψ. (All vectors here are assumed to have unit norms.) But it is not sufficient to merely assert that certain mathematical expressions are probabilities unless it can be shown that they satisfy the mathematical theory of probability. In particular, we must verify that such expressions obey the four axioms of probability theory and, in appropriate circumstances, the independence property (8).

The expression on the lefthand side of (13) can be read as "*the probability that the dynamical variable R has the value r_n, conditional on Ψ*". The latter portion of this statement requires some comment because the state vector Ψ is not a physical object. Its significance is twofold. First, it is an abstract mathematical object from which the probability distributions of observable quantities can be calculated. Second, to assert that the state vector is Ψ can be regarded as implying that the system has undergone a corresponding state preparation procedure, which could be described in more detail but all of the relevant information is contained in the specification of Ψ.

It is clear that (13) satisfies axiom (1), this being a direct consequence of the Schwartz inequality. If the state vector is the eigenvector $|\Psi\rangle = |r_n\rangle$, then (13) becomes $P(R=r_n|r_n)=1$, which is the equivalent of axiom (2). Axiom (3) follows from the more general additivity rule (7). The events described by $R=r_1$, $R=r_2$, etc. are mutually exclusive, and the additivity rule, $P\{(R=r_1)\vee(R=r_2)|\Psi\} = |\langle r_1|\Psi\rangle|^2 + |\langle r_2|\Psi\rangle|^2$, holds almost by definition. We shall defer consideration of axiom (4) because it can be better treated by a more general formalism. If the system consists of two independent non-interacting components, which are only formally regarded as a single system, then the state vector can be written in the form $|\Psi\rangle = |\psi\rangle^{(a)} \otimes |\psi\rangle^{(b)}$, where the superscripts refer to the two components. The joint probability distribution for the dynamical variables $R^{(a)}$ and $R^{(b)}$, belonging to components a and b, respectively, is

$$P(R^{(a)}=r_m \& R^{(b)}=r_n | \Psi) = |\langle r_m|\psi\rangle^{(a)}|^2 \; |\langle r_n|\psi\rangle^{(b)}|^2, \tag{14}$$

in agreement with (8). We have now verified that the quantum mechanical postulate (13) satisfies all the ingredients of probability theory that are needed to derive the *law of large numbers*. Indeed if we interpreted the components a and b above as systems in the sequence of measurements E^n, described in Sec. 4, we could recapitulate the derivation of the law of large numbers in the language of quantum mechanics.

The most general state description in quantum mechanics is by means of the *state operator* ρ, which has unit trace, is self-adjoint, and is non-negative definite:

$$\text{Tr } \rho = 1, \tag{15}$$
$$\rho = \rho^\dagger, \tag{16}$$
$$\langle u|\rho|u\rangle \geq 0 \quad \text{for all vectors u.} \tag{17}$$

In the special case of a pure state, represented above by the vector $|\Psi\rangle$, the state operator is $\rho=|\Psi\rangle\langle\Psi|$. Associated with the any dynamical variable R is a family of projection operators $M_R(\Delta)$ which are related to the eigenvalues and eigenvectors of (12) as follows:

$$M_R(\Delta) = \sum_{r_n \in \Delta} |r_n\rangle\langle r_n| \qquad (18)$$

where the sum is over all eigenvectors (possibly degenerate) whose eigenvalues lie in the subset Δ. The probability that the value of R will lie within Δ is postulated to be

$$P(R \in \Delta | \rho) = \text{Tr}\{\rho\, M_R(\Delta)\}. \qquad (19)$$

It is easily verified that the general form (19) reduces to (13) in the appropriate special case.

We must now verify that (19) satisfies the axioms of probability theory. Axiom (1) follows directly from (15) and the fact that $M_R(\Delta)$ is a projection operator. The analogue of (2) is obtained if we chose a state prepared in such a manner that the value of R is guaranteed to lie within Δ. This will be so for those states which satisfy $\rho = M_R(\Delta)\, \rho\, M_R(\Delta)$, for which (19) is identically equal to 1. Axiom (3) follows from the additivity rule (7). To verify it we consider two disjoint sets, Δ_1 and Δ_2, the union of which is denoted $\Delta_1 \cup \Delta_2$. Now $(R \in \Delta_1) \vee (R \in \Delta_2)$ is equivalent to $R \in (\Delta_1 \cup \Delta_2)$. Since the two sets, Δ_1 and Δ_2, are disjoint it follows that $M_R(\Delta_1)\, M_R(\Delta_2) = 0$, and the projection operator corresponding to the union of the sets is just the sum of the separate projection operators, $M_R(\Delta_1 \cup \Delta_2) = M_R(\Delta_1) + M_R(\Delta_2)$. Hence it is clear that (7) is satisfied. The factorization property (8) follows, as did (14), from the factorization of the state function for two independent uncorrelated systems, $\rho = \rho^{(a)} \otimes \rho^{(b)}$. We have now verified that the general statistical postulate (19) satisfies all of those parts of probability theory that are needed to derive the key theorem, the law of large numbers.

The remaining axiom (4) is not essential for most of the applications of probability in quantum mechanics (provided that (7) replaces (3) as an axiom), but it must be considered for completeness. Let R and S be two dynamical variables, represented by the operators R and S whose eigenvalues and eigenvectors are given by $R|r_n\rangle = r_n |r_n\rangle$ and $S|s_n\rangle = s_n |s_n\rangle$. The corresponding projection operators are denoted $M_R(\Delta_a)$ and $M_S(\Delta_b)$. Finally, let A denote the event of R taking on a value within the set Δ_a, and let B denote the event of S taking on a value within the set Δ_b. We must now evaluate each of the three probabilities in (4) with the conditional event C being the preparation of a general state represented by ρ.

The joint probability on the left-hand side of (4), $P(A \& B | \rho)$, can be evaluated from the formalism of quantum mechanics only if the operators R and S are commutative (the corresponding projection operators M_R and M_S then also being commutative). In that case the product $M_R(\Delta_a) M_S(\Delta_b)$ is also a projection operator, and the desired joint probability is given by (19) to be $P(A \& B | \rho) = \text{Tr}\{\rho\, M_R(\Delta_a) M_S(\Delta_b)\}$. But there is no accepted formula in quantum mechanics for a joint probability distribution for dynamical variables whose operators do not commute.

On the righthand side of (4), the second factor $P(A|\rho)$ is given directly by

(19) with $\Delta = \Delta_a$. However the first factor $P(B|A\&\rho)$ requires careful interpretation. The second argument of the probability function, which we have called "the conditional event", must describe the actual physical conditions to which the probability (or "propensity") refers. It does *not* denote mere subjective information or personal belief (as would be the case in a subjective interpretation of probability). Therefore, just as ρ signifies that the system has undergone a certain state preparation, so $A\&\rho$ implies that it has been subjected to additional filtering interactions that ensure the value of R lies within Δ_a. In principle one should analyze the dynamics of this process in detail in order to compute the resulting state function. But if this filtering process does nothing but remove unacceptable values of R then it is reasonable to represent its result by the projected and renormalized state operator,

$$\rho' = M_R(\Delta_a)\rho M_R(\Delta_a)/Tr\{M_R(\Delta_a)\rho M_R(\Delta_a)\}.$$

One then obtains the following result for the righthand side of (4):

$$P(B|A\&\rho) P(A|\rho) = Tr\{\rho' M_S(\Delta_b)\} Tr\{\rho M_R(\Delta_a)\}$$
$$= Tr\{M_R(\Delta_a)\rho M_R(\Delta_a) M_S(\Delta_b)\}$$

If M_R commutes with M_S, then this expression further reduces to

$$P(B|A\&\rho) P(A|\rho) = Tr\{\rho M_R(\Delta_a) M_S(\Delta_b)\}$$
$$= P(A\&B|\rho),$$

in agreement with axiom (4). Thus we see that this last axiom of probability theory is obeyed by the formalism of quantum mechanics provided the probability of the joint event A&B is defined in the formalism. The restriction, in this case, to dynamical variables whose operators commute is not a restriction on the applicability of "classical" probability theory to quantum mechanics. It is rather a limitation of the formalism of quantum mechanics, in that it does not assign meaning to the conjunction of arbitrary events. (See the discussion at the end of Sec. 3.)

6. Erroneous Applications of Probability Theory in QM

6.1 The Double Slit

This example has been repeated many times in slightly differing versions, the first of which may be that due to Feynman[4]. The experiment consists of a particle source, a screen with two slits (labeled #1 and #2), and a detector. By moving the detector and measuring the particle count rate at various positions, one can measure the probability of a particle passing through the slit system and arriving at the point X. If only slit #1 is open the probability of detection at X is $P_1(X)$. If only slit #2 is open the probability of detection at X is $P_2(X)$. If both slits are open the probability of detection is $P_{12}(X)$. Now passage through slit #1 and passage through slit #2 are certainly exclusive events, so one might expect, from (7), that $P_{12}(X)$ should be equal to $P_1(X) + P_2(X)$. But experiment clearly shows that this is not true, hence it might be concluded that the rule (7) of probability theory does not hold in quantum mechanics.

In fact the above argument draws its radical conclusion from an incorrect application of probability theory. One is well advised to beware of probability statements expressed in the form P(X) instead of P(X|C). The second argument may be safely omitted only if the conditional event or information is clear from the context, and is *constant* throughout the problem. This is not the case in the double slit example. The probability of detection at X in the first case (only slit #1 open) should be written as $P(X|C_1)$, where the conditional information C_1 includes (at least) the state function Ψ_1 for the particle beam and the screen state S_1 (only slit #1 open). In the second case (only slit #2 open) the probability should be written as $P(X|C_2)$, where C_2 includes the state function Ψ_2 for the particle beam and the screen state S_2 (only slit #2 open). In the third case (both slits open) the probability is of the form $P(X|C_3)$, where C_3 includes the state function Ψ_{12} (approximately equal to $\Psi_1+\Psi_2$, but this fact plays no role in our argument) and the screen state S_3 (both slits open). We observe from experiment that $P(X|C_3) \neq P(X|C_1) + P(X|C_2)$. This fact, however, has no bearing on the validity of rule (7) of probability theory. Essentially this counter argument to Feynman was given by Koopman[5].

6.2 The Superposition Fallacy

The following argument is taken from sec. 2.2 of the textbook by Trigg[6], although other versions of it exist.

"Classical probabilities are compounded according to the relation

$$P(B'|A') = \sum_{C'} P(B'|C') P(C'|A') \qquad (20?)$$

where the summation is over all members C' of a set of non-overlapping states connecting A' and B'."

It is then noted that in quantum mechanics this relation is satisfied by amplitudes,

$$\langle B'|A'\rangle = \sum_{C'} \langle B'|C'\rangle\langle C'|A'\rangle.$$

Since the probabilities are the squares of the amplitudes,

$$P(B'|A') = |\langle B'|A'\rangle|^2,$$

it follows that the equation (20?) of the "classical theory" can hold only if the quantum interference terms are negligible.

There is no doubt that (20?) fails to hold as written, but we must examine more closely its status with respect to probability theory. We may presume from the context that A',B', and C' denote events which can be characterized by unique values for certain corresponding dynamical variables, and moreover that the set of possible values of C', say, is a mutually exclusive and exhaustive set. That is to say, no more than one such value can occur at a time (exclusive), and there are no other possible outcomes in the relevant class of events than one of the values from this set (exhaustive). Put yet another way, if the set $\{C_1,C_2,C_3,\cdots\}$ contains all possible values of C' then the disjunction of all those possibilities, $C_1 \vee C_2 \vee C_3 \vee \cdots$, is a certainty. In attempting to derive (20?) we make use of (4) to obtain P(B'&C'|A') = P(B'|C'&A') P(C'|A'), and then sum over all possible values of C'. Since each of $B'\&C_1$, $B'\&C_2$, $\cdots$ are exclusive, it follows from (7) that

$$\Sigma_{C'} P(B'\&C'|A') = P\{B'\&(C_1 \vee C_2 \vee C_3 \vee \cdots)|A'\} = P(B'|A').$$

Therefore the correct deduction from "classical" probability theory is

$$P(B'|A') = \Sigma_{C'} P(B'|C'\&A') P(C'|A'), \tag{21}$$

rather than the questionable (20?). It is now apparent that the quantum mechanical superposition principle for amplitudes is in no way incompatible with the formalism of probability theory, and that the contrary claim was based on an incorrect application of probability theory. The error in this example is very similar to that in 6.1. In the former case the conditional argument of the probability function was omitted, leading to an erroneous conclusion. In this case only a part of the relevant conditional information was included by writing $P(B'|C')$ instead of $P(B'|C'\&A')$ in (20?). That would be permissible only if it could be shown that the additional information was not relevant, which is evidently not the case.

6.3 The Reciprocity Fallacy

This example is taken from sec. 2.3 of the same textbook by Trigg[6].

> "If the times involved in the specification of the two states are the same, the probability $P(B'|A')$ *[here I alter Trigg's notation slightly in order to conform to that used in this paper]*, is actually the probability that the system satisfy two conditions simultaneously. This cannot depend on the order in which the conditions are stated, so we require
> $$P(B'|A') = P(A'|B')." \tag{22?}$$

A probability theorist will immediately recognize that the author of the above quotation has confused the *conditional* probability $P(B'|A')$ with a *joint* probability. "The probability that the system satisfy two conditions simultaneously" (under some unspecified prior condition C), should be denoted as $P(A'\&B'|C)$, which has nothing to do with (22?).

The spurious equation (22?) draws its superficial plausibility from the amplitude relation, $P(B'|A') = |\langle B'|A'\rangle|^2$, which appears to support (22?), though not, of course, the specious argument that preceded it. But even that apparent connection is misleading.

To see more clearly the subtlety which is involved, let us rewrite the relation between probability and amplitude in a more explicit form,

$$P(R=r_n|\Psi) = |\langle r_n|\Psi\rangle|^2. \tag{13}$$

This is interpreted as the probability that the dynamical variable R has the value r_n, conditional on the state being Ψ. But what about the inverse probability, $P(\Psi|R=r_n)$, which is the probability that the state was Ψ on the condition that R was found to have the value r_n? Does it have the same value (13)?

One can easily show, by means of a simple example, that the inverse probability does not have the value (13). Suppose that space is divided into cells and that R is a discrete position variable. If Ψ is chosen to be a wavefunction localized within the n'th cell, then one will have $P(R=r_n|\Psi)=1$. From a knowledge of Ψ one can predict with certainty that the particle lies in the n'th cell. But suppose, on the other hand, that Ψ is unknown and one has only the the single measurement result $R=r_n$. Then all that one can infer about Ψ is that it must have been non-zero in the n'th cell. There is no assurance that the (definite but unknown) state preparation procedure which led to the state Ψ would, if repeated, yield the same value for R, and so there is no reason to believe Ψ to be localized. Therefore $P(\Psi|R=r_n)$ will definitely be less than one.

The relation between the direct and inverse probabilities can be deduced by observing that since the lefthand side of (4) is symmetrical in A and B, so must be the righthand side. Hence we obtain

$$P(A|B\&C) = P(B|A\&C) P(A|C)/P(B|C), \qquad (23)$$

which is known as *Bayes theorem*. Applied to the above example, it yields

$$P(\Psi|R=r_n\&C) = P(R=r_n|\Psi\&C) P(\Psi|C)/P(R=r_n|C). \qquad (24)$$

Here C denotes any other relevant prior information about Ψ, and it may be ignored if there is none. At this point we have left the domain of events and the propensity interpretation of probability. This is so because the occurrence of the state Ψ is not an event which can be causally influenced by the subsequent determination that $R=r_n$. We are instead engaging in inductive inference, attempting to infer what the state might have been, on the basis of information which is not adequate to determine it uniquely. The value of the first factor on the righthand side of (24) is simply $P(R=r_n|\Psi\&C) = P(R=r_n|\Psi) = |<r_n|\Psi>|^2$, since no further information C about Ψ is relevant if Ψ is given. The last factor, $P(R=r_n|C)$, is called the *prior* probability that $R=r_n$. It might be a uniform distribution over the portion of space that can possibly be occupied by the particle. The second factor, $P(\Psi|C)$, is the *prior* probability, conditional on whatever information C may be available, that the state should be Ψ. It is rather difficult to evaluate, and must be regarded as only a degree of reasonable belief. It should now be apparent that we are very unlikely to obtain $P(\Psi|R=r_n) = P(R=r_n|\Psi)$, contrary to the spurious equation (22?).

7. Conclusions

In this paper I have briefly reviewed axiomatic probability theory and its interpretation, with emphasis on those aspects that are most relevant to quantum mechanics. By demonstrating explicitly that the axioms of probability theory are satisfied by the formalism of quantum mechanics, I have refuted any and all claims that "classical" probability theory is not valid in quantum mechanics. The only anomaly is the fact that joint probability distributions for two or more dynamical variables are not conventionally defined unless the corresponding operators are commutative. But quantum mechanics can hardly be said to contradict (literally, "speak against"), probability theory on this point, since the accepted formalism of quantum mechanics is simply silent here.

Some examples of erroneous applications of probability theory in quantum mechanics have been exposed and analyzed. In view of the fact that these errors were committed by well educated physicists, one is led to the conclusion that probability theory needs greater emphasis in our curriculum. This is especially so in relation to quantum mechanics, where probability enters at a fundamental level. I hope that this paper will be helpful in achieving this goal.

Appendix A. – Derivation of (6), (based on sec. 5 of Cox[2])

To evaluate (5) by means of axioms (1)-(4) we require a lemma:
$$P(X\&Y|C) + P(X\&\sim Y|C) = P(X|C)P(Y|X\&C) + P(X|C)P(\sim Y|X\&C)$$
$$= P(X|C) \{P(Y|X\&C) + P(\sim Y|X\&C)\}$$
$$= P(X|C). \tag{A.1}$$
Here we have used (4) and (3). Using (A.1) with $X=\sim A$ and $Y=\sim B$, we obtain
$$P(\sim A\&\sim B|C) = P(\sim A|C) - P(\sim A\&B|C).$$
In the first term we use now (3), and in the second term we use (A.1) with $X=B$, $Y=A$. This yields
$$P(\sim A\&\sim B|C) = 1 - P(A|C) - \{P(B|C) - P(B\&A|C)\}.$$
Upon substitution of this result into (5) we obtain (6).

Appendix B. – Derivation of axiom (3) from (7)

By noting that A and ~A are mutually exclusive, and substituting $B=\sim A$ in (7), we obtain $P(A\vee\sim A|C) = P(A|C) + P(\sim A|C)$. Now $A\vee\sim A$, ("A or not A"), is intuitively a certainty, and if we set its probability equal to 1 then we immediately obtain (3). The only gap in this proof lies in the fact that *certainty* is defined by axiom (2), and we should relate our intuitive notion of certainty to that definition. A formal proof that $P(A\vee\sim A|C)=1$ is to be found on p.17 of the book by Cox[2].

References

1. T.L.Fine, *Theories of Probability, and Examination of Foundations* (1973, Academic Press, New York).
2. R.T.Cox, *The Algebra of Probable Inference* (1961, Johns Hopkins Press, Baltimore).
3. K.R.Popper, "The Propensity Interpretation of Calculus of Probability, and the Quantum Theory", pp. 65-70 in *Observation and Interpretation*, edited by S.Korner (1957, Butterworths, London).
4. R.P.Feynman, "The Concept of Probability in Quantum Mechanics", pp. 533 -541 in *Proceedings of the second Berkeley Symposium on Mathematical Statistics and Probability*, (1951, University of California Press, Berkeley, California).
5. B.O.Koopman, "Quantum Theory and the Foundations of Probability", pp. 97-102 in *Applied Probability*, edited by L.A.MacColl (1955, McGraw-Hill).
6. G.L.Trigg, *Quantum Mechanics* (1964, Van Nostrand, Princeton, N.J.).

MEASUREMENTS AND AMPLITUDES

Stanley Gudder
Department of Mathematics
and Computer Science
University of Denver
Denver CO 80208

ABSTRACT. We present a framework for quantum mechanics based on the concepts of measurements and amplitudes. The measurements are represented by functions and therefore commute. The interference effects characteristic of quantum mechanics are caused by the fact that probabilities are computed using amplitude functions. Moreover, in order to consider inexact measurements we define fuzzy (or nonsharp) amplitudes. We also study measurements that satisfy a positivity condition and that are covariant relative to a symmetry group. It is shown that the framework includes stochastic quantum mechanics which in turn includes traditional Hilbert space quantum mechanics.

1. INTRODUCTION

Most formulations of quantum mechanics are based on the concepts of states and observables [1, 4, 7, 8, 9, 12]. The states correspond to preparation procedures of a physical system and observables correspond to quantities that are observed after the system is suitably prepared. For example, in the Hilbert space formulation of quantum mechanics, the (pure) states are represented by unit vectors in a Hilbert space H and the observables by self-adjoint operators (or equivalently, spectral measures) on H. It is often thought that the interference effects that are characteristic of quantum mechanics result from the fact that certain observables do not commute. Unlike classical statistical theories we now have a "noncommutative" probability theory. In the present work we consider refinements of these concepts that we call measurements and amplitudes.

In this framework, the measurements are not operator-like objects that form a noncommutative system, but are functions that always commute. The interference effects are not the result of a noncommutativity of observables but are caused by the fact that probabilities are computed using amplitude functions. This framework provides a deeper description of a physical system and gives a way of including an underlying objective reality that is entirely missed by most quantum mechanical formulations. This formulation allows us to include hidden variables, traditional Hilbert space quantum mechanics, stochastic quantum mechanics, and classical probability theory within a single framework. Our theory is based on guidelines motivated by the path integral formalism for quantum mechanics [2, 3, 11].

The guidelines are given as follows.

(1) An outcome of a measurement is the result of various interfering alternatives and each of these alternatives has an amplitude for occurring.

(2) The amplitude of an outcome is the "sum" of the amplitudes of the alternatives that result in that outcome.

(3) The probability of an outcome is the modulus squared of its amplitude.

(4) The probability of an event for a measurement is the "sum" of the probabilities of the outcomes composing it.

We now amplify on the meaning of the above guidelines. We first assume that a physical system can be in precisely one of a set of possible configurations (alternatives, potentialities) and that each configuration has a probability amplitude of occurring. When the system interacts with a measuring apparatus, an outcome results. In general, an outcome may result from many interfering configurations. By interfering we mean that the configurations can only be distinguished by disturbing the system; that is, by executing at least one different measurement. The amplitude of an outcome is found by summing (in the discrete case) or integrating (in the continuum case) the amplitudes of the configurations that result in that outcome upon executing the measurement. An event for a measurement is a collection of outcomes for that measurement. The probability of an event is computed in the usual way by summing (integrating) probabilities (probability densities) over the outcomes composing the event.

These guidelines provide three structural levels. At the first level (Guideline 1), the set of configurations represent an underlying objective physical reality. This could be called a hidden variables level. At the second level (Guidelines 2 and 3), the outcomes represent the results of physical measurements and provide the quantum probability level. Guideline 4 is the usual basis for classical probability theory and the set of events provide the classical probability level.

This paper is a continuation of our work begun in [5]. In this previous work we only considered exact measurements and sharp amplitudes. We now extend those methods to include inexact measurements and fuzzy (nonsharp) amplitudes. The resulting framework then includes stochastic quantum mechanics [10] (which in turn includes traditional Hilbert space quantum mechanics). We shall only outline the relevant theory and shall not include proofs in the present paper. Details and proofs will appear elsewhere [6].

2. MATHEMATICAL FRAMEWORK

We now present a rigorous mathematical framework based on the previous guidelines. Let Ω be a nonempty set called a *sample space* and whose elements we call *sample points*. The sample points correspond to the possible configurations of a physical system S. A *measurement* is a map F from Ω onto its range $X_F = F(\Omega)$ satisfying the following conditions.

(M1) X_F is the base space of a σ-finite measure space (X_F, Σ_F, ν_F).

(M2) For every $x \in X_F$, $F^{-1}(x)$ is the base space of a σ-finite measure space $(F^{-1}(x), \Sigma_x, \mu_x)$.

We call $F^{-1}(x)$ the *fiber over* x, the elements of X_F are called *F-outcomes* and the sets in Σ_F are called *F-events*. A measurement F corresponds to a laboratory procedure or experiment that can be performed on S. For every $\omega \in \Omega$, $F(\omega)$ denotes the outcome resulting from executing F, using a perfectly accurate measuring apparatus, when S has configuration ω. For $x \in X_F$, the fiber $F^{-1}(x)$ is the set of sample points that result in outcome x using a perfect apparatus. The measure ν_F is an a priori weight for the F-events that is independent of the state of S and the state of the measuring apparatus. In case of total ignorance, ν_F is a uniform measure such as Lebesgue measure, Haar measure, or the counting measure in the discrete case. Similarly, μ_x is an a priori weight on the fiber $F^{-1}(x)$.

Let $F: \Omega \to X_F$ be a measurement. A function $f: X_F \times \Omega \to \mathbf{C}$ is a *fuzzy amplitude density* for F (*F-fad*) if f satisfies the following conditions.

(F1) For every x', $x \in X_F$,

$$f(x', \cdot) \mid F^{-1}(x) \in L^1(F^{-1}(x), \Sigma_x, \mu_x).$$

(F2) As a function of x

$$f_{x'}(x) \equiv \int_{F^{-1}(x)} f(x', \omega) d\mu_x(\omega) \in L^1(X_F, \Sigma_F, \nu_F).$$

(F3) Defining $H_F = L^2(X_F, \Sigma_F, \nu_F)$ we have

$$F(f)(x') \equiv \int_{X_F} f_{x'}(x) d\nu_F(x) \in H_F.$$

(F4) $\|F(f)\|_{H_F} = 1$.

An F-fad f represents a state for the combined system $S + M$ where M is a measuring apparatus for F. Since a perfect (sharp point) measuring apparatus is usually impossible to achieve, we assume that M has an intrinsic inaccuracy which is included in the description of f and is responsible for its fuzzyness. If another measuring apparatus M' is used to execute F, then the corresponding F-fad f' represents a state for $S + M'$ and in general f' would be different than f. From another point of view, we may think of M as a probe of S (for example, a scattering experiment) involving a quantum test particle α. In this case, f would contain a proper wave function or exciton state of α [10]. We denote the set of F-fads by fad(F).

For $f \in$ fad(F) we interpret $f(x', \omega)$ as the amplitude density that S has configuration ω and a subsequent execution of F using a fixed measuring apparatus M results in outcome x'. Notice that the result using a perfect apparatus would be the outcome $F(\omega)$ which, in general, differs from x'. Applying Guideline 2 we conclude that $F_{x'}(x)$ is the amplitude density that the result of measurement F

using M is x' when the result using a perfect apparatus is x. Applying Guideline 2 again, $F(f)(x')$ gives the amplitude density of the outcome x' when F is executed using M. We call $F(f)$ the (F, f) *wave function*. Applying Guideline 3 we interpret $|F(f)(x')|^2$ as the probability density for F at the outcome x'. Conditon F4 is motivated by Guideline 4 and ensures that $|F(f)(x')|^2$ is indeed a probability density.

We now give a general method for computing probabilities of events. Let $F: \Omega \to X_F$ be a measurement and let $f \in \text{fad}(F)$. A set $A \subseteq \Omega$ is a *generalized (F, f) event* if it satisfies the following conditions.

(E1) For every $x \in X_F$, $A \cap F^{-1}(x) \in \Sigma_x$.

(E2) As a function of x
$$f_F(A)(x', x) \equiv \int_{A \cap F^{-1}(x)} f(x', \omega) d\mu_x(\omega) \in L^1(X_F, \Sigma_F, \nu_F).$$

(E3) $f_F(A)(x') \equiv \int_{X_F} f_F(A)(x', x) d\nu_F(x) \in H_F$.

We denote the set of generalized (F, f) events by $\mathcal{E}(F, f)$. Notice that $\emptyset, \Omega \in \mathcal{E}(F, f)$ and $f_F(\emptyset) = 0$, $f_F(\Omega) = F(f)$. The elements of $\mathcal{E}(F, f)$ are the subsets of Ω for which a reasonable amplitude can be defined. In fact, $f_F(A)(x)$ is the "sum" of the amplitudes over the configurations in A that result in x upon execution of F. We interpret $f_F(A)(x)$ as the amplitude of A given that x occurs. Since $F(f)(x)$ is the amplitude that x occurs, we interpret $F^*(f)(x) f_F(A)(x)$ as the "probability" of A *and* x (where $*$ denotes the complex conjugate). Notice that this is a generalization of Guideline 3 which states that

$$|F(f)(x)|^2 = F^*(f)(x) F(f)(x) = F^*(f)(x) f_F(\Omega)(x)$$

is the probability that x ocurs. Motivated by these considerations, we define the (F, f) *pseudo-probability* of A as

$$P_{F,f}(A) = \int_{X_F} f_F(A) F^*(f) d\nu_F = \langle f_F(A), F(f) \rangle.$$

In its present generality, $P_{F,f}(A)$ may be complex so it cannot always be interpreted as a probability. However, we shall later specialize to measurements for which $P_{F,f}(A) \geq 0$.

We can extend pseudo-probabilities to pseudo-expectations in a natural way. We call a function $h: \Omega \to \mathbf{R}$ an $L^2(F, f)$ function if h satisfies the following conditions.

(L1) Defining $hf: X_F \times \Omega \to \mathbf{C}$ by $(hf)(x, \omega) = h(\omega) f(x, \omega)$, for every $x', x \in X_F$,
$$hf(x', \cdot) \mid F^{-1}(x) \in L^1(F^{-1}(x), \Sigma_x, \mu_x).$$

(L2) As a function of x
$$f_F(h)(x', x) \equiv \int_{F^{-1}(x)} (hf)(x', \omega) d\mu_x(\omega) \in L^1(X_F, \Sigma_F, \nu_F).$$

(L3) $f_F(h)(x') \equiv \int_{X_F} f_F(h)(x', x) d\nu_F \in H_F$.

For $h \in L^2(F, f)$, the (F, f) *pseudo-expectation* of h is

$$E_{F,f}(h) = \langle f_F(h), F(f) \rangle.$$

Notice that for $A \in \mathcal{E}(F, f)$, if χ_A denotes the characteristic function of A, we have $\chi_A \in L^2(F, f)$ and $E_{F,f}(\chi_A) = P_{F,f}(A)$. It follows that $E_{F,f}$ is the unique linear extension of $P_{F,f}$ to $L^2(F, f)$.

In the present generality, $\mathcal{E}(F, f)$ need not be a σ-algebra and $P_{F,f}$ need not be countably additive. However, weaker regularity conditions hold. A nonempty collection $\mathcal{S}$ of subsets of Ω is an *additive class* if $\mathcal{S}$ is closed under the formation of complements and finite disjoint unions.

Theorem 1. If $F: \Omega \to X_F$ is a measurement and $f \in \text{fad}(F)$, then $\mathcal{E}(F, f)$ is an additive class and $P_{F,f}$ is an additive complex-valued set function on $\mathcal{E}(F, f)$ with $P_{F,f}(\Omega) = 1$.

3. REGULAR AND COVARIANT MEASUREMENTS

We now specialize the treatment of Section 2 to measurements with certain regularity properties. If $F: \Omega \to X_F$ is a measurement, then the wave function map $f \mapsto F(f)$ is a transformation from $\text{fad}(F)$ into the unit sphere S_F of H_F. If there exists a set $\mathcal{F} \subseteq \text{fad}(F)$ such that $F: \mathcal{F} \to S_F$ is bijective, the F is $\mathcal{F}$-*regular*. For an $\mathcal{F}$-regular F we use the notation

$$\mathcal{E}(F, f) = \cap \{\mathcal{E}(F, \mathcal{F}): f \in \mathcal{F}\}.$$

If $\mathcal{B} \subseteq \mathcal{E}(F, \mathcal{F})$ is a σ-algebra and $P_{F,f}(A) \geq 0$ for every $A \in \mathcal{B}$ and $f \in \mathcal{F}$, then F is $(\mathcal{B}, \mathcal{F})$ *positive*. If F is $(\mathcal{B}, \mathcal{F})$ positive, it follows from the additivity of $P_{F,f}$ that

$$0 \leq P_{F,f}(A) \leq P_{F,f}(\Omega) = 1$$

for every $A \in \mathcal{B}$ and $f \in \mathcal{F}$. Hence, $P_{F,f}(A)$ can be interpreted as a probability and $P_{F,f}$ is a finitely additive probability measure on $\mathcal{B}$.

A function $f: X_F \times \Omega \to \mathbb{C}$ is a *fuzzy amplitude function* for F (F-*faf*) if f satisfies conditions (F1), (F2) and (F3). Thus, an F-fad is a normalized F-faf. Denoting the set of F-fafs by $\text{faf}(F)$ we see that $\text{faf}(F)$ is a complex linear space. If F is $\mathcal{F}$-regular, then the map $F: \mathcal{F} \to S_F$ extends to a linear bijection from the linear hull $\bar{\mathcal{F}} \subseteq \text{faf}(F)$ of $\mathcal{F}$ onto H_F which we also denote by F. For $f \in \bar{\mathcal{F}}$, $A \in \mathcal{E}(F, \mathcal{F})$, it is easy to check that A satisfies (E1), (E2) and (E3) so $f_F(A) \in H_F$ is well-defined. For $A \in \mathcal{E}(F, \mathcal{F})$, define $\hat{F}(A): H_F \to H_F$ by $\hat{F}(A)\psi = (F^{-1}\psi)_F(A)$. Then $\hat{F}(A)$ becomes a linear operator satisfying $\hat{F}(A)F(f) = f_F(A)$ for all $f \in \mathcal{F}$. Hence, for all $f \in \mathcal{F}$ we have

$$P_{F,f}(A) = \langle f_F(A), F(f) \rangle = \langle \hat{F}(A)F(f), F(f) \rangle. \tag{1}$$

Moreover, if F is $(\mathcal{B}, \mathcal{F})$ positive, then $\langle \hat{F}(A)\psi, \psi \rangle \geq 0$ for every $\psi \in H_F$, $A \in \mathcal{B}$, and we conclude that $\hat{F}(A)$ is a positive bounded linear operator on H_F. Since

$\hat{F}(\Omega) = I$, it follows that $\hat{F}$ is a finitely additive normalized POV (positive operator-valued) measure on $\mathcal{B}$.

Let G be a group of bijections from Ω onto Ω. If $F: \Omega \to X_F$ is a measurement, G is an *F-symmetry group* if there exists an irreducible, projective, unitary representation $g \mapsto U_g$ of G on H_F. We may think of $F(f) \mapsto U_g F(f)$ as the change of the wave function due to the symmetry transformation g on Ω. Moreover, if F is $\mathcal{F}$-regular, we can transfer the action of U_g to $\mathcal{F}$ by defining $\hat{U}_g(f) = F^{-1}U_g F(f)$ for every $f \in \mathcal{F}$, $g \in G$. Then $f \mapsto \hat{U}_g f$ gives the change in the fad f due to g on Ω. We say that a $(\mathcal{B}, \mathcal{F})$ positive measurement F is *covariant* with respect to an F-symmetry group G if $gA \equiv \{g\omega: \omega \in A\} \in \mathcal{B}$ for every $A \in \mathcal{B}$, $g \in G$ and

$$P_{F,f}(g^{-1}A) = P_{F,\hat{U}_g f}(A) \tag{2}$$

for every $f \in \mathcal{F}$, $A \in \mathcal{B}$, $g \in G$. Both sides of (2) give the probability of A after the symmetry transformation g. Applying (1) and (2), we have for all $f \in \mathcal{F}$, $A \in \mathcal{B}$

$$\begin{aligned}\langle \hat{F}(g^{-1}A)F(f), F(f)\rangle &= P_{F,f}(g^{-1}A) = P_{F,\hat{U}_g f}(A) \\ &= \langle \hat{F}(A)F(\hat{U}_g f), F(\hat{U}_g f)\rangle \\ &= \langle \hat{F}(A)U_g F(f), U_g F(f)\rangle \\ &= \langle U_g^* \hat{F}(A) U_g F(f), F(f)\rangle.\end{aligned}$$

It follows that

$$\hat{F}(gA) = U_g \hat{F}(A) U_g^* \tag{3}$$

for all $g \in G$, $A \in \mathcal{B}$. In a similar way, (3) implies (2) so (2) and (3) are equivalent.

We now consider an important class of $(\mathcal{B}, \mathcal{F})$ positive measurements. Let $F: \Omega \to X_F$ be a measurement and let $\{\xi_\omega : \omega \in \Omega\}$ be a set of vectors in H_F satisfying $\|\xi_\omega\| = \|\xi_{\omega'}\|$ for all $\omega, \omega' \in \Omega$. We call $\{\xi_\omega : \omega \in \Omega\}$ an *F-resolution of the identity* if for every $\psi, \phi \in H_F$, the function $(\psi, \phi)_\xi(\omega) = \langle \psi, \xi_\omega\rangle\langle\xi_\omega, \phi\rangle$ satisfies the following conditions.

(R1) $(\psi, \phi)_\xi \mid F^{-1}(x) \in L^1(F^{-1}(x), \Sigma_x, \mu_x)$ for $x \in X_F$.

(R2) $(\psi, \phi)_\xi(x) \equiv \int_{F^{-1}(x)} (\psi, \phi)_\xi(\omega) d\mu_x(\omega) \in L^1(X_F, \Sigma_F, \nu_F)$.

(R3) $\int_{X_F} (\psi, \phi)_\xi(x) d\nu_F(x) = \langle \psi, \phi\rangle$.

We thus have for every $\psi, \phi \in H_F$

$$\begin{aligned}\langle \psi, \phi\rangle &= \int_{X_F} \int_{F^{-1}(x)} \langle\psi, \xi_\omega\rangle\langle\xi_\omega, \phi\rangle d\mu_x(\omega) d\nu_F(x) \\ &\equiv \int_\Omega \langle\psi, \xi_\omega\rangle\langle\xi_\omega, \phi\rangle d\omega.\end{aligned} \tag{4}$$

For $\psi \in H_F$, define

$$(W_\xi \psi)(\omega) = \langle\psi, \xi_\omega\rangle = \tilde{\psi}(\omega)$$

and define the positive operators P_ω^ξ on H_F by

$$P_\omega^\xi(\psi) = \langle \psi, \xi_\omega \rangle \xi_\omega = \tilde{\psi}(\omega)\xi_\omega.$$

Equation (4) shows that the following Bochner integral satisfies

$$\int_\Omega P_\omega^\xi d\omega = I.$$

For $\psi \in H_F$, define $f_\psi: X_F \times \Omega \to \mathbf{C}$ by $f_\psi(x,\omega) = \xi_\omega(x)\tilde{\psi}(\omega)$ and let $\mathcal{F}_\xi = \{f_\psi: \psi \in S_F\}$. The next result shows that F is $\mathcal{F}_\xi$-regular.

Theorem 2. Let $\{\xi_\omega: \omega \in \Omega\}$ be an F-resolution of the identity. (a) For any $\psi \in S_F$, $f_\psi \in \text{fad}(F)$ and F is $\mathcal{F}_\xi$-regular. For any σ-algebra $\mathcal{B} \subseteq \mathcal{E}(F, \mathcal{F}_\xi)$, F is $(\mathcal{B}, \mathcal{F}_\xi)$ positive. (b) For any $\psi \in S_F$, $F(f_\psi) = \psi$. (c) If $A \in \mathcal{E}(F, \mathcal{F}_x)$ and $A_x = A \cap F^{-1}(x)$, then for any $\psi \in S_F$ we have

$$P_{F,f_\psi}(A) = \int_{X_F} \int_{A_x} |\tilde{\psi}(\omega)|^2 d\mu_x(\omega) d\nu_F(x)$$

and

$$\hat{F}(A) = \int_{X_F} \int_{A_x} P_\omega^\xi d\mu_x(\omega) d\nu_F(x).$$

Corollary 3. Suppose an F-resolution of the identity $\{\xi_\omega: \omega \in \Omega\}$ exists and $\mathcal{B}$ is a σ-algebra in $\mathcal{E}(F, \mathcal{F}_\xi)$. (a) Then $P_{F,f}$ is a (σ-additive) probability measure on $\mathcal{B}$ for every $f \in \mathcal{F}_\xi$ and $\hat{F}$ is a normalized (σ-additive) POV measure on $\mathcal{B}$. (b) If $h: \Omega \to \mathbf{R}$ is an $L^2(F, \mathcal{F}_\xi)$ function that is measurable with respect to $\mathcal{B}$, then for any $f = f_\psi \in \mathcal{F}_\xi$ we have

$$E_{F,f}(h) = \int_\Omega h(\omega) P_{F,f}(d\omega) = \int_\Omega h(\omega) \langle \hat{F}(d\omega)\psi, \psi \rangle.$$

We thus see that in this case, $P_{F,f}$ and $E_{F,f}$ have the desirable properties of a probability measure and an expectation. It also now follows that the present framework includes stochastic quantum mechanics [10] as a special case. For details and proofs, we refer the reader to the forthcoming paper [6].

4. REMARKS

The author would like to thank Professor T. Brody for his careful reading of this manuscript and his subsequent comments. Professor Brody inquired whether the present framework can include a non-Hilbert space formalism for quantum mechnics. We now respond to this inquiry.

Let Ω be a sample space describing the possible configurations of a physical system S. A measurement $F: \Omega \to X_F$ gives a view of the physical reality Ω as seen by the measurement F. The Hilbert space H_F describes this view in the following sense. If $f \in \text{fad}(F)$, then the wave function $F(f) \in H_F$ gives the probability distribution for the values of F for the "state" f. A different measurement G results in another view of Ω and another Hilbert space H_G. In this way we have a collection of Hilbert spaces $H_F, H_G, \ldots$, each giving a "projection" of physical reality. However, in general, no single Hilbert space completely describes the whole physical reality.

Even though the results of a single measurement are represented by a Hilbert space, the sample space Ω can be much more general and may not have a Hilbert space structure. For example, the points in Ω might be the set of trajectories for a quantum particle. If the particle evolves in accordance to a Brownian motion, then the trajectories will have unbounded variation and hence will not be square integrable. As another example, the sample points in Ω may represent a scattering experiment. In such a situation the sample points frequently describe incoming plane waves and outgoing spherical waves. Such waves are usually not square integrable and cannot be elements of a Hilbert space.

REFERENCES

1. E. Beltrametti and G. Cassinelli, *The Logic of Quantum Mechanics*, Addison-Wesley, Reading, 1981.
2. R. Feynman, "Space-time approach to non-relativistic quantum mechanics", *Rev. Mod. Phys.* 20, 367–398, 1948.
3. R. Feynman and A. Hibbs, *Quantum Mechanics and Path Integrals*, McGraw-Hill, New York, 1965.
4. S. Gudder, *Stochastic Methods in Quantum Mechanics*, North Holland, New York, 1979.
5. S. Gudder, "A theory of amplitudes", (to appear).
6. S. Gudder, "Fuzzy amplitude densities and stochastic quantum mechanics", *Found. Phys.* (to appear).
7. J. Jauch, *Foundations of Quantum Mechanics*, Addison-Wesley, Reading, 1968.
8. G. Mackey, *Mathematical Foundations of Quantum Mechanics*, Benjamin, New York, 1963.
9. C. Piron, *Foundations of Quantum Mechanics*, Benjamin, New York, 1976.
10. E. Prugovecki, *Stochastic Quantum Mechanics and Quantum Spacetime*, Reidel, Dordrecht, 1984.
11. L. Schulman, *Techniques and Aplications of Path Integration*, Wiley (Interscience), New York, 1981.
12. V. Varadarajan, *Geometry of Quantum Theory vol. I*, Van Nostrand, Princeton, 1968.

REPRESENTATIONS OF QUANTUM LOGICS AND TRANSITION PROBABILITY SPACES

Sylvia Pulmannova
Mathematics Institute
Slovak Academy of Sciences
814 73 Bratislava
Czechoslovakia

ABSTRACT. Two axiomatic approaches to quantum mechanics are compared : the quantum-logic approach and the transition-probability spaces approach. It is shown that the quantum-logic approach is more general than the transition-probability spaces approach. Necessary and sufficient conditions for the equivalence of these approaches are found. As a generalization of the above approaches, a notion of an orthogonality space is introduced. Conditions under which an orthogonality space can be represented in a generalized Hilbert space are investigated.

1. INTRODUCTION

The "quantum logic" approach is related to the classical paper of Birkhoff and von Neumann [2] . In this approach, the set of all experimentally verifiable propositions about a physical system is considered as fundamental. The well-known system of axioms introduced by Mackey [6] consist of a triple $(\mathcal{S}, \mathcal{O}, p)$ where $\mathcal{S}$ and $\mathcal{O}$ are two sets with elements representing the states and the observables of the physical system, respectively, while p is a function defined on $\mathcal{O} \times \mathcal{S} \times \mathcal{B}(\mathcal{R})$ ($\mathcal{B}(\mathcal{R})$ denoting the family of Borel subsets of the real line $\mathcal{R}$), with values in the interval $[0,1]$, and represents the probability $p(A,\alpha,E)$ that the measurement of the observable A when the system is prepared in the state α gives a result belonging to the Borel set E. Any observable A associates to each Borel set E a proposition $A(E)$, which can be interpreted as an observable giving only two possible values : 1 if the outcome of A belongs to E and 0 otherwise. Under a few easily interpretable axioms the set of all propositions possesses the algebraic structure of a σ -orthomodular poset which is called a "quantum logic". The set of all states defines an ordering on the quantum logic as follows : every proposition $A(E)$ can be considered as a function defined on the set $\mathcal{S}$ by $A(E)(\alpha) = p(A,\alpha,E)$, and the propositions are ordered by the natural ordering of functions, i.e. $A(E) \leq B(F)$ iff $A(E)(\alpha) \leq B(F)(\alpha)$ for all $\alpha \in \mathcal{S}$. On the other hand, for every $\alpha \in \mathcal{S}$, the map $A(E) \to p(A,\alpha,E)$ defines a probability measure on the quantum logic of all propositions.

The "transition probability" approach has also been initiated by von Neumann [13]. In an abstract form it has been introduced by Mielnik [11], [12]. In this approach, the set $\mathscr{S}$ of all pure states of a physical system is considered as an abstract set with a geometry prescribed by a transition probability. A transition probability is defined as a function $p: \mathscr{S} \times \mathscr{S} \to [0,1]$ which separates the states, i.e $p(\alpha,\beta) = 1$ iff $\alpha = \beta$, is symmetric, i.e. $p(\alpha,\beta) = p(\beta,\alpha)$ and complete (in the sense that

$$\sum_{\beta \in R} p(\alpha,\beta) = 1$$

for every maximal set R of pairwise orthogonal elements of $\mathscr{S}$, where α, β are said to be orthogonal if $p(\alpha,\beta) = 0$). The number $p(\alpha,\beta)$ is considered as the probability of spontaneous transition of the physical system prepared in the state α to the state β.

In the conventional Hilbert-space approach, the quantum logic is represented by the complete atomistic orthomodular lattice $\mathscr{L}(\mathscr{H})$ of all closed linear subspaces of a (complex, separable) Hilbert space $\mathscr{H}$. By the spectral theorem, observables are in one-to-one correspondence with self-adjoint operators on $\mathscr{H}$, and by the Gleason theorem, pure states on $\mathscr{L}(\mathscr{H})$ are represented by unit vectors in $\mathscr{H}$. Transition probabilities are defined by

$$p(\alpha,\beta) = |(\alpha,\beta)|^2 ,$$

where $(.,.)$ denotes the inner product in $\mathscr{H}$.

It is a natural question, under what (physically interpretable) conditions the quantum-logic approach and the transition-probability space approach lead to the conventional Hilbert-space formulation. In the present paper, we obtain a partial solution of this problem. We formulate conditions under which a representation in a generalized Hilbert space is possible. By a "generalized Hilbert space" we mean a vector space $\mathscr{V}$ over a division ring $\mathscr{D}$ with an involutive anti-automorphism θ and a hermitian form $f: \mathscr{V} \times \mathscr{V} \to \mathscr{D}$. Our theorem is an analogue of the well-known Piron [14] and MacLaren [7] representation theorems and slightly generalizes them. It is also a generalization of the result obtained in [15] for transition probability spaces.

2. QUANTUM LOGICS AND TRANSITION-PROBABILITY SPACES

Recall that a quantum logic (or equivalently a σ-orthomodular poset) is a partially ordered set $\mathscr{L}$ containing 0 and 1 and with the orthocomplementation $': \mathscr{L} \to \mathscr{L}$, which satisfies the following conditions:

(i) $(a')' = a$,
(ii) $a \leq b$ iff $b' \leq a'$,
(iii) $a \vee a' = 1$, $a \wedge a' = 0$,
(iv) if $\{a_i\} \subset \mathscr{L}$ and $a_i \leq a_j'$ for $i \neq j$, $i,j = 1,2,\ldots$, then $\bigvee_{i=1}^{\infty} a_i$ exists in $\mathscr{L}$,
(v) orthomodularity: $a \leq b$, $a' \wedge b = 0$ imply $a = b$.

We shall call $\mathscr{L}$ a complete orthomodular poset if the condition (iv) is replaced by (iv') if $\{a_i \mid i \in I\} \subset \mathcal{L}$, $a_i \leq a_j'$ for $i \neq j$, $i,j \in I$, where I is any set, then $\bigvee_{i \in I} a_i$ exists in $\mathcal{L}$.

We say that the elements a,b in $\mathcal{L}$ are orthogonal (written $a \perp b$), if $a \leq b'$.

A state on a quantum logic $\mathcal{L}$ is a map $m : \mathcal{L} \to [0,1]$ such that
(i) $m(1) = 1$
(ii) if $\{a_i\}$ is any sequence of pairwise orthogonal elements in $\mathcal{L}$, then $m(\bigvee_{i=1}^{\infty} a_i) = \sum_{i=1}^{\infty} m(a_i)$.

A set $\mathcal{M}$ of states on $\mathcal{L}$ is full if $m(a) \leq m(b)$ for all $m \in \mathcal{M}$ implies $a \leq b$ $(a,b \in \mathcal{L})$.

A set $\mathcal{M}$ of states on $\mathcal{L}$ is quite full if $\{m \in \mathcal{M} \mid m(a) = 1\} \subset \{m \in \mathcal{M} \mid m(b) = 1\}$ implies $a \leq b$. It is easily seen that if $\mathcal{M}$ is quite full, then it is full.

Now let $\mathcal{S}$ be any set. We say that a subset $\mathcal{L} \subset [0,1]^{\mathcal{S}}$ satisfies the orthogonality postulate if
(i) $0 \in \mathcal{L}$,
(ii) $a \in \mathcal{L}$ implies $1 - a \in \mathcal{L}$,
(iii) calling a and b orthogonal if $a+b \leq 1$ $(a,b \in \mathcal{L})$, we have $\sum a_i \in \mathcal{L}$ for any sequence $\{a_i\}$ of pairwise orthogonal elements of $\mathcal{L}$.

We shall say that $\mathcal{L} \subset [0,1]^{\mathcal{S}}$ satisfies the complete orthogonality postulate if instead of (iii) we have
(iii') $\sum_{i \in I} a_i \in \mathcal{L}$
for any set $\{a_i \mid i \in I\}$ of pairwise orthogonal elements of $\mathcal{L}$.

Let $\mathcal{L} \subset [0,1]^{\mathcal{S}}$ satisfy the orthogonality postulate. Put $a \leq b$ if $a(x) \leq b(x)$ for all $x \in \mathcal{S}$ and put $a' = 1-a$. The set $\mathcal{L}$ with the above order and orthocomplementation is a σ-orthomodular poset. If $\mathcal{L}$ satisfies the complete orthogonality postulate, then it becomes a complete orthomodular poset. Every point $u \in S$ induces a state on $\mathcal{L}$, where $m_u(a) = a(u)$ for all $a \in \mathcal{L}$, and the family of measures $\{m_u \mid u \in \mathcal{S}\}$ is full.

Conversely, if $(\mathcal{L}, \leq, \perp)$ is a σ-(complete) orthomodular poset with a full set $\mathcal{M}$ of states, then each $a \in \mathcal{L}$ induces a function $\underline{a}: \mathcal{M} \to [0,1]$, where $\underline{a}(m) = m(a)$ for all $m \in \mathcal{M}$. The set of all such functions $\underline{\mathcal{L}} = \{\underline{a} \mid a \in \mathcal{L}\}$ satisfies the (complete) orthogonality postulate and $(\underline{\mathcal{L}}, \leq, ')$ is isomorphic to $(\mathcal{L}, \leq, ')$.

The set $\mathcal{L} \subset [0,1]^{\mathcal{S}}$ satisfying the orthogonality postulate is called a functional logic.

Let $\mathcal{L} \subset [0,1]^{\mathcal{S}}$ be a functional logic. For each $a \in \mathcal{L}$, the set $S_a = \{u \in \mathcal{S} \mid a(u)=1\}$ is called the characteristic subset of $\mathcal{S}$ corresponding to a. The set of all characteristic subsets is denoted by $\mathcal{L}_{\mathcal{S}}$. The logic $\mathcal{L}$ is quite full if $a \leq b$ whenever $a(u) = 1$ implies

$b(u) = 1$. If $\mathscr{L}$ is quite full, the map $': \mathscr{L}_\mathscr{S} \to \mathscr{L}_\mathscr{S}$ defined by $S'_a = S_a$, is well defined, it is an orthocomplementation and $(\mathscr{L}_\mathscr{S}, \leq, ')$ is isomorphic with $(\mathscr{L}, \leq, ')$.

For the proofs of the above statement see Mączyński [8],[9].
Let $\mathscr{S}$ be any set and let $p: \mathscr{S} \times \mathscr{S} \to [0,1]$ satisfy the following requirements:
  (i) $p(\alpha, \beta) = 1$ iff $\alpha = \beta$,
  (ii) $p(\alpha, \beta) = p(\beta, \alpha)$,
  (iii)   calling α and β orthogonal if $p(\alpha, \beta) = 0$, we have
          $\sum_{\beta \in R} p(\alpha, \beta) = 1$

for any $\alpha \in \mathscr{S}$ and for every maximal set R of pairwise orthogonal elements of $\mathscr{S}$.

A function p satisfying (i),(ii),(iii) is called a (symmetric) transition probability and the couple $(\mathscr{S}, p)$ is called a transition--probability space (Mielnik [11],[12]. In what follows, we shall weaken the symmetry condition (ii) to a "weak symmetry":
  (ii') $p(\alpha, \beta) = 0$ iff $p(\beta, \alpha) = 0$.
and we shall call a function p satisfying (i),(ii'),(iii) a transition probability and the couple $(\mathscr{S}, p)$ a transition-probability space.

Let $(\mathscr{S}, p)$ be a (weakly symmetric) transition-probability space. Let $\mathscr{B}$ be the family of all orthogonal subsets of $\mathscr{S}$. We suppose that $\emptyset \in \mathscr{B}$ and $\{\alpha\} \in \mathscr{B}$ for all $\alpha \in \mathscr{S}$. For $A, B \in \mathscr{B}$ put $A \sim B$ if there exists a subset $C \in \mathscr{B}$ such that $A \cup C$ and $B \cup C$ are both maximal (pairwise) orthogonal subsets of $\mathscr{S}$. (By the Zorn lemma, every orthogonal subset is contained in a maximal one). Then $\sim$ is an equivalence relation. Denote by $\widetilde{\mathscr{B}}$ the set of all equivalence classes in $\mathscr{B}$. Let us define a map $f_B: \mathscr{S} \to [0,1]$ for $B \in \widetilde{\mathscr{B}}$ as follows: $f_\emptyset = 0$,
and if $\emptyset \neq B$, then
$$f_B(\alpha) = \sum_{\beta \in B} p(\alpha, \beta),$$
where $B \in \widetilde{B}$ is any representant of the equivalence class B. It is easy to show that f_B is well-defined. Denote by $\mathscr{L}$ the set of all functions f_B, i.e. $\mathscr{L} = \{f_B | B \in \widetilde{\mathscr{B}}\}$.

THEOREM 2.1.: Let $(\mathscr{S}, p)$ be a transition-probability space. The set $\mathscr{L} = \{f_B | B \in \widetilde{\mathscr{B}}\}$ is a complete quite full atomic quantum logic.

Proof: We show that the complete orthogonality postulate is satisfied:
  (i) $0 = f_\emptyset \in \mathscr{L}$,
  (ii) $1 - f_B = f_C$, where $B \cup C$ is a maximal orthogonal subset of $\mathscr{S}$,
  (iii) if $B_i \in \widetilde{\mathscr{B}}$, $i \in I$, are such that $f_{B_i} + f_{B_j} \leq 1$ for all $i \neq j$,
$i, j \in I$, then for any representants $B_i \in B_i$, $B_j \in B_j$ we have $B_i \perp B_j$ (in the sense that $\alpha \perp \beta$ for all $\alpha \in B_i, \beta \in B_j$) and $B = \bigcup_{i \in I} B_i$ belongs to $\mathscr{B}$. Then

$$\sum_{i \in I} f_{B_i} = f_B \in \mathscr{L}.$$

Every $\alpha \in \mathscr{S}$ induces a state on $\mathscr{L}$ if we put $\alpha(f_B) = f_B(\alpha)$.

For $B \in \mathscr{B}$ put $S_B = \{\alpha \in \mathscr{S} \mid f_B(\alpha) = 1\}$, i.e. S_B is the characteristic subset for f_B. To prove that $\mathscr{L}$ is quite full, i.e. that $f_{B_1} \leq f_{B_2}$ iff $S_{B_1} \subseteq S_{B_2}$, we need the fact that if A is a maximal pairwise orthogonal subset in S_B, then $A \in B$. Indeed, let $B \in \mathscr{B}$ be any representant and let C be such that $B \cup C$ is a maximal orthogonal subset of $\mathscr{S}$. Let $\delta \in \mathscr{S}, \delta \perp A \cup C$. Then

$$\sum_{\alpha \in A} p(\delta, \alpha) + \sum_{\gamma \in C} p(\delta, \gamma) = 0$$

Therefore,

$$1 = \sum_{\beta \in B} p(\delta, \beta) + \sum_{\gamma \in C} p(\delta, \gamma) = \sum_{\beta \in B} p(\delta, \beta) = f_B(\delta),$$

hence $\delta \in S_B$. But $\delta \perp A$ contradicts the maximality of A. This proves that $A \cup C$ is maximal orthogonal subset of $\mathscr{S}$, i.e. $A \in B$.

Now let $S_{B_1} \subset S_{B_2}$. We have $f_{B_1}(\alpha) = \sum_{\beta \in B_1} p(\alpha, \beta)$, where $B_1 \in \mathscr{B}_1$. If $\alpha \in B_1$, then $f_{B_1}(\alpha) = 1$, therefore $f_{B_2}(\alpha) = 1$, hence $B_1 \subset S_{B_2}$. Let B be a maximal orthogonal subset of S_{B_2} containing B_1. Then $B \in \mathscr{B}_2$, and $B_1 \subset B$ implies that $f_{B_1} \leq f_{B_2}$. The opposite implication is straightforward. It is easily seen that $\{\alpha\} \sim \{\beta\}$ iff $\alpha = \beta$, so that the set of all characteristic subsets contains all one-point sets. This implies that $\mathscr{L}$ is atomic with the atoms f_α, $\alpha \in \mathscr{S}$.

A result in the opposite direction is obtained in the following theorem, which has been proved in [16].

THEOREM 2.2. Let $\mathscr{S}$ be any set. Let $\mathscr{L} \subset [0,1]^\mathscr{S}$ be a quite full functional logic satisfying the complete orthogonality postulate. Let the set of all characteristic subset contain all one-point sets. Define $p(\alpha, \beta) = u_\beta(\alpha)$ for $\alpha, \beta \in \mathscr{S}$, where u_β is the (unique) element of $\mathscr{L}$ such that the characteristic subset of u_β is equal to $\{\beta\}$. Then $p(\alpha, \beta)$ is well-defined, and the couple $(\mathscr{S}, p)$ is a transition-probability space.

We note that the elements of $\mathscr{S}$ necessarily must be interpreted as pure states.

3. VECTOR-SPACE REPRESENTATIONS

Let $\mathscr{S}$ be any set and let $\perp$ be a symmetric, antireflexive binary relation on $\mathscr{S}$. Suppose that the following condition is satisfied:

(i) if $\beta \perp \gamma$ whenever $\beta \perp \alpha$ then $\gamma = \alpha$.

A symmetric, antireflexive binary relation $\perp$ satisfying the above condition (i) will be called an orthogonality relation and the couple $(\mathscr{S},\perp)$ will be called an orthogonality space.

Let us consider the following examples. Let $(\mathscr{S},p)$ be a transition probability space. We put $\alpha \perp \beta$ iff $p(\alpha,\beta) = 0$. Then $\perp$ is an orthogonality relation. Indeed, we have only to check the condition (i). Let R be a maximal pairwise orthogonal subset of $\mathscr{S}$ such that $\alpha \in R$ (such a set exists by Zorn's lemma). Let $\beta \perp \gamma$ whenever $\beta \perp \alpha$. Then $\gamma \perp \beta$ for all $\beta \in R - \alpha$. Therefore,
$$1 = \sum_{\beta \in R} p(\gamma,\beta) = p(\gamma,\alpha), \quad \text{i.e } \gamma = \alpha.$$

Let $\mathscr{L}$ be an atomic quantum logic. Let $\mathscr{A}$ be the set of all atoms in $\mathscr{L}$. Put $a \perp b$ iff $a \leq b'$ $(a,b \in \mathscr{A})$. If $b \perp c$ whenever $b \perp a$, then $a' \perp a$ implies that $c \leq a$, i.e. $c = a$. This proves that $(\mathscr{A},\perp)$ is an orthogonality space.

Let $(\mathscr{S},\perp)$ be an orthogonality space. For $M \subset \mathscr{S}$ put
$$M^{\perp} = \{\beta \in \mathscr{S} \mid \beta \perp \alpha \text{ for all } \alpha \in M\}$$
$$\bar{M} = (M^{\perp})^{\perp}.$$

The map $M \to \bar{M}$ is a closure operation, and therefore the set
$$\mathscr{F}(\mathscr{S}) = \{M \subset \mathscr{S} \mid M = \bar{M}\}$$
of all closed subsets of $\mathscr{S}$ is a complete lattice with the lattice operations
$$\wedge_i M_i = \bigcap_i M_i \quad \text{and} \quad \wedge_i M_i = \overline{(\bigcup_i M_i)},$$
where $\bigcap$ and $\bigcup$ are the set-theoretical intersection and union respectively. Moreover, for every $\alpha \in \mathscr{S}$ we have $\overline{\{\alpha\}} = \{\alpha\}$ by the property (i) of the orthogonality relation. Therefore, $\mathscr{F}(\mathscr{S})$ is atomic. It is also easily seen that the map $M \mapsto M^{\perp}$ is an orthocomplementation on $\mathscr{F}(\mathscr{S})$ (see [1]).

To simplify the notations, we shall write α instead of $\{\alpha\}$ and $\overline{\alpha\beta}$ instead of $\overline{\{\alpha,\beta\}}$. Put
$$P(\mathscr{S}) = \{\overline{\alpha\beta} \mid \alpha,\beta \in \mathscr{S}, \alpha \neq \beta\}$$

To obtain an embedding into a vector space, we shall consider $\mathscr{S}$ as a set of "points" and the set $P(\mathscr{S})$ as the set of "lines". We shall try to find conditions under which $\mathscr{S}$ and $P(\mathscr{S})$ form a projective space. Recall that a set P whose elements are called points is called a projective space if there exists a family of subsets of P called lines satisfying the following conditions:

(P1) Every line contains at least two points, and two different points p,q are contained in just one line denoted by $\overline{pq}$

(P2) Let p,q,r be three non-collinear points. If s,t are different points such that $s \in \overline{pq}$ and $t \in \overline{qr}$, then there exists a point u in $\overline{pr} \cap \overline{st}$. (See [10]).

We shall try to replace the conditions (P1) and (P2) by conditions having more physical sense. Let $(\mathscr{S},\perp)$ be an orthogonality space. We shall say that α is a superposition of A $(A \subset \mathscr{S})$ if $\alpha \in \bar{A}$. We shall say that α is a minimal superposition of A if $\alpha \in \bar{A}$ and

ON QUANTUM LOGICS AND TRANSITION PROBABILITY SPACES

$\alpha \notin \bar{B}$ for any proper subset B of A.

We note that the above definition of superposition agrees with Varadarajan's definition of superpositions of states on quantum logics (see [18]) : a state s is a superposition of a set M of states on a quantum logic $\mathscr{L}$ if $m(a) = 0$ for all $m \in M$ implies $s(a) = 0$ ($a \in \mathscr{L}$). Let $(\mathscr{S},p)$ be a transition-probability space and let $\mathscr{L}$ be a corresponding quantum logic by Th.2.1. If $A \subset \mathscr{S}$ and $\alpha \in \bar{A}$, then for any $f_B \in \mathscr{L}$, $\gamma(f_B) = 0$ for all $\gamma \in A$ implies $\beta \perp \gamma$ for all $\beta \in B$ ($B \in \tilde{B}$ is any representant). Therefore, $B \subset A^\perp$ and hence $\alpha \perp \beta$ for all $\beta \in B$, i.e. $\alpha(f_B) = 0$. On the other hand, if $\gamma(f_B) = 0$ for all $\gamma \in A$ ($A \subset \mathscr{S}$) implies $\alpha(f_B) = 0$ ($\alpha \in \mathscr{S}$), then $\gamma(f_\beta) = 0$ for all $\gamma \in A$ implies $\alpha(f_\beta) = 0$ ($\beta \in \mathscr{S}$). Hence $\alpha \in \bar{A}$.

Following Gudder [3], we shall say that the orthogonality space $(\mathscr{S},\perp)$ satisfies the postulate of minimal superposition (PMS) if for every finite subset F of $\mathscr{S}$ and every maximal superposition α of F we have

$$\overline{(\alpha \cup F_1)} \cap \overline{F_2} \neq \emptyset, \text{ where } \{F_1, F_2\} \text{ is any partition of } F.$$

We shall need the following weakening of the postulate of minimal superposition. We shall say that PMS of order n ($n \geq 2$) is satisfied if for every $\{\alpha_1, \alpha_2, \ldots, \alpha_k\}$ ($k \leq n$) and for every minimal superposition α of $\{\alpha_1, \alpha_2, \ldots, \alpha_k\}$ we have $\overline{\alpha \alpha_1} \cap \overline{\{\alpha_2, \alpha_3, \ldots, \alpha_k\}} \neq \emptyset$.

THEOREM 3.1 : Let $(\mathscr{S},\perp)$ be an orthogonality space. If the postulate of minimal superposition of order 3 is satisfied, then $\mathscr{S}$ with the lines $P(\mathscr{S})$ forms a projective space.

Before the proof, let us remark that PMS of order 3 has a simple geometric interpretation : let $F = \{\alpha_1, \alpha_2\}$ and let α be a minimal superposition. Then $\overline{\alpha \alpha_i} \cap \overline{\alpha_j} \neq \emptyset$ means that $\alpha_j \in \overline{\alpha \alpha_i}$ ($i,j \in \{1,2\}$). This shows that "lines" are well-defined. Now let $F = \{\alpha_1, \alpha_2, \alpha_3\}$ and let α be a minimal superposition. Then $\overline{\alpha \alpha_i} \cap \overline{\alpha_j \alpha_k} \neq \emptyset$ ($i,j,k \in \{1,2,3\}$) imply that two different lines have a nonempty intersection.

Proof of Th.3.1 : To prove (P1), let the points p,q ($p \neq q$) be contained in a line $\overline{rs}$. Suppose first that $p = r$, $q \neq s$. Then $q \in \overline{rs}$ is a minimal superposition. By PMS of order 2, $s \in \overline{qr} = \overline{pq}$, i.e. $\overline{rs} = \overline{pq}$. Now let p,q,r,s be all different. Then p,q are minimal superpositions of $\{r,s\}$. This means that $r \in \overline{ps}$, $r \in \overline{qs}$, hence $\overline{rs} = \overline{ps} = \overline{qs}$. Now $p \in \overline{qs}$ implies $\overline{pq} = \overline{qs} = \overline{rs}$.

To prove (P2), let p,q,r be non-collinear. Let $s \in \overline{pq}$ and $t \in \overline{qr}$, $s \neq t$. We have to prove that $\overline{st} \cap \overline{pr} \neq \emptyset$. We may assume that $s \notin \overline{pr}$

and $t \not\in \overline{pr}$. If $s = q$, we may suppose that $t \in \overline{qr}$ is a minimal superposition. But then $r \in \overline{qt} = \overline{st}$. If $s \neq q$, then $s \in \overline{pq}$ implies that $q \in \overline{ps}$, and hence $t \in \overline{qr} \subset \{p,r,s\}$. If $t \in \overline{sr}$ then $r \in \overline{ts}$ and if $t \in \overline{ps}$ then $p \in \overline{ts}$, so that in both cases $\overline{st} \cap \overline{pr} \neq \emptyset$.
In the remaining case, t is a minimal superposition of $\{p,r,s\}$, so that by PMS of order 3, $\overline{ts} \cap \overline{pr} \neq \emptyset$, and the proof is complete.

We note that also the opposite statement to Th.3.1 can be proved (see [5]). As a consequence we obtain that PMS is satisfied iff PMS of order 3 is satisfied.

By a generalized Hilbert space we shall mean a vector space $\mathcal{V}$ over a division ring $\mathcal{D}$ with an involutive anti-automorphism $\theta : \mathcal{D} \to \mathcal{D}$ and a hermitian form $f : \mathcal{V} \times \mathcal{V} \to \mathcal{D}$ (see e.g. [10] for the definitions). For $M \subset \mathcal{V}$ put $M^\circ = \{v \in \mathcal{V} \mid f(u,v) = 0 \text{ for all } u \in M\}$. A subset M of $\mathcal{V}$ is called f-closed if $M = M^{\circ\circ}$. Denote by $\mathcal{L}_f(V)$ the set of all f-closed subsets of $\mathcal{V}$. Then $\mathcal{L}_f(\mathcal{V})$ is a complete atomic lattice, which is orthocomplemented by the map $M \to M^\circ$.

The following theorem can be proved (see [5] for the details).

THEOREM 3.2 : Let $(\mathcal{S}, \perp)$ be an orthogonality space satisfying a postulate of minimal superposition of order 3. Let every line in $P(\mathcal{S})$ contain at least three points and let there be at least four independent points in $\mathcal{S}$ (in the sense that none of them is a superposition of the remaining three). Then there is a generalized Hilbert space $(\mathcal{V}, \mathcal{D}, \theta, f)$ such that the orthocomplemented complete lattice $\mathcal{F}(\mathcal{S})$ of all closed subsets of $\mathcal{S}$ is ortho-isomorphic with the lattice $\mathcal{L}_f(\mathcal{V})$ of all f-closed subsets of $\mathcal{V}$.

We note that the supposition that every line contains at least three points can be interpreted as a kind of a superposition principle. The division ring $\mathcal{D}$ is in general not specified. If we consider a transition-probability space which corresponds to a transition-amplitude space introduced in [4] (i.e. $p(\alpha, \beta) = |A(\alpha, \beta)|^2$ where A is a complex amplitude), then the division ring $\mathcal{D}$ is a subfield of the complex field $\mathbb{C}$ ([17]).

REFERENCES

1. Birkhoff, G. (1967) Lattice theory. Providence, Rhode Island.
2. Birkhoff, G., von Neumann, J. (1936) 'The logic of quantum mechanics.' Annals of mathematics 37, 823-843.
3. Gudder, S. (1970) 'Projective representations of quantum logics.' Int. J. Theor. Phys. 3, 99-108.
4. Gudder, S., Pulmannova, S. (1987) 'Transition amplitude spaces.' J. Math. Phys. 28, 376-385.

5. Hedlikova,J.,Pulmannova,S.:'Orthogonality spaces and atomic ortho-complemented lattices.' In preparation.
6. Mackey,G. (1963) *The mathematical foundations of quantum mechanics.* Benjamin, New York.
7. Mac Laren,D. (1964) 'Atomic orthocomplemented lattices.' *Pac. J. Math.* **14**, 597-612.
8. Mączyński,M. (1973) 'The orthogonality postulate in axiomatic quantum mechanics.' *Int. J. Theor. Phys.* **8**, 353-360.
9. Mączyński,M. (1974) 'Functional properties of quantum logics.' *Int.J Theor. Phys.* **11**, 149-156.
10. Maeda,F.,Maeda,S. (1970) *Theory of symmetric lattices.* Springer, Berlin
11. Mielnik,B. (1968) 'Geometry of quantum states.' *Commun. Math. Phys.* **9**, 55-80.
12. Mielnik,B. (1969) 'Theory of filters.' *Commun. Math. Phys.* **15**, 1-46.
13. von Neumann,J. (1981) 'Continuous geometries with a transition probability.' *Mem.Amer.Math.Soc.* **34**, No 252.
14. Piron,C. (1976) *Foundations of quantum physics.* Benjamin, Reading Mass.
15. Pulmannova,S. (1986) 'Transition probability spaces.' *J.Math.Phys.* **27**, 1791-1795.
16. Pulmannova,S. 'Functional properties of transition probability spaces.' *Rep.Math.Phys.* , to appear.
17. Pulmannova,S.,Gudder,S. (1987) 'Geometric properties of transition amplitude spaces.' *J.Math.Phys.* **28**, 2393-2399.
18. Varadarajan,V. (1968) *Geometry of quantum theory.* Van Nostrand, Princeton New Jersey.

PROBABILITY IN QUANTUM MECHANICS

G.T. Rüttimann
University of Berne
Institute of Mathematical Statistics
Sidlerstrasse 5
CH-3012 Berne
Switzerland

ABSTRACT. Non-commutative measure theory serves as a mathematical basis to compare the classical and the quantum mechanical probability scheme.

Probability theory and statistics are supposed to provide the tools to analyse the aleatory nature of physical phenomena. One of the main goals of any probabilistic assessment consists of making predictive judgments the correctness of which determines the success of the model considered.
Classical mechanical systems are governed by Hamiltonian equations which yield position and momentum of the particles involved as functions of time. This setting conveys a determinism, i.e. predictability of future events in terms of the present ones, in its purest form. A first departure from this scheme of certainty is provided by statistical mechanics where uncontrolled variables interfere with the variables being studied and where the former account for the random character of the predictions; although uncontrolled, they are still recognized as variables of the system. Classical mechanics and classical probability theory are built upon the same conceptual basis as far as the observation of possible properties of a system are concerned. It therefore comes at no surprise that statistical mechanics in its original form appears to be a first order descendant of both. The idea of a Laplacian determinism is rooted in the assumption that it is possible, at least in principle, to obtain a complete knowledge of the observed system per se, at any given time, an assumption which in turn requires a definite and sharp separation between observer and the system under consideration.
Quantum mechanics the realm of which is at the atomic and subatomic level not only brought us the quantization of certain physical quantities in agreement with the experiments, but also made the existence of a grand-canonical observation, which yields a complete knowledge of the system, highly questionable. The discovery that submicroscopic systems are not just scaled down versions of macroscopic ones does not so much concern our observational possibilities as it does our conceptual basis of perceiving physical phenomena. Heisenberg's uncertainty principle gives the cognitive act a new accent: rather than having it intrinsically, a system

carries a property potentially, decided precisely when being "questioned" in terms of suitable experiments.

Quantum mechanical reasonings and predictions are of a probabilistic nature and we may ask what are its bearings upon probability theory. Kolmogorov's postulational scheme of probability claims to have the criteria for an assessment of random phenomena. However, at least in terms of successful models for quantum mechanical systems, there is sufficient evidence that quantum phenomena need not obey these postulates:

"The quantum mechanical laws of the physical world approach very closely the laws of Laplace as the size of the objects involved in the experiments increases. Therefore, the laws of probability which are conventionally applied are quite satisfactory in analyzing the behaviour of the roulette wheel but not the behaviour of a single electron or a photon of light". (R. Feynmann).

The aim of this paper is to compare Kolmogorov's probability theory with the probability scheme emanating from the Hilbert space model of quantum mechanics. A generalized measure theory will be used to provide the basis for this comparison both at a mathematical as well as at a conceptual level.

Assessing a physical system from a classical point of view amounts to determine a triple $(\Omega, \mathcal{A}, P)$ called a <u>probability space</u> where Ω is a non-empty set of <u>elementary events</u>, $\mathcal{A}$ is a σ-field of subsets of Ω called <u>events</u>, and P is a σ-additive measure with $P(\Omega) = 1$ on the σ-field $\mathcal{A}$, called a <u>probability functional</u>. For an event A, P(A) is interpreted as the likelihood for the event A to occur while a measurement of A is performed.

Observable physical quantities are represented by random variables. A (real-valued) <u>random variable</u> is understood to be a map $X: \Omega \to \mathbb{R}$ such that for all $\Delta \in \mathcal{B}(\mathbb{R})$, $\mathcal{B}(\mathbb{R})$ denotes the class of Borel sets of the set $\mathbb{R}$ of real numbers, the set

$$X^{-1}(\Delta) := \{\omega \in \Omega \mid X(\omega) \in \Delta\}$$

is an event of the probability space. The composition $P \circ X^{-1}$ is a probability functional on the σ-field $\mathcal{B}(\mathbb{R})$; $(P \circ X^{-1})(\Delta)$ $(= P(X^{-1}(\Delta)))$ is the the probability that the random variable X attains a value in the subset $\Delta \in \mathcal{B}(\mathbb{R})$. The <u>expectation value</u> of the random variable X with respect to the probability functional P is defined to be

$$\int t \, d(P \circ X^{-1}) \in \mathbb{R} \cup \{\pm \infty\}$$

provided the integral exists.

It is quite easy to determine the probability space that corresponds to the experiment of drawing balls from an urn. In other instances it is a non-trivial problem. Aside from their use in the conceptual foundations, probability spaces play an insignificant rôle in the further development of the theory. Auxiliary quantities such as random variables and their distribution functions become the dominant objects. However, notice that for every distribution function $F: \mathbb{R} \to \mathbb{R}$ (i.e.: $0 \leq F(t) \leq 1$ $\forall t \in \mathbb{R}$, F is non-decreasing, right-continuous, $\lim_{t \to -\infty} F(t) = 0$ and $\lim_{t \to +\infty} F(t) = 1$) there exists a probability space $(\Omega, \mathcal{A}, P)$ and a random variable X such that

$$P(\{\omega \in \Omega \mid X(\omega) \leq t\}) = F(t)$$

for all $t \in \mathbb{R}$. This is a consequence of a powerful theorem of Kolmogorov which, in fact, involves multi-dimensional stochastic processes.

The probability functionals on a σ-field of subsets of a set may be considered as the possible <u>stochastic models</u> for the system under consideration as they describe its long-run random behaviour. The problem of statistical inference is to decide on the basis of incoming data which probability functional represents the appropriate stochastic model.

Let us now focus our attention on the quantum probability model. An assessment of a physical system from a quantum mechanical standpoint consists of choosing a Hilbert space H as the basic descriptional frame, e.g. a two-dimensional Hilbert space if it is agreed upon that the system under consideration is a spin 1/2-particle moving in a magnetic field.

By a <u>state</u> of the system we understand a, possibly encoded, maximal knowledge of it. In the conventional Hilbert space model states are represented by positive trace-class operators on H of norm one. Observable physical quantities or <u>observables</u>, for short, such as energy, spin, momentum, angular momentum etc., are represented by (not necessarily bounded) self-adjoint operators on H. The <u>expectation value</u> of an observable A with respect to a state W is defined by

$$\text{Tr}(AW) \in \mathbb{R} \cup \{\pm\infty\}$$

provided that the trace exists.

Here we have two seemingly different probability schemes. The measure-theoretic ingredients of the former are not at all apparent in the latter. On the other hand, quantum mechanical systems produce interference patterns and within the Hilbert space model we have a fairly accurate idea of what the superposition of two states should look like. None of this is present in statistical mechanics which, as we recall, is based on Kolmogorov's postulational scheme of probability theory.

A quantum theory is supposed to be at a level of generality which includes the description of classical and microscopic systems as well. However, in many concrete examples, predictions concerning macroscopic systems are only obtained as a result of complicated approximations the justification of which are seldom rigorous.

At the probabilitic level <u>non-commutative measure theory</u> gives us a mathematical frame which is general enough to contain both theories as special cases. Moreover, it serves as a basis for exchanging and comparing classical and quantum-mechanical concepts. Let us make the following definitions:

An <u>orthomodular poset</u> is a triple $(L, \leq, ')$ where $(L, \leq)$ is a partially ordered set with a least element (0) and a greatest element (1) and ' denotes a map from L into L such that, for all elements p and q in L,

(i) $p \leq q \Longrightarrow q' \leq p'$,
(ii) $p'' = p$,
(iii) if $p \leq q'$, then the supremum of $\{p, q\}$, denoted by $p \vee q$,

exists in $(L, \leq)$,
(iv) $p \vee p' = 1$,
(v) $p \leq q'$ and $p \vee q = 1 \Longrightarrow p = q'$

holds true. A pair (p,q) of elements of L is said to be <u>orthogonal</u>, denoted by $p \perp q$, provided that $p \leq q'$. A subset N of L is said to be <u>orthogonal</u>, if $p \perp q$ for all elements p and q in N with $p \neq q$. An orthomodular poset is called σ-<u>orthocomplete</u> if suprema of countable orthogonal subsets of L exist in the partially ordered set $(L, \leq)$. Let $(L, \leq, ')$ and $(P, \leq, ')$ be orthomodular posets. A map $z : L \to P$ is said to be an <u>orthomorphism</u> from $(L, \leq, ')$ into $(P, \leq, ')$ if

(i) $z(1) = 1$,
(ii) if $p \perp q$, then $z(p) \perp z(q)$,
and $z(p \vee q) = z(p) \vee z(q)$.

If z is an orthomorphism from $(L, \leq, ')$ into $(P, \leq, ')$ then it follows that, for element p and q in L,

(iii) $z(p') = (z(p))'$,
(iv) $z(0) = 0$,
(v) $p \leq q \Longrightarrow z(p) \leq z(q)$.

Let $(L, \leq, ')$ and $(P, \leq, ')$ be σ-orthocomplete orthomodular posets. An orthomorphism $z : L \to P$ is said to be a σ-<u>orthomorphism</u> if

(vi) $z(\vee N) = \vee z(N)$ for every non-empty countable orthogonal subset N of L.

Let $(L, \leq, ')$ be an orthomodular poset. A map $\mu : L \to \mathbb{R}_+$ where $\mathbb{R}_+$ denotes the set of positive real numbers, is said to be a <u>measure</u> on L if

$$\mu(p \vee q) = \mu(p) + \mu(q) \text{ for all } p, q \in L \text{ with } p \perp q.$$

A measure μ is said to be <u>normalized</u> if $\mu(1) = 1$.

If $(L, \leq, ')$ is a σ-orthocomplete orthomodular poset then a measure μ an L is said to be σ-<u>additive</u> provided that for every non-empty countable orthogonal subset N of L and every strict enumeration $i \to p_i$ of N, $\sum_{i=1}^{\infty} \mu(p_i)$ equals $\mu(\vee N)$.

Let X be a non-empty set and let A be a σ-algebra of subsets of X. Then the triple $(A, \subseteq, {}^c)$, where $\subseteq$ denotes set-theoretical inclusion and ^{c} denotes the formation of the set-theoretical complement is a σ-orthocomplete orthomodular poset. Notice that a pair of elements of A is orthogonal if and only if they are set-theoretically disjoint. As a consequence, the triple $(B(\mathbb{R}), \subseteq, {}^c)$, where $B(\mathbb{R})$ denotes the class of Borel sets of the set $\mathbb{R}$ of real numbers, is a σ-orthocomplete orthomodular poset.

Let $(L, \leq, ')$ be any σ-orthocomplete orthomodular poset. For a σ-orthomorphism z from $(B(\mathbb{R}), \subseteq, {}^c)$ into $(L, \leq, ')$ and a σ-additive normalized measure μ on $(L, \leq, ')$ we define the <u>average value</u> $E(z)(\mu)$, of z with respect to μ as

$$E(z)(\mu) = \int t\, d(\mu \circ z) \in \mathbb{R} \cup \{\pm\infty\}$$

provided the integral exists. Notice that the map $\Delta \in \mathcal{B}(\mathbb{R}) \to \mu(z(\Delta)) \in \mathbb{R}$ is a σ-additive normalized measure on the orthomodular poset $(\mathcal{B}(\mathbb{R}), \subseteq, ^c)$. Such a σ-orthomorphism is said to be <u>bounded</u> provided that there exists a closed interval $[s,r]$ of $\mathbb{R}$ such that

$$z([s,t]) = 1.$$

A <u>projection</u> P on a Hilbert space H is a bounded selfadjoint (P=P*) and idempotent (P²=P) operator on H. We denote the collection of projections on H by $\mathcal{P}(H)$ and define for elements P and Q of $\mathcal{P}(H)$

$$P \leq Q : \iff PQ = P,$$
$$P^\perp := 1 - P,$$

where 1 is the unit operator on H. Then the triple $(\mathcal{P}(H), \leq, ^\perp)$ is a σ-orthocomplete orthomodular poset. Notice that a pair of elements of $\mathcal{P}(H)$ is orthogonal if and only if PQ = 0, where 0 denotes the null operator on H.

The two examples of orthomodular posets, $(\mathcal{A}, \subseteq, ^c)$ and $(\mathcal{P}(H), \leq, ^\perp)$, are of a particular interest. The former pertains to classical probability whereas the latter involves the basic descriptional frame of a quantum mechanical system. The question arises as to the specialization or rather the implementation of the notions of a positive normalized σ-additive measure and a σ-orthomorphism, respectively:

	Classical Probability Scheme	Quantum Mechanical Probability Scheme
	X a non-empty set $\mathcal{A}$ a σ-field of subsets of X	H a separable Hilbert space, dim H ≥ 3. $\mathcal{P}(H)$ collection of projections on H.
orthomodular poset $(L, \leq, ')$	$(\mathcal{A}, \subseteq, ^c)$	$(\mathcal{P}(H), \leq, ^\perp)$
σ-additive normalized measures on $(L, \leq, ')$	every <u>probability functional</u> on the σ-field $\mathcal{A}$ is a σ-additive normalized measure on $(\mathcal{A}, \subseteq, ^c)$ and vice versa.	Let W be a <u>state</u>. Then the map $$P \in \mathcal{P}(H) \to \mu_W(P) := \text{Tr}(WP) \in \mathbb{R}$$ is a σ-additive normalized measure on $(\mathcal{P}(H), \leq, ^\perp)$.

μ			**Gleason's theorem**: To each σ-additive normalized measure μ there exists a unique state W such that $$\mu(P) = \mu_W(P) \; \forall P \in \mathcal{P}(H).$$ Therefore, the map $$W \to \mu_W$$ is a bijection between the collection states and the σ-additive normalized measures on $(\mathcal{P}(H), \leq, ')$.
σ-orthomorphisms from $(\mathcal{B}(\mathbb{R}), \subseteq, ^c)$ into $(L, \leq, ')$ z	Let X be a <u>random variable</u>. Then the map $$\Delta \in \mathcal{B}(\mathbb{R}) \to z_X(\Delta) := X^{-1}(\Delta) \in \mathcal{A}$$ is a σ-orthomorphism from $(\mathcal{B}(\mathbb{R}), \subseteq, ^c)$ into $(\mathcal{A}, \subseteq, ^c)$. **Sikorsky's theorem**: To each σ-orthomorphism z from $(\mathcal{B}(\mathbb{R}), \subseteq, ^c)$ into $(\mathcal{A}, \subseteq, ^c)$ there exists a unique random variable X such that $$z(\Delta) = z_X(\Delta) \; \forall \Delta \in \mathcal{B}(\mathbb{R}).$$ Therefore, the map $$X \to z_X$$ is a bijection, between the collection of random variables and σ-orthomorphisms from $(\mathcal{B}(\mathbb{R}), \subseteq, ^c)$ into $(\mathcal{A}, \subseteq, ^c)$.		Let z be a σ-orthomorphism from $(\mathcal{B}(\mathbb{R}), \subseteq, ^c)$ into $(\mathcal{P}(H), \subseteq, \perp)$. Then the subset $$D_z := \{\varphi \in H \mid \int t^2 \, d<\varphi, z_t \varphi> < \infty\},$$ where $z_t = z(-\infty, t]$, is dense in H and there exists a unique (not necessarily bounded) self-adjoint operator A_z of H with D_z as its domain of definition such that $$<\varphi, A_z \varphi> = \int t \, d<\varphi, z_t \varphi> \; \forall \varphi \in D_z$$ **Spectral theorem**: To each self-adjoint operator (A,D) of H there exists a unique σ-orthomorphism z from $(\mathcal{B}(\mathbb{R}), \subseteq, ^c)$ into $(\mathcal{P}(H), \leq, \perp)$ such that $$(A,D) = (A_z, D_z).$$ Therefore, the map $$z \to (A_z, D_z)$$ is a bijection between

		the collection of σ-orthomorphism from $(\mathcal{B}(\mathbb{R}),\subseteq,^c)$ into $(\mathcal{P}(H),\leq,^\perp)$ and the collection of <u>observables</u>.
average <u>value</u> <u>of</u> z <u>with</u> <u>respect</u> <u>to</u> μ (provided it exists) $E(z)(\mu) =$	$\int t \, d(\mu \circ X^{-1})$ where the random variable X satisfies the condition $z = z_X$.	$\mathrm{Tr}(W A_z)$ where the state W satisfies the condition $\mu = \mu_W$.

The two seemingly unrelated probability schemes outlined at the beginning appear as special cases of a generalized measure theory. The procedure of comparison and exchange of notions and concepts consists of their generalization in the context of non-commutative measure theory, a process which is by no means unique, followed by a specialization.

The perfect analogy between the orthomodular poset of events, a Boolean lattice, in the classical probability scheme and the orthomodular poset of projections, a non-Boolean lattice, in the quantum mechanical case is suited to conceptualize quantum mechanics in terms of an alternative logic.

Literature

E.G. Beltrametti, G. Cassinelli: <u>The Logic of Quantum Mechanics</u>. Encyclopedia of Mathematics and Its Applications, Vol. 15. Addison-Wesley Publishing Company, Reading Mass., 1981.

G. Birkhoff, J. von Neumann: 'The Logic of Quantum Mechanics'. Ann. of Math. <u>37</u> (1936), 823-842.

W. Feller: <u>An Introduction to Probability Theory and Its Applications</u>, Vol. 1. Wiley § Sons, New York, 1950.

B. de Finetti: <u>Theory of Probability</u>, <u>Vol</u>. 1. Wiley § Sons, Inc., New York. 1974.

D.J. Foulis, C.H. Randall: 'Operational Statistics I'. J. Math. Phys. <u>13</u> (1972), 1667-1675.

S.P. Gudder: <u>Stochastic Methods in Quantum Mechanics</u>. North Holland, New York, 1979.

J.M. Jauch: <u>Foundations of Quantum Mechanics</u>. Addison-Wesley, Reading Mass., 1968.

J. von Neumann: <u>Mathematical Foundations of Quantum Mechanics</u>, Princeton University Press, 1955.

C. Piron: <u>Foundations of Quantum Physics</u>. W.A. Benjamin, Reading Mass.,

1976.
C.H. Randall, D.J. Foulis: 'A Mathematical Setting for Inductive Reasoning'. In Foundations of Probability Theory, Statistical Inference, and Statistical Theories of Science, Vol. III. Ed. C.A. Hooker. Reidel Publishing Company, Dordrecht-Holland, 1976, pp. 169-205.
G.T. Rüttimann: Logikkalküle der Quantenphysik. Duncker und Humblot, Berlin, 1977.
G.T. Rüttimann: 'Non-commutative Measure Theory'. Habilitationsschrift, Universität Bern, 1980.
G. T. Rüttimann: 'Expectation Functionals of Observables and Counters'. Reports on Mathematical Physics 21 (1985), 213-222.
V.S. Varadarajan : Geometry of Quantum Theory. Von Nortrand, Princeton N.J., 1968.

Part 3

Aspects of the arguments on nonlocality,
Bell's theorem and EPR correlations

BELL'S THEOREM: A COUNTEREXAMPLE THAT AGREES WITH THE QUANTUM FORMALISM

Thomas D. Angelidis

University College London
Gower Street
London WC1E 6BT, England.

ABSTRACT

A family $\{A^\mu\}$ of models is here constructed whose members satisfy *all* the postulates of locality due to Clauser and Horne (CH), and whose members *converge uniformly* to a unique *limit function* identical with the function of the quantum formalism (QF) model for the Einstein-Podolsky-Rosen-Bohm (EPRB) *ideal* experiment. This renders *invalid* Bell's theorem. My construction establishes my proposed *local explanatory theory* of the EPRB experiment from more basic postulates of a structural character as Einstein had in mind. The theory explains, in purely local terms, *the characteristic trait of the EPRB experiment where the directions of polarization of the single photon pairs are chosen at random* by the process of annihilation from the singlet state, and the directions of the polarizer settings are *chosen at random* (or nearly so) by the switches as in the Aspect experiment. Moreover, a bona fide *specified form* of the generalized CH inequality, known as CH(4), is here constructed which is *satisfied by* the QF model itself. This *directly* demonstrates the *consistency* of CH(4) with the quantum formalism.

1. INTRODUCTION

Einstein's arguments against the Copenhagen interpretation of the quantum formalism (QF) have one common point: They attribute to it an action at a distance which, as Einstein[1] put it, "*contradicts the postulate of relativity.*" Einstein never accepted the spooky character of this action at a distance or, as it is now called, non-locality.
    Even though Einstein claimed that the Copenhagen interpretation was non-local, he nowhere appears to have claimed that the quantum formalism itself was non-local. Thus, Einstein did *not* identify the quantum formalism with its Copenhagen interpretation, as is clear from his argument, based on his *principle of locality*,[2] that if the (unproved) Copenhagen interpretation of the quantum formalism is true, then there must be action at a distance.
    With Bell's[3] reformulation of the Einstein-Podolsky-Rosen-Bohm[4]

(EPRB) argument, the issue went beyond the conclusions drawn in the EPR paper.[4a] For Bell's theorem asserts that the quantum formalism itself is non-local, since the theorem claims to put the QF predictions *outside* the bounds of the range within which lie the predictions of *all* local theories, and therefore of special relativity. This is *the* essence of what I have elsewhere[5] called the Universality Claim (UC) associated with Bell's theorem.

The issue reached its climax when recent experiments, and especially those of Aspect et al.,[6] supported the QF predictions, and thereby appeared to refute, at one fell swoop, *all* local theories and Einstein's most attractive principle of locality, and even to shatter our ideas of spacetime and of physical reality.[7,8]

More precisely, it is claimed that in EPRB type experiments local action is *not* sufficient to explain all that the quantum formalism predicts, and therefore action at a distance is needed, according to the claim, in order to explain the Aspect experiment with the switches. Now action at a distance is *incompatible with* Einstein's special theory of relativity, and even more so with the *weaker* principle of locality due to Einstein, since special relativity *implies* the principle of locality but not vice versa.

I shall here argue in defence of Einstein's principle of locality, and show that Bell's theorem, sharpened and made testable by the inequality of Clauser and Horne [9] (CH), cannot sustain the experiments that have been based on it and the shattering conclusions that have been drawn from it.

According to Bell, his theorem asserts that NO model exists whose postulates satisfy the CH formal conditions of locality (L1) to (L3) (see below), and whose function is, as Bell[3] writes, "*either accurately or arbitrarily closely*" equal to the functions:[8,9]

$$[p_{12}(\alpha,\beta)]_{QF} = \tfrac{1}{4}[1+ \cos 2(\alpha-\beta)] \tag{1}$$

$$[p_1]_{QF} = [p_2]_{QF} = \tfrac{1}{2}, \tag{2}$$

of the quantum formalism (QF) model for the EPRB *ideal* experiment.

In EPRB type experiments, single photon pairs (γ_1,γ_2) are being born by the annihilation process from the singlet state with different directions of polarization *chosen at random*. Each correlated photon γ_1 and γ_2, respectively, proceeds in opposite directions along, say, the $(+z,-z)$ axes towards the polarizers (P_1,P_2) and the detectors (D_1,D_2). The values $\alpha_1, \alpha_2,\ldots$ and $\beta_1,\beta_2,\ldots$ of the variables α,β are respectively the angles at which the polarizers P_1 and P_2 may be set *at random*; these angles are taken with respect to an arbitrary (x) axis lying in a plane orthogonal to the z axis. Each *value* of the function $[p_{12}(\alpha,\beta)]_{QF}$ corresponds to the conditional probability of the coincidence detection of *the single photon pair* (γ_1,γ_2) by the detectors D_1 and D_2 after passing the polarizers P_1 and P_2 respectively. And each *value* of the (constant) functions $[p_1]_{QF}$ and $[p_2]_{QF}$ corresponds to the single detection probability of γ_1 and γ_2 respectively.

My problem now is: Can a model be constructed whose postulates satisfy (L1) to (L3), and whose constructed function is, as Bell[3] writes, "*either accurately or arbitrarily closely*" equal to the functions (1) and (2) of the QF model ? The answer is yes: *A local model can be constructed whose function is the unique limit function of the family* $\{A^\mu\}$ *of the* A^μ *models, and this limit function is*:

$$\lim_{\mu,N} \{[p^\mu_{12}(\alpha,\beta)]\} = [p_{12}(\alpha,\beta)]_{QF} = \tfrac{1}{4}[1+ \cos2(\alpha-\beta)] \text{ for all } \alpha,\beta, \qquad (3)$$

which is *identical with* the function of the QF model.

The construct of *deleted neighbourhoods* N together with the *uniform convergence* of the family of the A^μ models are sufficient to ensure that the postulates of the A^μ models and their unique *limit function* (3) satisfy (L1) to (L3) meticulously. *The family of the* A^μ *models that converge uniformly to the QF model are not deprived of their local character (in the CH sense) at the deleted limit.*

What is meant by the *family*[10] $\{A^\mu\}$ of the A^μ models, and its *uniform convergence*[11,12] to the unique *limit function* (3) is explained in Section 3. The physical significance of *uniform convergence* is important. By definition, this kind of convergence does *not* depend upon any particular values, say, $\alpha_1, \alpha_2, \ldots$ or $\beta_1, \beta_2, \ldots$, of the variables α, β. So, *the uniform convergence to the limit function (3) does not depend upon any particular choices of the settings of the polarizers*.

I shall show, in Section 4, how my proof of the *limit function* (3) (see formula (13) of Theorem 2) establishes my proposed *local explanatory theory* of the EPRB experiment. More precisely, in the presence of the universal quantifiers ($\forall \mu \in M$) and ($\forall \alpha, \beta \in D$) prefixed to formula (13) of Theorem 2, the *choice* of the values of the variable μ is *independent of* the *choice* of the values of the variables α and β. This explains, in purely *local* terms, *the* characteristic trait of the EPRB experiment where the directions of polarization (given by values of μ) of the single photon pairs are *chosen at random* by the annihilation process from the singlet state, and the directions of the settings (given by values of α and β) are *chosen at random* (or nearly so) by the switches as in the Aspect[6] experiment. In addition, it explains that only the *local interaction* between each polarizer P_1 (P_2) and the corresponding *correlated* photon γ_1 (γ_2), contained in each photon pair (γ_1, γ_2), is involved where each polarizer, according to its setting, selects or rejects the correlated photon. This selection or rejection is a purely *local* and *probabilistic* affair according to my postulates (P1a) and (P1b) (see Section 3).

With this, formula (3) can be interpreted in purely *local* terms as the *chance* each single photon pair has, upon its birth at the source by the annihilation process, to be *selected* by the polarizers from the population of photon pairs emitted by the source *and* to cause a coincidence count with probability given by (3).

So my model, whose function is the unique *limit function* (3), is not only local in the CH sense but, more importantly, it is also local in the *physical sense*. It describes the probabilistic *local selection* of the single photon pairs (γ_1, γ_2) by the polarizers from the population of photon pairs emitted by the source which generates *all* the time the *same* flux of photon pairs with the *same* angular distribution of polarizations *irrespective* of the chosen settings of the polarizers.

In other words, *local action*, in the shape of my proposed probabilistic selection based on my formula (13), is *sufficient* to explain all that the quantum formalism predicts for the EPRB experiment, and much more, from more basic postulates of a structural character as Einstein had in mind.

Moreover, in Section 5, I shall derive a bona fide *specified form* of the generalized CH inequality, known as CH(4), and then show that it is *satisfied by* the QF model itself. This *directly* demonstrates the *consistency* of CH(4) with the quantum formalism. I have shown elsewhere[13] that what *is* violated is some other inequality *unrelated by derivation* to CH(4) and to the CH Lemma. This helped me to locate the invalid step in the CH proof of Bell's theorem.

The argument of the present paper is also directed against the Universality Claim[5] (UC) of Bell and CH that, for EPRB experiments, *all* local theories give predictions that significantly *differ* from those of the QF model. My Theorems 2 and 3 disprove (UC): The "infinite tail" of the family $\{A^\mu\}$ is the class of counterexamples to Bell's theorem and to (UC). Moreover, since CH(4) is satisfied by the QF model, the QF predictions lie *inside* the bounds within which, it is claimed, lie the predictions of *all* local theories, thereby showing again the *falsity* of (UC).

But my present argument, giving a *construction* of counterexamples, is completely independent of the argument in my earlier papers.[5] However, in view of the objections raised against one of my earlier papers,[5a] and of the misunderstandings and confusion that ensued from their mutually contradictory character, some of my critics claimed that (UC) is true and some others that (UC) is false,[14] I have elsewhere[15] clarified a little the outcome of the earlier debate on (UC). There, I also reinstated some important facts which had been omitted, inadvertently or otherwise, from that debate. Here, I shall confine myself to stating the salient point of that debate: Since Horne and Shimony *twice* admitted the *falsity* of (UC), and since this was the issue in my earlier papers,[5] the issue is settled, at least between us.

Incidentally, from my local model for the EPRB *ideal* experiment local models for the non-ideal photon-cascade experiments are easily obtained by introducing measures of the inefficiency of the apparatus. I have shown elsewhere[15] that all my local models, for the ideal and the non-ideal experiments, satisfy the *ad hoc* "no-enhancement assumption" (NEA) of CH. Therefore, it *cannot* be claimed that local models *must* exhibit enhancement in order to match the predictions of the QF model. In addition, I showed there that the NEA is *compatible with* both locality *and* non-locality as defined by CH, and therefore the NEA *cannot possibly* be used even as a criterion of locality in the CH sense. It seems that, as far as the problem of locality is concerned, the NEA has virtually nothing to say.

2. THE CLAUSER AND HORNE POSTULATES OF LOCALITY

The Clauser and Horne[9] formulae CH(2), CH(2'), CH(3) together with their note 13 (formulae from Ref. 9 are here prefixed by the letters CH) intend to give a characterization of physical locality. I shall make this more explicit by stating the three formal postulates of locality which, according to CH and Bell's[16] later endorsement of them, characterize physical locality (but see paragraph (J) of Section 4):

(L1) The definition of the function p_{12} is given by $p_{12}(\alpha,\beta) := \int_\Lambda d\lambda \rho(\lambda) \hat{p}_{12}(\lambda,\alpha,\beta) = \int_\Lambda d\lambda \rho(\lambda) \hat{p}_1(\lambda,\alpha) \hat{p}_2(\lambda,\beta)$. (The factorized form $\hat{p}_{12}(\lambda,\alpha,\beta) = \hat{p}_1(\lambda,\alpha)\hat{p}_2(\lambda,\beta)$ is known as the factorizability condition.)

(L2) The definition of the domain Λ of the variable λ must not depend either upon the variable α or upon the variable β.

(L3) The definition of the normalised distribution function $\rho(\lambda)$ of the variable λ must not depend either upon the variable α or upon the variable β.

Note that (L3) does *not* exclude the possibility that the function $\rho(\lambda)$ may depend upon some other variable, say μ, provided μ is *independent of* both the variables α and β.

3. CONSTRUCTION OF A COUNTEREXAMPLE (LOCAL MODEL)

I shall here explain the construction of my counterexample (local model) in four steps:

STEP 1. I postulate here the following functions $\hat{p}_1(\lambda,\alpha)$ and $\hat{p}_2(\lambda,\beta)$ which describe the standard probabilistic (local) interaction between photons and polarizers.

Postulate 1.

(P1a) $\qquad \hat{p}_1(\lambda,\alpha) = \tfrac{1}{2}[1 + \cos 2(\lambda - \alpha)]$

(P1b) $\qquad \hat{p}_2(\lambda,\beta) = \tfrac{1}{2}[1 + \cos 2(\lambda - \beta)]$.

These postulated functions satisfy the CH factorizability condition (L1). Clearly, the values of $\hat{p}_1(\lambda,\alpha)$ and $\hat{p}_2(\lambda,\beta)$ are bounded by 0 and 1, as probabilities should be. This answers what Feynman[17] calls the *"fundamental problem."*

STEP 2. I show here that I can construct a normalised distribution function $\rho_p(\lambda)$ characterizing the states λ of polarization of the *single photon pairs* (γ_1, γ_2) and a domain Λ of the λ's such that $\rho_p(\lambda)$ and Λ do not depend either upon the variable α or upon the variable β, as required by (L2) and (L3). [The subscript p in $\rho_p(\lambda)$ stands for "photon pairs".]

Within QF, predictions (1) and (2) are obtained[8] with the help of a projection operator for linear polarization. For any direction κ there are *two* discrete eigenvalues each corresponding to one of two *orthogonal* states of linear polarization: one state for the κ direction and the other state for the $\kappa \pm \tfrac{1}{2}\pi$ direction.

The following postulate (P2a) translates this characteristic of QF as closely as possible into the definition of $\rho_p(\lambda)$: $\rho_p(\lambda)$ is characterized precisely by the two states, λ and $\lambda \pm \frac{1}{2}\pi$, for each value of λ. I postulate (or construct) the following normalised distribution function $\rho_p(\lambda)$ and domain Λ:

Postulate 2.

(P2a) $\quad \int_\Lambda d\lambda \rho_p(\lambda) = \frac{1}{2} \int_\Lambda d\lambda [\delta(\lambda) + \delta(\lambda \pm \frac{1}{2}\pi)] = 1$

(P2b) $\quad \Lambda = \{\lambda | -\infty < \lambda < +\infty\}$,

where $\delta(\ )$ is the Dirac[18] distribution (or "function") and the domain Λ includes all possible states λ of polarization of the single photon pairs emitted by the source.

Since the symbols α, β *nowhere* occur in the postulates (P2a) and (P2b), it is manifestly obvious that the postulated function $\rho_p(\lambda)$ of the variable λ and the postulated domain Λ of the variable λ do *not* depend either upon α or upon β, and therefore the CH conditions (L2) and (L3) are *meticulously satisfied.*

Postulates (P2a) and (P2b) define $\rho_p(\lambda) := \frac{1}{2}[\delta(\lambda) + \delta(\lambda \pm \frac{1}{2}\pi)]$. This postulated distribution $\rho_p(\lambda)$ characterizes the *single photon pairs,* each different photon pair (γ_1, γ_2) being characterized by a different value of λ. Since each value of λ corresponds to the azimuthal angle of polarization of each photon pair, λ is clearly *not* a hidden variable. Also, the postulated distribution $\rho_p(\lambda)$ imposes a condition, or a structure, on the variable λ such that λ characterizes, as Bell[16] suggests, the common *"causal factors."* Obviously, for each value of λ, my $\rho_p(\lambda)$ gives each single photon pair the mark of its birth by the annihilation process from the singlet state: As Wheeler[19] writes, *"... if one of these photons is linearly polarized in one plane (say, of azimuthal angle λ), then the photon which goes off in the opposite direction with equal momentum is linearly polarized in the perpendicular plane (of azimuthal angle $\lambda \pm \frac{1}{2}\pi$)."*

STEP 3. I introduce here the *independent* variable μ distinct from the variables α and β. From postulates (P2a) and (P2b) I can *deduce,* as Dirac[20] says, the following formula as a theorem:

$$\int_\Lambda d\lambda \rho_p(\lambda - \mu) = \frac{1}{2} \int_\Lambda d\lambda [\delta(\lambda - \mu) + \delta(\lambda - \mu \pm \frac{1}{2}\pi)] = 1. \qquad (4)$$

Postulate (P2b) and formula (4) define the *shifted distribution:*

$$\rho_p(\lambda - \mu) := \frac{1}{2}[\delta(\lambda - \mu) + \delta(\lambda - \mu \pm \frac{1}{2}\pi)], \qquad (5)$$

where the range (domain) of values of the independent variable μ is the set M of real numbers.

Since the symbols α, β *nowhere* occur in formulae (4) and (5), it is manifestly obvious that my chosen definition of the distribution function $\rho_p(\lambda - \mu)$ satisfies (L3) *meticulously.*

This definition of the function $\rho_p(\lambda - \mu)$ depends upon the variable μ. Since in a *construction* I am entitled to use any permissible construct to satisfy the conditions of the problem, I am entitled to use the construct of *deleted neighbourhoods N* (see definition 1 below). *This enables me to choose the range of values of the variable* μ *to be different from the range of values of the variables* α *and* β. *This ensures, as Church*[21] *says, that the independent variable* μ *is kept distinct from the variables* α *and* β.

More precisely, I employ the *strict* inequality $0<|\mu-\alpha|<\delta=2\varepsilon$ as the *antecedent* in the quantified formula (13) of my proof of Theorem 2. This ensures that the values of the variable μ, including those values of μ required for convergence to the *limit function* (3), as shown by the *consequent* in formula (13), are *any* real numbers *except* equal to the values of the variable α (and β), and therefore the independent variable μ is *kept distinct* from the variables α and β by construction.

To sum up: The definition (5) of the function $\rho_p(\lambda-\mu)$ satisfies (L3) *meticulously* since the proof of Theorem 2, in the shape of my formula (13), ensures that the range of values of μ is *different from* the range of values of α and β. So, the identification of the variable μ with either the variable α or the variable β is impossible in my construction.

Incidentally, any two *distinct* variables must be kept distinct even if they happen to have the same range. Otherwise, as Church[21] says, we would be faced with the absurdity that any two variables x and y whose range is, say, the real numbers R must be identical !

STEP 4. I construct here the family $\{A^\mu\}$ of models whose members satisfy *all* the postulates (L1) to (L3) of locality due to Bell and CH, and whose members *converge uniformly* to the *limit function* (3) identical with the function of the QF model for the EPRB *ideal* experiment.

THEOREM 1: For every admissible real value of μ*, the following functions can be constructed using (L1) together with the postulates (P1a), (P1b), (P2b) and formula (5) of the* A^μ *models:*

$$[p_{12}^\mu(\alpha,\beta)] = \int_\Lambda d\lambda \rho_p(\lambda-\mu)\hat{p}_1(\lambda,\alpha)\hat{p}_2(\lambda,\beta)$$

$$= \tfrac{1}{4}[1+\cos2(\mu-\alpha)\cos2(\mu-\beta)] \tag{6}$$

$$[p_1^\mu(\alpha)] = \int_\Lambda d\lambda \rho_p(\lambda-\mu)\hat{p}_1(\lambda,\alpha) = \tfrac{1}{2} \tag{7}$$

$$[p_2^\mu(\beta)] = \int_\Lambda d\lambda \rho_p(\lambda-\mu)\hat{p}_2(\lambda,\beta) = \tfrac{1}{2}. \tag{8}$$

Comment: For every admissible real value of $\mu \in M$, I obtain an A^μ model, defined by the function $[p_{12}^\mu(\alpha,\beta)]$ etc., *which is unquestionably local in the CH sense:* The postulates (P1a), (P1b), (P2b), formula (5) and the integration over λ meticulously satisfy *all* the postulates of locality (L1) to (L3) due to Bell and CH.

Let $D = \{(\alpha_i,\beta_j) \in R^2\}$ be the range of values of the variables α,β. Let M be the range of values of the variable μ. Since a *family*[10] is itself a *function*, what I consider here is the function p_{12}: $M \times D \to R$ defined by:

$$p_{12}(\mu,\alpha,\beta) := \{[p_{12}^\mu(\alpha,\beta)]\}. \tag{9}$$

The value of the function p_{12} at an index μ, called a *member* of the family, is denoted by $[p_{12}^\mu(\alpha,\beta)]$. Since to every admissible value of $\mu \in M$ corresponds an A^μ model defined by the function $[p_{12}^\mu(\alpha,\beta)]$, the family $\{A^\mu\}_{\mu \in M}$ of the A^μ models has domain the *index set* M and codomain the *indexed set* $\{[p_{12}^\mu(\alpha,\beta)]\}$ of functions.

Definition 1: Let T be a subset of a topological space. With any accumulation point ξ of T, let N be the family of all sets of the form $N = T \cap U \smallsetminus \{\xi\}$ where N is any set obtained by taking any basic neighbourhood U of the point ξ and *removing* ξ from it. Then N is said to be a *directed* (by downward inclusion) family of *deleted basic neighbourhoods* of ξ or that N is a *direction in* T.(12)

The notion of *uniform convergence* is not limited to sequences and series of functions, but can be validly extended to a family of functions since a family is a generalization of the notion of a sequence to an index set other than the set of natural numbers. I now define it more formally.

Definition 2: Let D be a subset of R^2. Let M be a subset of R, and let N be a direction in M. Let $f: M \times D \to R$ be a function. To *each* value of μ in M, let there correspond a function $f^\mu: D \to R$ given by $f^\mu(x) = f(\mu,x)$. Let $g: D \to R$ be a function. The family $\{f^\mu\}$ is said to *converge uniformly* to g on D if for every $\varepsilon > 0$ there exists a $\delta > 0$ (with $\delta(\varepsilon)$ depending *only* on ε) corresponding to a deleted neighbourhood $N_{\delta(\varepsilon)}$ in N such that for *all* $\mu \in M$ and for *all* $x \in D$ whenever the values of μ are in $N_{\delta(\varepsilon)}$, then $|f^\mu(x) - g(x)| < \varepsilon$ holds. In symbols, $\lim_{\mu, N} \{f^\mu(x)\} = g(x)$ uniformly on D. The function $g: D \to R$ is said to be the *limit function* of the family $\{f^\mu\}$, and it is *uniquely* defined.(12)

This definition requires that for each ε one single $\delta(\varepsilon)$, depending *only* on ε, can be found which serves at *every* value of x in D. In other words, the corresponding $N_{\delta(\varepsilon)}$ can freely move about the *whole* range D of the variable x since uniformity is with respect to *all* the values of x in D. So, $N_{\delta(\varepsilon)}$ does not depend upon any particular value of x.(12) [This should be even more clear from the quantified formula (13); see also Comment 2 on Theorem 2 below.]

To avoid certain misunderstandings, some reminders seem in order here. (a) The *limit function* should not be confused with the limit of a function at a point. (b) When it exists, the value of a function at any point of its domain has *no* influence on the existence or the value of the *deleted limit* of the function at that point. (c) Continuity is defined in terms of convergence and expressed in terms of limit operations and not vice versa. (d) The notion of convergence should *not* be misconstrued as "dependence". For, by the axioms of the real number system (complete *ordered* field), *any* two real numbers are either the same or different. So, no matter how large or small is the *distance* $|p-q|$ between any two real numbers p and q, it asserts no "dependence" between them.(11,12)

THEOREM 2: *The family* $\{[p_{12}^\mu(\alpha,\beta)]\}$ *of functions of the* A^μ *models converges uniformly to the function* $[p_{12}(\alpha,\beta)]_{QF}$ *of the QF model for* all α,β and its unique limit function is:

$$\lim_{\mu,N} \{[p_{12}^{\mu}(\alpha,\beta)]\} = [p_{12}(\alpha,\beta)]_{QF} = \tfrac{1}{4}[1+\cos 2(\alpha-\beta)] \qquad (10)$$

$$\lim_{\mu,N} \{[p_1^{\mu}(\alpha)]\} = [p_1]_{QF} = \tfrac{1}{2}; \quad \lim_{\mu,N} \{[p_2^{\mu}(\beta)]\} = [p_2]_{QF} = \tfrac{1}{2}. \qquad (11)$$

Proof: Using the identity that, for every value of μ,

$$\cos 2(\alpha-\beta) = \cos 2(\mu-\alpha)\cos 2(\mu-\beta) + \sin 2(\mu-\alpha)\sin 2(\mu-\beta), \qquad (12)$$

I obtain $|[p_{12}^{\mu}(\alpha,\beta)] - [p_{12}(\alpha,\beta)]_{QF}| = \tfrac{1}{4}|\sin 2(\mu-\alpha)\sin 2(\mu-\beta)|$. Using the inequality $|\sin 2(\mu-\alpha)\sin 2(\mu-\beta)| \leq |\sin 2(\mu-\alpha)|$, valid for *all* α,β and *all* μ, together with the inequality $|\sin z| < |z|$, valid for $z \neq 0$, it follows that $|[p_{12}^{\mu}(\alpha,\beta)] - [p_{12}(\alpha,\beta)]_{QF}| = \tfrac{1}{4}|\sin 2(\mu-\alpha)\sin 2(\mu-\beta)| \leq \tfrac{1}{4}|\sin 2(\mu-\alpha)| < \tfrac{1}{2}|\mu-\alpha|$ for *all* α,β and *any* μ except $\mu=\alpha$. It suffices to choose $\delta = 2\varepsilon$ (δ depending *only* on ε) to establish that for *all* α,β and every μ in $N_{2\varepsilon}$: $0 < |\mu-\alpha| < \delta = 2\varepsilon$ *implies* $|[p_{12}^{\mu}(\alpha,\beta)] - [p_{12}(\alpha,\beta)]_{QF}| < \varepsilon$, and that the limit function of the A^{μ} models is given by (10) and (11). [Note that the same result follows if the inequality $|\sin 2(\mu-\alpha)\sin 2(\mu-\beta)| \leq |\sin 2(\mu-\beta)|$ was used instead; the situation is completely symmetrical between α and β, as it should be.]

The following comments are intended to elucidate some important features of this proof. They may also help prevent certain misunderstandings. But first I should introduce and explain some necessary notation and terminology borrowed from Church.[21]

The symbols $(\forall x)$ and $(\exists x)$ stand for the *universal* and *existential* quantifiers (when, e.g., the operator variable is x), and may be read respectively as "for all x" (or "for every x") and "there is an x such that". The symbol $\supset$ stands for the (truth-functional) *conditional* which, with some caution, may be read as "If..., then..." (or "implies"). The sign $\vdash$, when prefixed to a formula, is used to express that the formula which follows it is a theorem (true proposition).

Comment 1: Using ε and δ as variables whose range is the positive real numbers, and since μ, α, β are distinct variables whose ranges are respectively subsets of the real numbers, Theorem 2 may be expressed by the following quantified formula:

$$\vdash (\forall \varepsilon > 0)(\exists \delta > 0)(\forall \mu \in M)(\forall \alpha, \beta \in D) \; [0 < |\mu-\alpha| < \delta = 2\varepsilon \supset$$
$$|[p_{12}^{\mu}(\alpha,\beta)] - [p_{12}(\alpha,\beta)]_{QF}| < \varepsilon \;]. \qquad (13)$$

Here, in the presence of the *antecedent* $0 < |\mu-\alpha| < \delta = 2\varepsilon$ of formula (13), the values of μ are *any* real numbers *except* equal to the values of α (and β), and therefore the independent variable μ is kept *distinct* from the variables α and β by construction. Moreover, in the presence of the universal quantifiers $(\forall \mu \in M)$ and $(\forall \alpha,\beta \in D)$, the *choice* of values of μ is *independent of* the *choice* of values of α and β. So, my proof of Theorem 2 formally establishes that the definition (5) of $\rho_p(\lambda-\mu)$ satisfies (L3) *meticulously*, including the values of μ required for convergence to the *limit function* (10) as shown by the *consequent* of formula (13) (see also Step 3 above).

It certainly would be a mistake to suggest that, whenever the values of μ and α are chosen to satisfy the antecedent $0<|\mu-\alpha|<\delta=2\varepsilon$, this somehow implies a functional dependence of μ upon α (or upon β) which, if it were true, would violate (L3). This suggestion misconstrues "closeness" as "dependence". But, once again, no matter how large or small is the distance $|p-q|$ between any two real numbers, it asserts no "dependence" between them ! That this suggestion is *absurd* can be seen at once from what was said above: In the presence of the universal quantifiers ($\forall \mu \in M$) and ($\forall \alpha, \beta \in D$) the choice of values of μ is *independent of* the choice of values of α and β. So, there *cannot possibly* be any functional dependence, *nor any kind of dependence of* μ *upon* α *and* β. *Nothing violates (L3)*.

At this point, it should be understood that the antecedent $0<|\mu-\alpha|<\delta=2\varepsilon$ is a *propositional form*. Its associated *propositional function* (21) is a function from ordered triples of values of μ, α, δ (in that order) as arguments to *truth-values*, one of them being *truth* and the other one *falsehood*. So, what we have here is *not* any kind of mutual dependence between the arguments themselves, but rather a functional dependence between the mutually *independent* arguments of this propositional function and truth values. [A propositional function is said to be *satisfied by* an ordered system of arguments if its value for that ordered system of arguments is *truth*.]

Comment 2: Since in formula (13) the existential quantifier ($\exists \delta > 0$) *precedes* the universal quantifier ($\forall \alpha, \beta \in D$), what we have here is *uniform convergence* and the choice of δ depends only upon ε and NOT upon any particular values, say $\alpha_1, \alpha_2, \ldots$ or $\beta_1, \beta_2, \ldots$, of the variables α and β. Thus, as chosen above, the *same* $\delta = 2\varepsilon$ serves at *every* point of D since uniform convergence has been proved with respect to the *whole* domain D of the variables α and β (see Definition 2). The physical significance of this result is very important: *The uniform convergence on D of the family* $\{[p_{12}^{\mu}(\alpha,\beta)]\}$ *to the limit function (10) does not depend upon any particular choices of the settings of the polarizers.*

The familiar distinction between uniform convergence and pointwise convergence[11,12] is based on the difference made by the *order* in which the quantifiers are applied. For example, should Theorem 2 be valid only for the case where the universal quantifier ($\forall \alpha, \beta \in D$) preceded the existential quantifier ($\exists \delta > 0$) in formula (13), convergence would be pointwise only and the choice of δ would depend upon ε and upon α and β, and thereby N_δ would depend upon particular values of α and β. This would in turn imply an implicit dependence of the convergence to the *limit function* (10) upon particular choices of the settings of the polarizers. But I must not have this: *The stronger condition of uniform convergence obtained here excludes any such implicit dependence by ensuring that* δ *is independent of* α *and* β.

It should be noted that choosing the values of μ to be *any* real numbers *except* equal to the values of α and β is *not* sufficient by itself to ensure that δ is independent of α and β. More is needed, and this is the additional condition of *uniform convergence*.

BELL'S THEOREM: A COUNTEREXAMPLE   81

Comment 3: The construct of *deleted neighbourhoods* N and the *uniform convergence* of the family of the A^μ models are sufficient to ensure that the postulates of the A^μ models and their *limit function* (10) are perfectly consistent with (L1) to (L3): *(L1) to (L3) are meticulously satisfied even at the limit,* and therefore the A^μ models are not deprived of their local character (in the CH sense) at **the** deleted limit.

Comment 4: My chosen *local model* for the EPRB *ideal* experiment corresponds to the unique *limit function* (10) of the family $\{A^\mu\}$ of the local A^μ models, and this unique *limit function* (10) is identical with the function of the QF model.

THEOREM 3: *Bell's theorem and the Universality Claim (UC) are invalid.*

Proof: By Theorem 2.

Comment 1: My construction of the *limit function* (10) *refutes* Bell's theorem, and the "infinite tail" of $\{A^\mu\}$ is the class of *counterexamples* to the Universality Claim[5] (UC) of Bell and CH that, for EPRB experiments, *all* local theories give predictions that significantly *differ* from those of the QF model.

Comment 2: By Bell's theorem I mean precisely what Bell means: That NO model exists whose postulates satisfy (L1) to (L3), and whose function p_{12} etc. is, as Bell[3] writes, "*either accurately or arbitrarily closely*" equal to the function $[p_{12}(\alpha,\beta)]_{QF} = \frac{1}{4}[1 + \cos 2(\alpha-\beta)]$ for *all* α, β.

4. CONSTRUCTION OF A LOCAL EXPLANATORY THEORY OF THE EPRB EXPERIMENT

Two formal facts from my proof of Theorem 2, in the shape of formula (13), play a prominent role here. One is that, in the presence of the universal quantifiers $(\forall \mu \in M)$ and $(\forall \alpha, \beta \in D)$, the choice of values of the variable μ is *independent of* the choice of values of the variables α and β. The other one is the antecedent $0 < |\mu - \alpha| < \delta = 2\varepsilon$. Together they establish my *local explanatory theory* of the EPRB experiment which *consistently* explains the following three physical facts, and much more.

(A) The directions of polarization (values of μ) of the single photon pairs are *chosen at random* by the annihilation process from the singlet state.
(B) The directions of the settings (values of α and β) of the polarizers are *chosen at random* (or nearly so) by the switches as in the Aspect[6] experiment.
(C) The source generates *all* the time the *same* flux of photon pairs with polarizations in all directions.

By these three physical facts, there are some photon pairs, amongst the *population* of photon pairs emitted by the source, whose directions of polarization *happen*, by pure chance, to be within what I may here call the "*coincidence sector*" associated with *every* chosen polarizer setting, thereby enabling such photon pairs to be *selected by* the polarizers from the population of photon pairs emitted by the source *and to cause* coincidence counts with probability given by (10). [When the polarizers are orthogonal, there are no photon pairs amongst the population of photon pairs with the property to cause coincidence counts, even though each *correlated* photon, contained in each photon pair (γ_1,γ_2), *still* impinges upon the corresponding polarizer causing single counts with probability given by (11).]

More precisely, in the presence of the universal quantifiers ($\forall \mu \in M$) and ($\forall \alpha, \beta \in D$), the *choice* of values of the variable μ is *independent of* the *choice* of values of the variables α and β, and therefore what the antecedent $0 < |\mu-\alpha| < \delta = 2\varepsilon$ of formula (13) says is that, for *every* chosen value of α, *whenever* a value of μ, characterizing the *random* direction of polarization of a single photon pair, *happens by pure chance* to fall inside the set:

$$S_\alpha^c = \{\mu \mid -2\varepsilon + \alpha < \mu < \alpha + 2\varepsilon\}, \qquad (14)$$

this single photon pair is *selected by* the polarizers *from* the population of photon pairs emitted by the source *and causes* a coincidence count with probability given by (10), as shown by the *consequent* of formula (13). [Note that the situation is completely symmetrical between the variables α and β, as it should be; see proof of Theorem 2. The superscript c in S_α^c stands for "coincidence counts".]

So, the formal fact that the *choice* of values of the variable μ is *independent of* the *choice* of values of the variables α and β consistently explains the physical facts (A) and (B) which describe *the* characteristic trait of the EPRB experiment designed by Aspect et al.

The same formal fact renders *invalid* any suggestion that the photon pairs, at the moment of their birth by the annihilation process at the source, somehow have to adjust their random directions of polarization to fit any randomly chosen direction of the polarizer settings, or that the source somehow knows anything about the polarizer settings. *Without any proof*, this suggestion asserts that there is a conspiratorial dependence of the choice of values of μ upon the choice of values of α and β by which the source knowingly adjusts the directions of polarization (values of μ) of the photon pairs to fit, or to be "very close" to, *any* chosen setting of the polarizers (values of α and β). But, apart from mistaking "closeness" for "dependence" (see Comment 1 on Theorem 2), this assertion is *disproved* by my proof of formula (13): Since the *choice* of values of μ is *independent of* the *choice* of values of α and β, the source *cannot "knowingly"* adjust the directions of polarization of the photon pairs being *chosen at random by* the annihilation process, and the source *cannot "know"* anything about the polarizer settings being *chosen at random by* the switches. Such *"spooky action at a distance,"* and any such conspiratorial adjustment are ruled out *ab initio* by my proof of formula (13).[22]

On the other hand, my proposed *local explanatory theory*, based on my proof of formula (13), shows that *everything that happens is due to the common birth of the single photon pairs, local action, pure chance, and probabilistic selection*. With this, my theory explains that:

(D) Choosing different settings of the polarizers (values of α and β) has *no* influence upon the directions of polarization (values of μ) of the single photon pairs issuing from the source.

(E) Only the probabilistic *local interaction* between each polarizer P_1 (P_2) and the corresponding *correlated* photon γ_1 (γ_2), contained in each photon pair (γ_1,γ_2), is involved. Each polarizer, according to its setting (values of α and β), selects or rejects the impinging *correlated* photons with random directions of polarization (values of μ). This selection or rejection is a purely *local* and *probabilistic* affair according to my postulates (P1a) and (P1b) (since integration over λ using a Dirac delta function amounts to substituting μ for λ).

(F) Formula (10) can be interpreted, in purely *local* terms, as the overall measure of the *chance* each single photon pair has, upon its birth by the annihilation process, to be *selected* by the polarizers from the population of photon pairs emitted by the source *and to cause* a coincidence count with probability given by (10), as shown by the *consequent* of formula (13). [It follows from this that formula (10) can also be interpreted, in purely *local* terms, as a statistical prediction for the population of photon pairs emitted by the source to be *selected* by the polarizers *and to cause* coincidence counts.]

(G) It follows from (D) that choosing different settings of the polarizers has *no* influence upon the angular distribution of the population of photon pairs emitted by the source which (population) remains *invariant* upon selection of the *different* sets $S^c_{\alpha_1}$ ($S^c_{\beta_1}$), $S^c_{\alpha_2}$ ($S^c_{\beta_2}$) etc. corresponding to *different* choices of the settings of the polarizers. This can be seen from the following which also explains the *axial invariance* of the population of photon pairs emitted by the source.

Let the source be fixed. Rotate both polarizers P_1 and P_2 about the z axis so that their *relative* setting $\phi = |\alpha - \beta|$ is fixed to some *arbitrary* value (I may fix both polarizers and rotate the source about the z axis; the situation is completely symmetrical). Since the choice of values of the variable μ, characterizing the different directions of polarization of the single photon pairs emitted by the source, is *independent of* the choice of values of the variables α and β, as the rotating polarizers *sweep different* directions, such that $|\alpha_1 - \beta_1| = |\alpha_2 - \beta_2| = \ldots = $ constant, *different* sets $S^c_{\alpha_1}$ ($S^c_{\beta_1}$), $S^c_{\alpha_2}$ ($S^c_{\beta_2}$) etc. of photon pairs are being selected from the population of photon pairs emitted by the source. But, according to the *limit function* (10) and (11), *nothing* changes: The number of coincidence and single counts remains the same which implies that the flux of photon pairs emitted by the source and their angular distribution of polarizations remain *invariant* upon selection of the different sets $S^c_{\alpha_1}$ ($S^c_{\beta_1}$), $S^c_{\alpha_2}$ ($S^c_{\beta_2}$) etc. of photon pairs. [Actually, this mode of selection is used as a test of the *axial invariance* of the population of photon pairs emitted by the source; and, in all experiments performed so far,[23] no deviation from it has been detected.]

Of course, to each different value of φ corresponds a different number of coincidence counts which, however, remains the *same* upon rotating both polarizers with their relative setting fixed to that value of φ. Note that the number of single counts is the *same* for *all* values of φ according to (11).

But actually more has been shown than said above. Since the choice of a value of ϕ (fixed relative setting) was arbitrary, the flux of photon pairs emitted by the source and their angular distribution of polarizations remain *invariant* upon selection of *any* set S_α^c (or S_β^c) of photon pairs. In other words, for *all* values of ϕ, the source generates the *same* flux of photon pairs with the *same* angular distribution of polarizations *irrespective of which set S_α^c (or S_β^c) happens to be selected from the population of photon pairs emitted by the source.*

The same arguments allow me to add a little to the *local* explanation of my formula (10) given in (F) above:

(H) Since the source generates the *same* flux of photon pairs with the *same* angular distribution of polarizations, and since the choice of values of μ is *independent of* the choice of values of α and β, choosing a *different* setting, say, α_2 ($\neq \alpha_1$) of polarizer P_1 simply means that a *different* set $S_{\alpha_2}^c$ ($\neq S_{\alpha_1}^c$) of photon pairs is selected from the *invariant* population of photon pairs emitted by the source to cause coincidence counts with a *different* probability given by my formula (10).

To sum up: There is *no* need whatever either to invoke or appeal to action at a distance (nonlocality) between the source and the polarizers, or between the polarizers themselves, in order to explain the EPRB experiment. *Local action suffices*. My local explanatory theory, based on my formula (13), is *sufficient* to explain all that the quantum formalism predicts for the EPRB experiment *especially* the physical facts (A) and (B) which describe *the* characteristic trait of the EPRB experiment as designed by Aspect *et al.*, and much more:

(I) Since the QF model is uniquely *derivable from* the postulates of the A^μ models, which describe the *structure* and polarization properties of the single photon pairs born by the annihilation process and their subsequent *local* interaction with the polarizers, what we have here is a *local explanatory theory* of the QF model itself from more basic postulates of a structural character; and since this local explanatory theory is *not* part of the quantum formalism, it can be described as an *extension* of the quantum formalism in the direction which Einstein had in mind when he called for a derivation or an explanation of the quantum formalism by a local theory of a structural character.

(J) My theory shows the *physical inadequacy* of (L3). Since the directions of polarization of the photon pairs born by the annihilation process are random, *nothing* prevents the birth of *some* photon pairs whose directions of polarization *happen by pure chance* to be equal to that of the settings of the polarizers. Obviously, it would be physically unreasonable to exclude *ad hoc* this perfectly local state of affairs. Yet *this is precisely what (L3) DOES exclude*. So, this local state of affairs *refutes* (L3), and it shows that formal locality in the CH sense is certainly *not* equivalent to *physical locality*.

Thus, even though my proof of formula (13) meticulously satisfies (L3), I *could* contend, on the physical grounds given in (J) above, that the *strict* inequality $0<|\mu-\alpha|<\delta=2\varepsilon$ may be replaced by $|\mu-\alpha|<\delta=2\varepsilon$ *without* violating locality in the physical sense. Note that, upon such replacement, the variable μ *remains* distinct from the variables α and β, and the choice of values of µ *remains* independent of the choice of values of α and β, even if the range of the distinct variable µ were to cover the range of α or β rather than some of their values[21] (see paragraph immediately before Step 4 in Section 3). With this, I conclude my proposed local explanation of the EPRB experiment.

5. THE QUANTUM FORMALISM IS CONSISTENT WITH CH(4)

Starting from the CH Lemma as a premise, using the postulates of the A^μ models which satisfy (L1) to (L3), and by meticulously following the CH derivation of the generalized CH inequality, known as CH(4), I shall derive a *bona fide specified form* of CH(4), and then I shall demonstrate its *consistency with* the quantum formalism by showing that it is satisfied by the function $[p_{12}(\alpha,\beta)]_{QF}$ of the QF model itself for *all* values of α,β.

The CH Lemma is the proposition (for a proof see Ref. 9) that if the values of the (numerical) variables A,B,C,D are in the range [0,+1], then

$$-1 \leq [AB+BC+CD-DA-C-B] \leq 0. \tag{15}$$

Next I meticulously follow the CH derivation of CH(4) and I substitute $\hat{p}_1(\lambda,\alpha_1)$ for A, $\hat{p}_2(\lambda,\beta_1)$ for B, $\hat{p}_1(\lambda,\alpha_2)$ for C and $\hat{p}_2(\lambda,\beta_2)$ for D in (15) to obtain:

$$-1 \leq [\hat{p}_1(\lambda,\alpha_1)\hat{p}_2(\lambda,\beta_1) - \hat{p}_1(\lambda,\alpha_1)\hat{p}_2(\lambda,\beta_2) + \hat{p}_1(\lambda,\alpha_2)\hat{p}_2(\lambda,\beta_1)$$
$$+ \hat{p}_1(\lambda,\alpha_2)\hat{p}_2(\lambda,\beta_2) - \hat{p}_1(\lambda,\alpha_2) - \hat{p}_2(\lambda,\beta_1)] \leq 0, \tag{16}$$

where all the terms of (16) must be evaluated at the *same* value of the variable λ for *every* choice of λ from the range Λ. Then, I multiply by the *same* value of the function $\rho_D(\lambda-\mu)$, evaluated at that value of λ, each and every term and the bounds -1 and 0 of inequality (16), and I integrate with respect to λ over the range Λ according to the postulates of the A^μ models to obtain the following *bona fide specified form* of CH(4):

$$-2 \leq [\cos 2(\mu-\alpha_1)\cos 2(\mu-\beta_1) - \cos 2(\mu-\alpha_1)\cos 2(\mu-\beta_2)$$
$$+ \cos 2(\mu-\alpha_2)\cos 2(\mu-\beta_1) + \cos 2(\mu-\alpha_2)\cos 2(\mu-\beta_2)] \leq +2. \tag{17}$$

Since inequality (17) is one proposition with *one* and the *same* variable µ occurring in its four terms, *each and every term* of (17) must be evaluated at the *same* value of µ for *every* choice of µ from the range M. Note that integration over λ using a Dirac delta function amounts to substituting µ for λ, as Dirac[18] says, and all the terms of inequality (16) must be evaluated at the *same* value of λ for *every* choice of λ ∈ Λ.

THEOREM 4: *The proposition* $-2 \leq ab+bc+cd-da \leq +2$ *is true for all the values of the (numerical) variables a,b,c,d in the range* [-1,+1].

Proof: Let V= ab+bc+cd-da= (c+a)b+ (c-a)d. For any fixed values of a and c, V is a linear function of b and d, and therefore the global maximum and minimum of V occur at the boundaries of the range $[-1,+1]$, at $b= \pm 1$ and $d= \pm 1$. There are four possible cases: if $b= 1$ and $d= 1$, then $V= 2c$; if $b= 1$ and $d= -1$, then $V= 2a$; if $b= -1$ and $d= 1$, then $V= -2a$; if $b= -1$ and $d= -1$, then $V= -2c$. Since the values of a, c are in $[-1,+1]$, it follows that $-2 \leq V \leq +2$ which proves Theorem 4.

THEOREM 5: *The inequality (17) is true for all the values* $\alpha_1, \alpha_2, \beta_1, \beta_2$, *and for all values of* μ.

Proof: Inequality (17) follows from the proposition $-2 \leq ab+bc+cd-da \leq +2$ of Theorem 4 by substituting $\cos2(\mu-\alpha_1)$ for a, $\cos2(\mu-\beta_1)$ for b, $\cos2(\mu-\alpha_2)$ for c, and $\cos2(\mu-\beta_2)$ for d. Obviously, the values of a,b,c,d are in $[-1,+1]$ for *all* the values $\alpha_1, \alpha_2, \beta_1, \beta_2$ and for *all* the values of μ. Conversely, $-2 \leq ab+bc+cd-da \leq +2$ follows from the inequality (17) by substituting a for $\cos2(\mu-\alpha_1)$, b for $\cos2(\mu-\beta_1)$, c for $\cos2(\mu-\alpha_2)$, and d for $\cos2(\mu-\beta_2)$. So, inequality (17) is *equivalent* to the proposition $-2 \leq ab+bc+cd-da \leq +2$ of Theorem 4.

Comment: By the proof of Theorem 5, inequality (17) *can be derived and defined* for *all* values of μ, since inequality (17) is true for *all* values of μ. This permits the extension of the range of the *distinct* variable μ from M to M* which may include the values of the variables α and β. Let $S(\mu, \alpha_1, \alpha_2, \beta_1, \beta_2)$ denote the expression within the square brackets of inequality (17). Then, Theorem 5 may be expressed by the following quantified formula:

$$\vdash (\forall \mu \in M^*) [-2 \leq S(\mu, \alpha_1, \alpha_2, \beta_1, \beta_2) \leq +2]. \tag{18}$$

It should be noted from formula (18) that μ is *not* fixed to any particular value chosen from the range M*. Also, in formula (18), $\alpha_1, \alpha_2, \beta_1, \beta_2$ are treated as constants since they are *values* of the variables α, β. However, should $\alpha_1, \alpha_2, \beta_1, \beta_2$ themselves be considered as variables (rather than as values of α, β), then the universal quantifier $(\forall \alpha_1, \alpha_2, \beta_1, \beta_2 \in D)$ must be prefixed to formula (18).

Next I shall show that the QF formula (1) can be transformed into an *equivalent* conditional formula whose *antecedent* satisfies CH(4) in the specified form (17). This *directly* demonstrates the *consistency* of CH(4) with the quantum formalism.

THEOREM 6: *For all* α, β *and for all* μ, *the QF formula (1) is logically and mathematically equivalent to the quantified formula:*

$$\vdash (\forall \mu \in M^*)(\forall \alpha, \beta \in D) [(\mu= \alpha) \vee (\mu= \beta) \supset$$
$$[P_{12}(\alpha,\beta)]_{QF} \equiv [P_{12}^{\mu}(\alpha,\beta)]_{QF} = \tfrac{1}{4}[1+ \cos2(\mu-\alpha)\cos2(\mu-\beta)]]. \tag{19}$$

Proof: That formula (1) follows from formula (19) by substituting for μ either α or β is trivial. That (19) follows from (1) can be shown in two steps: (i) Substitute the identity (12), valid for *all* α, β and *all* μ, for $\cos2(\alpha-\beta)$ in formula (1); (ii) Note that the *antecedent* of formula (19), that is, $\mu= \alpha$ or $\mu= \beta$, ensures that the product of the two sine terms occurring in (12) becomes zero.

Comment 1: The symbol v in the antecedent ($\mu= \alpha$) v ($\mu= \beta$) of (19) stands for the (truth-functional) inclusive *disjunction*,[21] and may be read as "OR".

Comment 2: The transformation of formula (1) with two variables α, β into the equivalent conditional formula (19) with three variables μ, α, β is established by the identity (12). This transformation is, of course, *logically independent of* the existence of the A^μ models.

THEOREM 7: The QF model itself satisfies CH(4) in the specified form (17) for all α, β.

Proof: Since inequality (17) is true for *all* values of μ by Theorem 5 (see formula (18)), it is true for the antecedent $\mu= \alpha$ or $\mu= \beta$ of (19). This *directly* demonstrates the *consistency* of CH(4) with the QF formula (1) for *all* values of α, β.

Comment 1: Theorem 7 has the following physical interpretation: *Each emitted single photon pair* (γ_1, γ_2) *and its coincident selection (count) by ANY ONE pair* (α_i, β_j) *of polarizer settings is ONE EXPERIMENT whose probability of occurrence is given by the* value *of the function* $[p_{12}(\alpha, \beta)]_{QF}$ *evaluated at* (α_i, β_j) *and CH(4) in the specified form (17) is satisfied by the antecedent* $\mu= \alpha_i$ *or* $\mu= \beta_j$ *of formula (19).* [Since Theorems 6 and 7 are true for *all* α, β they are true for *any* pair (α_i, β_j) of settings.]

Comment 2: This interpretation is consistent with the empirical fact noted by Feynman[17] who wrote (for i,j= 1,2): *"When a single photon pair comes in, one always finds that only ONE of the four counters is triggered off in coincidence."*
In other words, a single photon pair *cannot possibly be selected by more than one pair of settings at a time*, say (α_1, β_1) and (α_2, β_2). The coincidence counts are consecutive: one coincidence count after another.
For example, the experiments with the switches[6] merely move from *one* experiment with settings (α_1, β_1) to *another* experiment with settings (α_1, β_2) to *another* with settings (α_2, β_1) to *another* with settings (α_2, β_2) in some nearly random way; and each *one* of the four *different* experiments occurs with probability given by *one* value from the set:

$$\{p_{12}(\alpha_1,\beta_1), \ p_{12}(\alpha_1,\beta_2), \ p_{12}(\alpha_2,\beta_1), \ p_{12}(\alpha_2,\beta_2)\}_{QF} \tag{20}$$

of values obtained by evaluating the function $[p_{12}(\alpha,\beta)]_{QF}$ at the corresponding arguments (pairs of settings).

Comment 3: This physical interpretation explains the "OR" in the antecedent $\mu= \alpha_i$ or $\mu= \beta_j$ of formula (19), and it agrees with the formal fact that *there is permitted one and only one substitution for μ at a time in each and every term of inequality (17)* since (17) is one proposition with one and the same variable μ occurring in its four terms. That is, we cannot possibly have "AND" replace the "OR" in the antecedent

$\mu = \alpha_i$ or $\mu = \beta_j$ of formula (19), and we cannot possibly substitute two different values for μ in inequality (17) at a time, say, $\mu = c_1$ in the first two terms of (17) AND $\mu = c_2$ in the last two terms of (17). This would be both *invalid* and *contrary* to the physical fact that a single photon pair, characterized by a value of μ, *cannot possibly be selected by more than one pair of settings at a time*.

Comment 4: Surprising as it may seem to some people, the fact that the QF model itself satisfies CH(4) is as it should be for the following almost trivial logical reason: Since the postulates of the A^μ models and their unique *limit function* (10) satisfy CH(4), and since the function $[p_{12}(\alpha,\beta)]_{QF}$ of the QF model is *derivable from* the postulates of the A^μ models and is *identical with* their limit function (10), the QF model itself must satisfy CH(4) as shown by Theorem 7.

6. CONCLUSIONS

A proof can be shown to be invalid in several ways. One way is to show that the conclusion is mistaken. Bell and CH claim to have reached the conclusion that NO local model exists whose constructed function is *"either accurately or arbitrarily closely"* equal to the function of the QF model for the EPRB *ideal* experiment. My construction of the family $\{A^\mu\}$ of the meticulously *local* A^μ models and of their *limit function*, which is *identical with* the function of the QF model, shows that the conclusion of Bell and CH is mistaken.

The *consistency* of the quantum formalism with CH(4) has been here demonstrated by the construction of a *bona fide specified form* of CH(4) which is *satisfied by* the function of the QF model for *all* α, β.

The Aspect experiments, with and without the switches, cannot be considered as *crucial* between locality and non-locality since the theory (Bell's theorem) on which these experiments are based is mistaken. These experiments, even though they support (as expected) the QF predictions, *cannot possibly refute Einstein's principle of locality*.

Local action, in the shape of my proposed probabilistic selection based on my formula (13), is *sufficient* to explain all that the quantum formalism predicts for the EPRB *ideal* experiment, and much more, from more basic postulates of a structural character as Einstein had in mind.

The *"causal factors"*(16) involved here are the births of the single photon pairs by the annihilation process from the singlet state, a purely *random* and *local affair* confined at the source, and the subsequent *local interaction* between each correlated photon and the corresponding polarizer.

My proposed *local explanatory theory* explains *the* characteristic trait of the EPRB experiment where the directions of polarization of the single photon pairs are being *chosen at random by* the annihilation process from the singlet state, and the directions of the polarizer settings are being *chosen at random* (or nearly so) *by* the switches as in the experiments designed by Aspect et al.

ACKNOWLEDGEMENTS

I am indebted to Professor C.W. Kilmister for his helpful suggestions.

REFERENCES AND NOTES

1. M. Jammer, *The Philosophy of Quantum Mechanics* (John Wiley, New York, 1974), p. 116.
2. A. Einstein, *Dialectica* 2, 320 (1948).
3. J.S. Bell, *Physics (N.Y.)* 1, 195 (1964).
4. A. Einstein, B. Podolsky, and N. Rosen, *Phys. Rev.* 47, 777 (1935); D.J. Bohm, *Quantum Theory* (Prentice-Hall, Englewood Cliffs, N.J., 1951), pp. 611-623.
5. Th.D. Angelidis, *Phys. Rev. Lett.* 51, 1819 (1983); and in *Open Questions in Quantum Physics*, A. van der Merwe et al., eds. (Reidel, Dordrecht, Holland, 1985), pp. 51-62.
6. A. Aspect, J. Dalibard, and G. Roger, *Phys. Rev. Lett.* 49, 1804 (1982).
7. B. d'Espagnat, *In Search of Reality* (Springer-Verlag, London, 1983).
8. J.F. Clauser and A. Shimony, *Rep. Prog. Phys.* 41, 1881 (1978).
9. J.F. Clauser and M.A. Horne, *Phys. Rev. D* 10, 526 (1974).
10. P.R. Halmos, *Naive Set Theory* (Van Nostrand Reinhold, New York, 1960), p. 34.
11. A.M. Gleason, *Fundamentals of Abstract Analysis* (Addison-Wesley, Reading, Mass., 1966), pp. 77, 245-248.
12. E.J. McShane and T.A. Botts, *Real Analysis* (Van Nostrand, Princeton, London, 1959), pp. 32-33, 68-69, 81-82.
13. Th.D. Angelidis, in *Microphysical Reality and Quantum Formalism*, A. van der Merwe et al., eds. (Reidel, Dordrecht, Holland, 1988), pp. 457-478.
14. Th.D. Angelidis, *Phys. Rev. Lett.* 53, 1022 (1984); and *Phys. Rev. Lett.* 53, 1297 (1984).
15. Th.D. Angelidis, in *Causality and Locality: 50 Years of the Einstein-Podolsky-Rosen Paradox* (Hellenic Physical Society and Interdisciplinary Research Group, University of Athens, 1987).
16. J.S. Bell, CERN preprint TH-2926 (1980).
17. R.P. Feynman, *Intl. J. Theor. Phys.* 21, 467 (1982). Feynman wrote: "The only difference between a probabilistic classical world and the... quantum world is that somehow or other it appears as if the probabilities would have to go negative... that's the fundamental problem." Note 18 of Ref. 15 explains how I tackled this "fundamental problem."
18. P.A.M. Dirac, *The Principles of Quantum Mechanics* (Oxford University Press, Oxford, 1958), p. 59.
19. J.A. Wheeler, *Ann. N.Y. Acad. Sci.* 48, 219 (1946).
20. See Ref. 18; and I.N. Sneddon, *The Use of Integral Transforms* (Tata McGraw-Hill, New York, Delhi, 1974), p. 499. More formally, if $T(x)$ is a distribution defined over the set R of real numbers, and a $\in$ R, the *shifted distribution* $T(x-a)$ can be defined by the equantion $<T(x-a),\phi(x)> = <T(x),\phi(x+a)>$ for all test functions $\phi(x)$ defined on the space $C_0^\infty(R^n)$.
21. A. Church, *Introduction to Mathematical Logic* (Princeton University Press, Princeton, New Jersey, 1956), pp. 9-10, 28, 37, 41-44.
22. My local explanatory theory, based on my proof of formula (13), shows that a defender of local realism does not have to rely on the *absurd* loophole of the backward light cones. This loophole, used by those

who believe in non-locality to dismiss local realism, conjures up an utterly improbable conspiracy, which would leave even a Cartesian demon gaping, that physical events in the overlap of the backward light cones determine both the random directions of polarization of the photon pairs and the random directions of the settings of the polarizers. Although this loophole is an instance of what one could call "Bohr's principle of the *total* experimental conditions," and of the related notion of "Bohm's *undivided* wholeness," and although it it admitted by Bell, Shimony *et al.* [see Ref. 8] to be a loophole, my local explanatory theory shows that there is no such conspiracy. On the other hand, those who believe in non-locality do not seem to have noticed that, should the "spooky action at a distance" they uphold be true, *everything* in the physical world would be conspiratorial, not only the events in the backward light cones ! It would be a world whose very structure would prevent us from learning more about these spooky effects at *any* distance, and from relying on the significance of *any* experiment including that of Aspect *et al.*
23. C.A. Kocher and E.D. Commins, *Phys. Rev. Lett. 18*, 575 (1967). In this experiment, one polarizer [P_1] is fixed and the other [P_2] is movable. These authors state: *"We have made runs with different orientations [settings] of the fixed polarizer [P_1], obtaining in each case a correlation which depends only on the relative angle [ϕ]."* ; A. Aspect, P. Grangier, and G. Roger, *Phys. Rev. Lett. 47*, 460 (1981). These authors wrote: *"We never observed any deviation from rotational [axial] invariance."*

QUANTUM PROBABILITY AND QUANTUM POTENTIAL APPROACH TO QUANTUM MECHANICS

Anastasios Kyprianidis

Institut Henri Poincaré
Laboratoire de Physique Théorique
11, rue Pierre et Marie Curie
75231 PARIS CEDEX 05

ABSTRACT

Starting from a classical Hamilton-Jacobi theory the Schrödinger equation of non relativistic quantum mechanics is derived by introducing an averaging procedure over the free parameters of the classical formalism. The conceptual implications, the relation to the causal interpretation and possible extensions of the formalism are discussed.

1. THE HAMILTON-JACOBI EQUATION IN CLASSICAL MECHANICS

One of the possible ways to determine a solution of a given dynamical problem in classical mechanics is given in terms of the Hamilton-Jacobi theory. In this frame the dynamical problem is reduced to the determination of Hamilton's principal function W satisfying the Hamilton-Jacobi (H-J) differential equation.

$$H(q_r, \frac{\partial W}{\partial q_r}) + \frac{\partial W}{\partial t} = 0 \qquad r = 1, \ldots, n \qquad (1)$$

Any complete integral of this equation, i.e, an integral containing as many arbitrary constants as there are independant variables, yields the equations of motion of the system. The function W is the generating function producing a canonical transformation to new variables P and Q such that all new momenta P are constants of the motion. A complete solution to the equation (1) contains n constants of integration $\alpha_1 \ldots \alpha_n$, and the new constant canonical momenta P can be chosen as any n independent functions of the n constants of integration, i.e.,

$P_r = P_r(\alpha_1 \ldots \alpha_n)$. The complete integral to the H-J equation then reads as :

$$W = W(q_r, P_r, t)$$

The transformation equations have the following form [1] :

$$p_r = \frac{\partial W}{\partial q_r}(q,P,t) \qquad Q_r = \frac{\partial W}{\partial P_r}(q,P,t) \qquad (2)$$

They yield the old momenta p and the new coordinates Q in terms of the old coordinates q and the new momenta P.

It should be noted here that the transformation function W essentially depends on the new momenta P, and that it is not defined on a trajectory q(t) but represents a scalar field which serves to determine the trajectories : using the transformation equations and fixing the value of the constants P we obtain the particle trajectory. Furthermore, the function W is not identical to the classical action function. The action integral between any two points of an orbit can be obtained

a) if one fixes the values of P in W corresponding to the specific orbit and

b) if one takes the difference between the respective values of W at the two endpoints.

Finally, note that W is not a phase-space function since its arguments belong to different phase-space representations. However, a phase-space picture of the system's motions can be deduced from W. Namely, once the P's are fixed, say to a value $P_r^{(a)}$, the relation $p_r = \frac{\partial W}{\partial q_r}(q, P^{(a)})$ represents an n-dimensional surface $\Sigma^{(a)}$ in the 2n-dimensional phase-space which will contain all phase-space orbits of the classical system for different initial positions. A different choice, say $P_r^{(b)}$, fixes a $\Sigma^{(b)}$ with analogous properties. A continuous variation of the P-values induces a succession of n-dimensional surfaces Σ which continuously scan the whole of phase-space. Nevertheless, one should keep in mind that for a specific classical system only one surface Σ exists which contains the phase-space orbits of the system.

2. THE CAUSAL INTERPRETATION OF QUANTUM MECHANICS

We revisited in the first section the classical Hamilton-Jacobi theory and clarified the role of Hamilton's principal function W in this frame, because our aim is to investigate the validity of the above basic features in the causal interpretation of quantum mechanics [2]. The reason for this attempt lies in the fact that quantum particle dynamics are reproduced in the causal interpretation by means of a modified Hamilton-Jacobi theory. To this end one substitutes $\psi = R e^{i(S/\hbar)}$ in the Schrödinger equation

$$\hbar i \frac{\partial \psi}{\partial t} = - \frac{\hbar^2}{2m} \nabla^2 \psi + V \psi \tag{3}$$

and by decomposing into real and imaginary part one obtains a set of two equations

$$\frac{\partial S}{\partial t} + \frac{(\nabla S)^2}{2m} + V - \frac{\hbar^2}{2m} \frac{\nabla^2 R}{R} = 0$$

and (4)

$$\frac{\partial R^2}{\partial t} + \nabla (R^2 \frac{\nabla S}{m}) = 0$$

The first is the modified H-J equation governing the particle dynamics and the second is the continuity equation introducing a probability distribution of the particle positions, i.e., the probabilistic element in the theory.

The usual statement in the causal interpretation is then that the modification of the H-J equation essentially consists in the presence of the purely quantum-mechanical term, the quantum-potential $Q = - \frac{\hbar^2}{2m} \frac{\nabla^2 R}{R}$ which is determined by the quantum-Schrödinger field ψ, solution of equation (3). The probabilistic element is thought to be a consequence of the lack of knowledge of the particle positions, and manifests itself in the presence of the position probability density R^2. However, this is a secondary aspect of the theory, while the main novelty consists in the emergence in the dynamical equations of the quantum-potential term, which guides non-classically the motion of the particle.

However, there are some striking deviations in the quantum H-J equation (4) from the classical one, beyond the presence of the additional quantum-potential term [3]. The most basic one is that in classical dynamics the complete integral of the H-J equation is a function

$W(q,P,t)$ while in the quantum case the phase of the wawe function, which is supposed to play the role of W, is $S(x,t)$ and does not depend on any constant momenta. Consequently, the classical trajectory obtained from $m\dot{q}_r = P_r = \frac{\partial W}{\partial q_r}(q,P,t)$ depends on the specific choice of the constant momenta P and initial positions $q_r(t=0)$, while the quantum trajectory, which is caculated in the causal interpretation from $m\dot{x}_r = p_r = (\partial S/\partial x_r)(x,t)$, only depends on the initial position $x_r(t=0)$. Furthermore, as already remarked by Keller [4] the initial conditions allowed in the causal interpretation refer to the configuration-space fields, e.g. $S_0(x) = S(x,t=0)$, and the particle momenta $p_0 = \nabla S_0$ are herby fixed throughout space. In contrast to this, classical mechanics determines the trajectory simply by having $p_0 = \frac{\partial W_0}{\partial q_r}$ $(q = q_0, t = 0)$, i.e. by fixing the initial momentum of the particle. Finally let us remark that the quantum potential cannot be simply considered as any additional independent potential term in the H-J equation, because $R(x,t)$ is always coupled with $S(x,t)$, since only $\psi = Re^{i(S/\hbar)}$ satisfies the Schrödinger equation. In fact this coupling of the R and S fields is at the origin of the specific form of $S(x,t)$ appearing in the quantum H-J equation, because it is this specific form of S that satisfies simultaneously the continuity equation.

To illustrate this latter point we will present a simple example where the quantum potential is considered as an "external potential". We write down the H-J equation in one space-dimension

$$\frac{\partial W}{\partial t} + \frac{1}{2m}(\frac{\partial W}{\partial q})^2 + V(q) + Q(q) = 0 \qquad (6)$$

and choose the case of a free dispersive Gaussian wave-packet moving with velocity u along the +q-axis where we have [5]

$$V(q) = 0 \qquad Q(q) = -\frac{\hbar^2}{2m}\frac{1}{4\sigma_t^2}\left[(q-ut)^2 - 2\sigma_t^2\right]$$

with

$$\sigma_t^2 = \sigma_o^2(1 + \alpha^2 t^2) \text{ and } \alpha = \frac{\hbar}{2m\sigma_o^2}$$

Equation (6) can now be integrated to yield Hamilton's principal function W as :

$$W(q,t,v_o) = \frac{1}{2} m \frac{\alpha^2 t}{1 + \alpha^2 t^2} (q - ut)^2 + mu(q - \frac{ut}{2}) + mqv_o \frac{1}{\sqrt{1 + \alpha^2 t^2}}$$

(7)

$$- muv_o \frac{t}{\sqrt{1 + \alpha^2 t^2}} - \frac{1}{2} mv_o^2 \frac{1}{\alpha} \arctan(\alpha t) - m\alpha\sigma_o^2 \arctan(\alpha t)$$

where v_o is the initial velocity appearing as a constant of integration. Comparing Hamilton's principal function $W(q,t,v_o)$ with the phase of the wave-function

$$S(x,t) = \frac{1}{2} m \frac{\alpha^2 t}{1 + \alpha^2 t^2} (x - ut)^2 + mu(x - \frac{ut}{2}) - m\alpha\sigma_o^2 \arctan(\alpha t) \quad (8)$$

one readily deduces that $S(x,t)$ is one specific solution of the H-J equation, namely

$$S(x,t) = W(q,t,v_o) \Big|_{\substack{v_o = 0 \\ q \leftrightarrow x}} \qquad (9)$$

The same condition ($v_o = 0$) reduces the generic trajectories obtained from the H-J theory, i.e.,

$$q(t) = \sqrt{1 + \alpha^2 t^2} \left[q_o + \frac{v_o}{\alpha} \arctan(\alpha t) \right] + ut \qquad (10)$$

to the form of the hyperbolic trajectories of a quantum particle, as calculated in the causal interpretation from the phase of the wave function, i.e.,

$$x(t) = \sqrt{1 + \alpha^2 t^2} \, x_o + ut \qquad (11)$$

This simple example shows that the quantum potential cannot be considered as an independent "external potential" in the H-J equation (6) since the coupling of the amplitude and the phase of the wave-function imposes constraints on the complete integral of the H-J equation which amount to a specific choice of the integration constants. Furthermore it demonstrates the essential role played by the integration constants in H-J theory which influence significantly the form of the particle trajectories deduced from Hamilton's principal function W. In what follows we will focus our attention on the integration constants of the classical Hamilton-Jacobi theory because, as we will demonstrate they can play a major part in the comprehension of the transition from classical to quantum mechanics.

3. THE TRANSITION FROM CLASSICAL TO QUANTUM MECHANICS

In the preceding section we pointed out that one basic feature which distinguishes the classical from the quantum H-J equation is tied to the presence of the integration constants in the former and their absence in the latter. Furthermore, we indicated a specific procedure by means of which the integration constants can be eliminated in the frame of the quantum H-J equation. However, this equation had the insufficiency that the introduction of the quantum potential in the H-J equation (6) is entirely ad hoc and the elimination of the constants is achieved by taking into account the continuity equation (5) and fixing the value of the constant to be zero. In what follows we will attempt a different procedure for the elimination of the constants, which as will be shown represents a way of performing the transition from classical to quantum mechanics or a means to derive the quantum equations from a classical Hamilton-Jacobi equation.

To this end consider the phase-space relations characteristic of the classical H-J approach and compare them to the corresponding ones in the causal interpretation of quantum mechanics. If we retain the picture we advanced in the introduction, then, for a two-dimensional phase-space, the classical H-J theory introduces a relation $p = (\partial W/\partial q)(q,P,t)$, i.e. a curve in the (p,q) plane, that for different values of the constant P scans through the entire phase-space. Conversely, the postulate of the causal interpretation $p = (\partial S/\partial x)(x,t)$ simply fixes a single line in the (x,p) plane, since no parameters are present herein. The problem is now to find a well-determined prescription which maps the classical H-J system, with the additional degree of freedom manifested by P, on the quantum system which is uniquely constrained; and this without the ad hoc introduction of the quantum potential and of additional constraints as performed in the preceding section. In other words, we wish to start from the classical H-J theory and arrive at the quantum equation by means of a specific assumption.

It is now clear that the classical H-J system is going to be represented by the H-J equation

$$(\partial W/\partial t) + (1/2m)(\partial W/\partial q)^2 + V = 0 \qquad (12)$$

and the quantum system by the "H-J type" equation

$$\frac{\partial S}{\partial t} + \frac{1}{2m}\left(\frac{\partial S}{\partial x}\right)^2 + V + Q = 0 \tag{13}$$

plus the conntinuity equation (5). The two functions W and S will therefore not coincide, and a simple choice of the value of P in W will not produce the desired mapping.

In a previous publication [6] we examined for the sake of simplicity the interaction free-case V = 0 and we were able to make the transition from classical to quantum mechanics by introducing a statistical hypothesis into the H-J theory. We will now extend this approach to the case where a time-independent potential is present, i.e. treat the set of equations (12) and (13) with V = V(q). The basic idea is now to declare the parameter P as a random variable and admit a distribution of the P's in the frame of the classical H-J theory. In our phase-space picture this means that we do not fix the value of the parameter P, i.e. a surface $p = \frac{\partial W}{\partial q}$ (q, P = P_0, t), but instead admit a possible distribution of the P's, or a possible distribution of surfaces $p = \frac{\partial W}{\partial q}$ (q,P,t). Then the goal will be to represent the system possessing this distribution of surfaces in phase-space by a corresponding average surface $p_Q = \langle\frac{\partial W}{\partial q}\rangle_P$. The average $\langle ... \rangle_P$ now means some statistical average over the distribution of parameters P. To find an appropriate form for the distribution function of the P's we recall the procedure adopted in Ref. [6] for the V = 0 case : Given the form of Hamilton's principal function for the free-particle case, i.e. $W = Pq - \frac{P^2}{2m}t + C$, we can show that the classical H-J equation can be formally rewritten in the form of a free Schrödinger equation

$$\hbar i \frac{\partial \psi}{\partial t} = -\frac{\hbar^2}{2m}\frac{\partial^2 \psi}{\partial q^2} \tag{14}$$

where we introduced for ψ the expression $\psi = \exp\left(\frac{i}{\hbar} W(q,P,t)\right)$ which plays the role of a "classical wave-function". The distribution function of P, i.e. $\rho(P)$, acts on this "classical wave-function" and yields the quantum wave function as an averaged quantity, namely

$$\psi(q,t) = \int dP \rho(P) e^{(i/\hbar)W(q,P,t)} \tag{15}$$

which automatically satisfies the Schrödinger equation. However, this scheme, which is straightforward in the V = 0 case and has been explained in detail in Ref. [6] cannot be extended as it stands for the case where a potential is present. The reason is that the H-J equation (12) cannot be put in a Schrödinger form (as was done in equation (14) for V = 0) due the form of $W(q,P,t)$ which reads now :

$$W(q,P,t) = \pm \int dq\sqrt{P^2 - 2mV} - \frac{P^2}{2m} t + C \qquad (16)$$

The correct choice of sign is made by performing the $V \to 0$ limit and comparing with the free form, i.e. $W(q,P,t)|_{V=0} = P_q - \frac{P^2}{2m} t$: the (+) sign for P > 0 and the (−) sign for P < 0. Furthermore we cannot expect the distribution of the P's to be independent of the space coordinate since the presence of V(q) breaks the space isotropy. Taking these considerations into account we will introduce the quantum wave function in an analogous manner to equation (15) by writing

$$\psi(q,t) = \int dP\{H(P) e^{i\frac{W}{\hbar}(q,P)} + H(-P) e^{-i\frac{W}{\hbar}(q,P)}\} \rho(q,P) e^{-\frac{i}{\hbar}\frac{P^2}{2m} t} \qquad (17)$$

Here $\rho(q,P)$ is the coordinate-dependent generalized distribution function of the constants P, $W(q,P) = \int dq\sqrt{P^2 - 2mV}$ and H(P) is the Heavyside function. Since $\rho(q,P)$ is not a phase-space distribution function, no positivity conditions are imposed hereon. This generalized characteristic distribution of the random variable P, enables us to calculate the characteristic quantities of the fictitious average substitute. As a consequence we can introduce for the average quantity $\psi(q,t)$, the quantum wave function, two configuration space fields R and S $(\psi = R e^{i(S/\hbar)})$ which can be directly derived from classical quantities :

(a) The average Hamilton's principal function or phase of the wave function

$$S = \langle W \rangle = \frac{\hbar}{2i} \ln \left\{ \frac{\int dP\rho(q,P) [H(P) e^{i(W/\hbar)} + H(-P) e^{-i(W/\hbar)}] e^{-(i/\hbar)(P^2/2m)t}}{\int dP\rho(q,P) [H(P) e^{-i(W/\hbar)} + H(-P) e^{+i(W/\hbar)}] e^{(i/\hbar)(P^2/2m)t}} \right\} \qquad (18)$$

(b) The square of the amplitude of the wave function which has no direct analog in the classical theory :

$$R^2 = \int\int dPdP' \rho(q,P)\rho(q,P') [H(P) e^{i\frac{W}{\hbar}} + H(-P) e^{-i\frac{W}{\hbar}}][H(P') e^{-i\frac{W'}{\hbar}} + \qquad (19)$$
$$+ H(-P') e^{i(W'/\hbar)}] e^{-(i/\hbar)[(P^2-P'^2)/2m]t}$$

The average $<W>$ (or S) defines a single surface in the (p,q) phase-space by means of the relation $p = \frac{\partial <W>}{\partial q}$, i.e with $H_\pm = H(\pm P)$:

$$p = \left[\frac{\int dP \rho \frac{\partial W}{\partial q} \left[H_+ e^{i(W/\hbar)} - H_- e^{-i(W/\hbar)} \right] e^{-(i/\hbar)(P^2/2m)t}}{\int dP \rho \left[H_+ e^{i(W/\hbar)} + H_- e^{-i(W/\hbar)} \right] e^{-(i/\hbar)(P^2/2m)t}} + c.c. \right] \quad (20)$$

which is a unique average surface in contrast to the series of surfaces for different values of P of the classical system. If we now search for the equation that $<W>$ obeys, then, in analogy with the V = 0 case, we can immediately see that it does not obey the classical H-J equation since we can directly verify that

$$\frac{\partial <W>}{\partial t} + \frac{1}{2m} \left(\frac{\partial <W>}{\partial q} \right)^2 + V \neq 0 \quad (21)$$

For the sake of comparison we recall that in the V = 0 case we were able to show that independently of the choice of $\rho(P)$ the quantum H-J equation for $<W>$ (q,t), where the quantum potential appeared as a new term constraining the behaviour of the average system, is valid. Furthermore we showed that the average fields R and $<W>$ obey an additional relation, the continuity equation, so that the quantum Schrödinger equation is automatically satisfied by the quantum wave function $\psi(q,t)$.

In the present case, proceeding in an analogous manner to the free case and after lengthy but straightforward calculations we can show that the validity of the classical H-J equation (12) leads to the quantum H-J equation

$$\frac{\partial <W>}{\partial t} + \frac{1}{2m} \left(\frac{\partial <W>}{\partial q} \right)^2 + V - \frac{\hbar^2}{2m} \frac{1}{R} \frac{\partial^2 R}{\partial q^2} = 0 \quad (22)$$

and the continuity equation

$$\frac{\partial}{\partial t} (R^2) + \frac{\partial}{\partial q} \left(R^2 \frac{1}{m} \frac{\partial <W>}{\partial q} \right) = 0 \quad (23)$$

<u>provided</u> a subsidiary condition on the characteristic distribution $\rho(q,P)$ holds :

$$\int dP \left[\frac{\hbar}{i} \frac{\partial^2 \rho}{\partial q^2} A_+ + \rho \frac{\partial^2 W}{\partial q^2} A_- + 2 \frac{\partial \rho}{\partial q} \frac{\partial W}{\partial q} A_- \right] e^{-\frac{i}{\hbar} \frac{P^2}{2m} t} = 0 \quad (24)$$

where $A_\pm = H(P) e^{i(W/\hbar)} \pm H(-P) e^{-i(W/\hbar)}$ and $W = W(q,P)$, $\rho = \rho(q,P)$. We can satisfy the subsidiary condition by requesting, as a sufficient

condition, the vanishing of the integrand in the brackets which yields
a set of two equations, i.e. (with $\rho^{\pm} = \rho(q, \pm P)$)

$$2 \frac{\partial W}{\partial q} \sin \frac{W}{\hbar} \cdot \frac{\partial}{\partial q}(\rho^+ + \rho^-) + \frac{\partial^2 W}{\partial q^2} \sin \frac{W}{\hbar} \cdot (\rho^+ + \rho^-) - \hbar \cos \frac{W}{\hbar} \frac{\partial^2}{\partial q^2}(\rho^+ + \rho^-) = 0 \quad (25a)$$

$$2 \frac{\partial W}{\partial q} \cos \frac{W}{\hbar} \cdot \frac{\partial}{\partial q}(\rho^+ - \rho^-) + \frac{\partial^2 W}{\partial q^2} \cos \frac{W}{\hbar} (\rho^+ - \rho^-) + \hbar \sin \frac{W}{\hbar} \frac{\partial^2}{\partial q^2}(\rho^+ - \rho^-) = 0 \quad (25b)$$

Then by using the H-J equation (12) and reshaping we can put them into
a form of eigenvalue Schrödinger equations, namely

$$\left[-\frac{\hbar^2}{2m}\frac{\partial^2}{\partial q^2} + V\right](\rho^+ + \rho^-)\cos\frac{W}{\hbar} = \frac{P^2}{2m}(\rho^+ + \rho^-)\cos\frac{W}{\hbar} \quad (26a)$$

$$\left[-\frac{\hbar^2}{2m}\frac{\partial^2}{\partial q^2} + V\right](\rho^+ - \rho^-)\sin\frac{W}{\hbar} = \frac{P^2}{2m}(\rho^+ - \rho^-)\sin\frac{W}{\hbar} \quad (26b)$$

This shows explicitly that for $V \neq 0$ the distributions $\rho(q,P)$ that accomplish the transition from classical H-J theory to the quantum Schrödinger equation cannot be supposed as arbitrary but themselves obey the same type of eigenvalue equations. In fact all the properties characteristic of quantization which are usually derived from the eigenvalue problem for the wave function can be transposed in the frame of this approach to the eigenvalue problem for the $\rho(q,P)$. Since this essential point is going to be developed extensively in a future publication, we will content ourselves here with an illustration on the basis of a concrete example.

Suppose that we have a P-symmetric distribution ρ, namely $\rho(q,P) = \rho(q,-P)$ or $\rho^+ = \rho^-$. Then equation (26b) vanishes and equation (26a) becomes

$$(-\frac{\hbar^2}{2m}\frac{\partial^2}{\partial q^2} + V)2\rho \cos\frac{W}{\hbar} = \frac{P^2}{2m} 2\rho \cos\frac{W}{\hbar} \quad (27)$$

Suppose furthermore that we know a solution of the eigenvalue equation

$$(-\frac{\hbar^2}{2m}\frac{\partial^2}{\partial q^2} + V)f_\alpha(q) = \alpha f_\alpha(q) \quad (28)$$

Then $\rho(q,P)$ can be expressed in terms of this solution since for $P = \pm\sqrt{2m\alpha}$ the expression $2\rho \cos\frac{W}{\hbar}$ can be identified with $f_\alpha(q)$. Therefore we are able to write for a solution of equation (27)

$$2\rho \cos\frac{W}{\hbar} = f_\alpha \cdot \{\delta(P - \sqrt{2m\alpha}) + \delta(P + \sqrt{2m\alpha})\} \quad (29)$$

or equivalently for the characteristic distribution

$$\rho(q,P) = \frac{f_\alpha(q)}{2\cos\frac{W(q,P)}{\hbar}} \{\delta(P - \sqrt{2m\alpha}) + \delta(P + \sqrt{2m\alpha})\} \quad (30)$$

With this form of distribution $\rho(q,P)$ we can evaluate the corresponding wave function $\psi(q,t)$ by means of equation (17). We thus obtain

$$\psi(q,t) = \int dP \frac{f_\alpha(q)}{2\cos\frac{W(q,P)}{\hbar}} \{\delta(P - \sqrt{2m\alpha}) + \delta(P + \sqrt{2m\alpha})\} \cdot$$
$$\cdot \{H(P)e^{i(W/\hbar)} + H(-P)e^{-i(W/\hbar)}\} \cdot e^{-((i/\hbar)(P^2/2m))t} \quad (31)$$

which yields

$$\psi(q,t) = f_\alpha(q) e^{-i(\alpha/\hbar)t} \quad (32)$$

This is an eigensolution of constant energy α of the Schrödinger equation. The form of ψ is a consequence of the subsidiary condition, imposing the discretization of the P-spectrum of the characteristic distribution $\rho(q,P)$. As we could see in the above simple example the property of ψ being an energy eigenfunction with eigenvalue α is equivalent to $\rho \cdot \cos\frac{W}{\hbar}$ being an eigenfunction with eigenvalue $P^2/2m = \alpha$ and two allowed values of the random variable P, namely $P = \pm\sqrt{2m\alpha}$.

Apart from the important feature of obtaining the wave function of stationary states of the Schrödinger equation by means of the allowed values of the ρ-distribution, some further interesting aspects of the approach should be noted:

(a) While it is the distribution $\rho(q,P)$ that acts on the classical analog of the wave function $\psi_{cl} = H(P)e^{(i/\hbar)(W(q,P))} + H(-P)e^{-(i/\hbar)(W(q,P))}$, the effective quantized distributions are $[\rho(q,P) + \rho(q,-P)]\cos\frac{W}{\hbar}(q,P)$ and $[\rho(q,P) - \rho(q,P)]\sin\frac{W}{\hbar}(q,P)$. Therefore, these two effective distributions govern the observable dynamics of the system.

(b) The efffective distributions $(\rho^+ + \rho^-)\cos\frac{W}{\hbar}$, $(\rho^+ - \rho^-)\sin\frac{W}{\hbar}$ possess now the properties that are normally ascribed to the wave function. Therefore, the superposition principle for eigenstates is translated in the frame of this formalism to a superposition of effective distributions of a process. An eigenstate and a superposition of eigenstates can be shown to differ in the fact that the distribution function allows only a fixed magnitude P in the former and a variable one in the latter case.

(c) This approach explains quite naturally why equation (22) cannot be considered as a genuine H-J equation, since the quantum potential Q(R) and the function <W> are built by using the same distribution function ρ(q,P) and are interrelated. Therefore Q cannot be considered as an independent "external" potential. [see Ref. [6]].

CONCLUSION

It has been indicated elsewhere [6] how this approach can be connected with an assumption of an underlying stochastic process as well as how it can be brought into relation to the Feynman path integrals. Possible extensions of the formalism to the case of time-dependent potentials V(q,t), to the many particle system and to the phase-space representation of quantum mechanics will be presented elsewhere [7]. Concluding, we wish to stress that the above approach shows that it is indeed possible to analyze a quantum system beyond the limits of complementarity, and understand the origin of the quantum probability in terms of well defined classical concepts. It is not a pure coincidence that the guiding principle of this investigations was the effort to understand the origin and the role played by the quantum potential in the quantum H-J type equation. It merely shows that the causal interpretation of quantum mechanics supplies us with powerful tools in the struggle to grasp the quantum reality.

ACKNOWLEDGEMENTS : The author thanks J.P. Vigier for constant encouragement and P.R. Holland and N. Cufaro-Petroni for many helpful discussions, as well as the Institut Henri Poincaré for its hospitality.

REFERENCES

[1] H. Goldstein, Classical Mechanics, Addison-Wesley, Cambridge, 1956, Ch.9.

[2] D. Bohm, Phys. Rev. 85 (1952) 166 ; 181.

[3] See e.g. H. Freistadt, Suppl. Nuov. Cim. V, Sec. X (1957) 1.

[4] J.B. Keller, Phys. Rev., 89 (1953) 1040.

[5] See e.g. F.J. Belifante, A survey of hidden variable theories, Pergamon Press 1973, Part II, App. G.

[6] A. Kyprianidis : Hamilton-Jacobi theory and quantum mechanics, to appear in Proceedings of Conf. "Problems in quantum physics", Gdansk, 1987.

[7] A. Kyprianidis : Hamilton-Jacobi theory, classical phase-space and quantum mechanics, IHP Preprint, to be published.

TIME AND ENHANCEMENT: TWO POSSIBLE LOCAL EXPLANATIONS FOR THE EPR PUZZLE

S. Pascazio
Département de Mathématique, Université Libre de Bruxelles, Campus Plaine, CP 217, B-1050 Brussels, Belgium
and
Theoretische Natuurkunde, Vrije Universiteit Brussel, Pleinlaan 2, B-1050 Brussels, Belgium.

ABSTRACT. The alleged generality of the experimental results against the so-called local realist theories is challenged on two major grounds : first, the "naturalness" of the notorious no-enhancement hypothesis is questioned ; second it is disclosed that the possibility of subluminal information exchanges cannot be ruled out in atomic cascade tests if the emission lifetimes of both photons are taken into account.

1. MOTIVATIONS

The purpose of this note is twofold. First, it will be argued that the notorious no-enhancement hypothesis, whose validity has always been assumed when carrying out experimental tests of Bell's inequality with atomic cascades, is far from being obvious and, more to that, is even not a "natural" assumption. Second, we will show that the typical times involved in atomic cascade experiments are long enough in order to account for local information exchanges between the photons. Since every other test of Bell's inequality (polarization correlation of γ's produced by e^+e^- annihilation, spin correlation in pp scattering on carbon [1]) is easily explainable in terms of local models, we are compelled to admit that, at present, <u>no experimental evidence</u> exists against locality.

2. ENHANCEMENT

There is a drastic difference between the original Bell's inequality [2] and the Bell-type inequalities experimentally tested so far [3,4]. The former presupposes only Einstein's definition of locality and "physical reality", whilst the latter assumes the validity of untested (and sometimes untestable) extra-assumptions. One of these assumptions is the so-called no-enhancement hypothesis (NEH) [4], that can be stated in two different ways :

a) if a pair of photons emerges from two polarizers, the probability of their joint detection is independent of the polarisers' settings a and b.

b) for every state λ, the probability of a count with a polarizer in place is less than or equal to the probability with the polarizer removed :

$$0 \leq p_1(\lambda,a) \leq p_1(\lambda,\infty) \leq 1$$
$$0 \leq p_2(\lambda,b) \leq p_2(\lambda,\infty) \leq 1$$
for every λ (1)

In formula (1) λ is a parameter (hidden variable) that specifies the state of the particle under investigation, a and b are polarizer settings, ∞ denotes "absence of the polarizer" and p_i is the probability that the i-th photon be detected by a photomultiplier, given the polarizer's setting and the particle's state. We stress that NEH requires that eq.(1) be valid for every value of the parameter λ. Marshall [5] warned us against the widespread misuse of defining the statement a) as NEH. This is, in his opinion, historically misleading because no-enhancement was first introduced in [4], as eq.(1), and is conceptually different from a). We will follow here the usual description of NEH in terms of a) and b). The interested reader can refer to [6] for a deeper discussion of the differences between the two formulations.

If we assume that λ spans the space Λ with distribution function $\rho(\lambda)$, then the single and double count probabilities are :

$$p_1(a) = \int_\Lambda d\rho_\lambda p_1(a,\lambda)$$
$$p_2(a) = \int_\Lambda d\rho_\lambda p_2(b,\lambda)$$ (2)
$$p_{12}(a,b) = \int_\Lambda d\rho_\lambda p_{12}(a,b,\lambda) = \int_\Lambda d\rho_\lambda p_1(a,\lambda) p_2(b,\lambda)$$

respectively. The so-called locality condition has been assumed, which implies, following Bell, Clauser and Horne, that the elementary double count probability $p_{12}(a,b,\lambda)$ is factored as in the last formula (2). From equation (2) the following inequality can be derived [4] :

$$-1 \leq 3p_{12}(\theta) - p_{12}(3\theta) - p_1 - p_2 \leq 0$$ (3)

where $\theta = |a-b|$ and p_1, p_2 are defined in (2) and have been supposed polarizer-setting independent. We will refer to (3) as inhomogeneous Clauser-Horne inequality [7] ; eq. (3) involves double (p_{12}) as well as single (p_1, p_2) count probabilities. If one considers that the p's in (2) are counting probabilities, one easily realizes that they are proportional to the so-called detector efficiencies η_i, so that

$$p_i = \eta_i P_i \quad (i = 1,2)$$
$$p_{12}(\theta) = \eta_1 \eta_2 P_{12}(\theta) \quad (4)$$

where the P's are the transmission probabilities through the polarizers. Due to the low value of η_i (for present photomultipliers one has, at most, $\eta = .15$), inequality (3) can never be violated in experimental tests. The single count probabilities p_i's and the double count ones p_{12}'s are indeed affected differently by η. This is the reason for the "inhomogeneous" attribute given to (3).

In order to get an inequality that can be violated by realizable experiments, we must call NEH on the stage. From (1) and the definition (2) one can derive the following inequality

$$-p_{12}(\infty,\infty) \leq 3p_{12}(\theta) - p_{12}(3\theta) - p_{12}(a',\infty) - p_{12}(\infty,b) \leq 0 \quad (5)$$

which is very different from (3) because, by making the substitution (4) :

$$p_{12} = \eta_1 \eta_2 P_{12}$$

the efficiencies η_i can be factored away. Eq.(5) is the so-called homogeneous inequality [7] and is the only one that has been tested in atomic cascade experiments [8]. A through-out discussion of the differences between (3) and (5) goes far beyond the purpose of this note. In refs [6,7,9,10] the reach and importance of NEH is discussed, together with the consequences of its negation (enhancement models, variable detection probability). We will just show in the following section that NEH is much less "natural" than it seems at first sight.

3. ON THE "NATURALNESS" OF NEH

We have never really grasped the meaning of "naturalness" in physics. What was natural in last century's physics is not necessarily natural nowadays and in turn what is natural today will probably turn out to be wrong one day or another. NEH is natural because it refers to the "obvious" idea that <u>an obstacle cannot increase the probability of detection</u>. Substitute "polarizer" for "obstacle" in the previous sentence and you will get NEH. We give hereafter two counterexamples to the "obviousness" of NEH.

Consider the situation in fig.1.

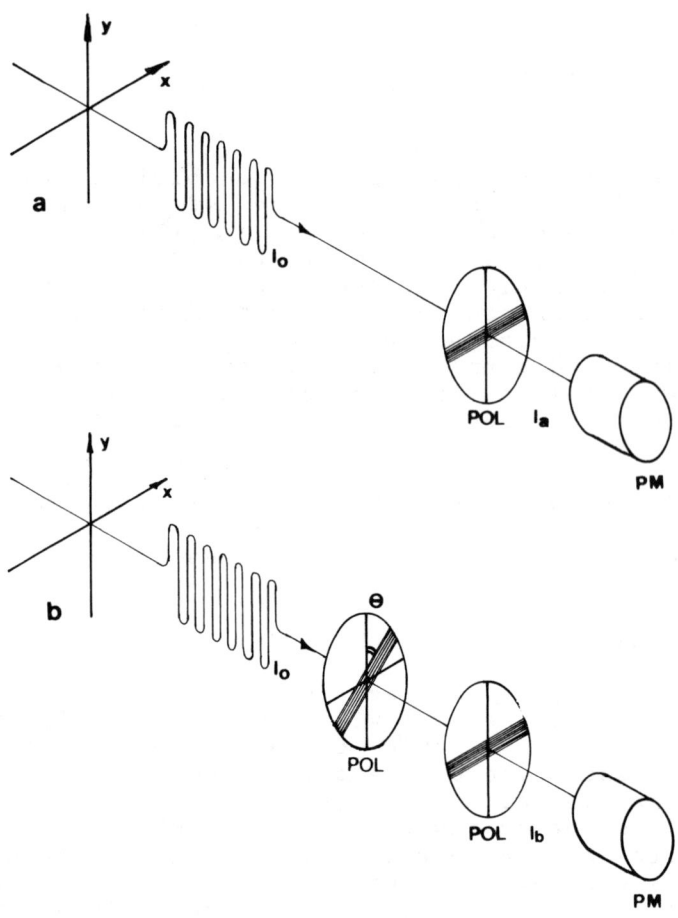

Fig.1. a) the intensity detected by the photomultiplier is $I_a = 0$
b) the intensity detected by the photomultiplier is $I_b = I_o/4$.
(POL = polarizer, PM = photomultiplier)

In fig.1a a y-polarized photon beam impinges upon a polarizer whose axis is set along the x-direction. The photomultiplier detects a null intensity $I_a = 0$. On the other hand, if another polarizer is inserted like in fig.1b, at an angle $\theta = 45°$ with respect to the y-direction, the photomultiplier detects an intensity $I_b = I_o/4$ (where I_o is the intensity of the incoming beam). This example is considered in [10] in connection with the first version of NEH (first introduced in [3] and denoted as a) in section 2).

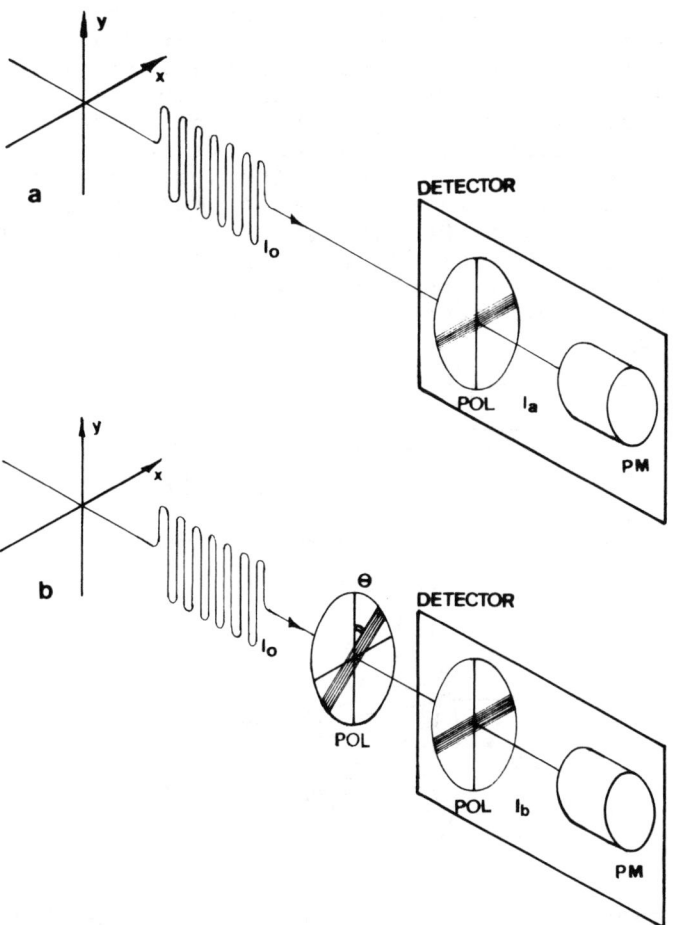

Fig.2. "Redefining" the detector

Consider now the situation in fig.2, in which everything is identical to fig.1, but the detector has been, so to speak, "redefined" : a black box contains both the x-polarizer and the photomultiplier. NEH is violated ! The obstacle (polarizer) does increase the detection probability.

A sensible objection has been put forward by Garuccio [11] against the "redefined" detector in fig.2 : the new detector is in fact polarization dependend and is therefore a "bad detector". In connection with Garuccio's criticism we think it is worth observing that polarization is a well-known property of photons, while a hidden variable is, by definition, unknown. We are facing here a curious situation: a careful analysis can point out if a detector is "bad" or "good" only if we know which variables are involved : it is always possible to test if a detector is polarization dependent just because we know what a polarization is. But if an unknown parameter is involved, we may not be able to test the detector's characteristics just because we have never observed the (new) effects due to the unknown parameter. In short : One cannot state : this polarizer is λ-independent, if one has never observed λ before. It is only in the light of these considerations that we can really understand the danger concealed in NEH.

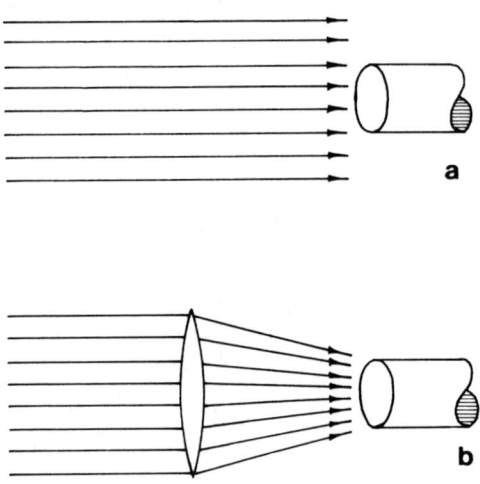

Fig.3. A convex lens increases the detection probability

A second counterexample is given in fig.3. A light beam is forwarded to a photomultiplier. If a convex lens is inserted, like in fig.3b, the detected intensity increases. Once more, an "obstacle" does increase the detection probability. Once more the philosophical ground on which NEH relies is blurred. With regards to this last example, we will make an objection ourselves : a lens is not a polarizer, and NEH concerns only polarizers. To answer this objection let us observe that there exist semiclassical radiation models in which a polarizer can "strengthen" some electromagnetic signals. In such a case, a polarizer would act as a lens for some photons, enhancing their detection probability, and NEH would be violated. Of course, on the average no effect would be observed . Incidentally, we stress that eq. (1) is supposed to be valid <u>for every</u> λ, so that averages are trivially included. For instance the general requirement

$$p_1(a) \leq p_1(\infty) \qquad (6)$$

(The condition $p_1(a) = \frac{1}{2} p_1(\infty)$ is valid only for an unpolarized incident beam. Eq.(6) holds in general.) is trivially satisfied by (1), because if

$$p_1(a,\lambda) \leq p_1(\infty,\lambda) \qquad \forall \lambda$$

then

$$p_1(a) = \langle p_1(a,\lambda) \rangle_\lambda \leq \langle p_1(\infty,\lambda) \rangle_\lambda = p_1(\infty) \qquad (7)$$

But one can easily require the condition (6) even if NEH is violated. Suppose that one has

$$p_1(a, \lambda^*) > p_1(\infty, \lambda^*) \qquad \text{for some } \lambda^* \in \Lambda$$

then nothing hinders that (7) holds true (or that $p_1(a) = \frac{1}{2} p_1(\infty)$ for an unpolarized incident beam). Plenty of such examples are known today [6,7,9].

In conclusion, we think that the very presence of NEH in the derivation of Bell-type inequalities blurs the clearness and simplicity of the philosophical ground upon which the original Bell inequality relied. The two examples just shown (fig.1-3) are just meant to "warn" us against the naturalness of some facile assumptions in connection with the EPR problem. We cannot know, at present, if no-enhancement is a property of nature. Its "naturalness" is not enough to guarantee its correctness and, more to that, its naturalness is unreliable if we think of the examples proposed in this section.

4. TIME

The search for a local explanation of the EPR puzzle led us to consider the consequences of time delays in Bell-inequality tests. Bell's inequality is in itself rather strange : it is indeed a bound for the so-called local realistic theories, but it makes use, in the very definition of correlation function, of a <u>Galilean Time</u>. Moreover, the experimental definition of coincidence in Bell-inequality experimental tests has got a <u>nonlocal flavour</u> [12]. But besides these qualitative arguments, a quantitative one can be given if the photons' lifetimes are considered. In [13] it has been shown that the possibility of subluminal information exchanges cannot be ruled out in atomic cascade tests of Bell inequality [8]. The argument is the following: if we assume that a photon's polarization is created together with the photon itself at the instant of its emission, then the probability that neither photons of a cascade can exchange any information with either polarizers is :

$$R = 1 - e^{-2L/c\tau_1}\frac{\tau_1}{\tau_1-\tau_2} + e^{-2L/c\tau_2}\frac{\tau_2}{\tau_1-\tau_2} \qquad (8)$$

where L is the source-polarizer distance, $\tau_1 (\tau_2)$ the first (second) photon's lifetime. A typical experimental setup is shown in fig. 4. The value of R is considerably low in all the atomic cascade experiments so far performed (table 1).

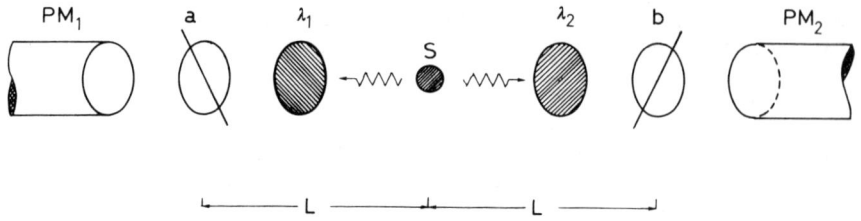

Fig.4. A typical test of Bell's inequality with atomic cascades. L is the source-polarizer distance, PM_i are photomultipliers, λ_i filters (i = 1,2)

If we consider that R is the percentage of "clean" photon couples (i.e. those couples for which no information-exchange mechanism could take place), we realize that, at most, 32.4% of the cascades cannot have been influenced by the polarizer's settings.
It is somehow incomprehensible the fact that the low value of R has not been formerly pointed out in the literature on Bell's inequality. The reason might be found in two sentences by Aspect, Dalibard and Roger [8] (referred to the famous experiment with polarizer switching) : "In this experiment, switching between the two channels occurs about each 10 ns. Since this delay, as well as the

Exp.	Atom	Cascade	τ_1(ns)	τ_2(ns)	L(m)	R
FC	Ca	$4p^2\ {}^1S_0 - 4s4p\ {}^1P_1 - 4s^2\ {}^1S_0$	89	5	.5	.010
HP	Hg	$9\ {}^1P_1 - 7\ {}^3S_1 - 6\ {}^3P_0$	–	48	.07	<.010*
C	Hg	$9\ {}^1P_1 - 7\ {}^3S_1 - 6\ {}^3P_0$	–	48	.5	<.067*
FT	Hg	$7\ {}^3S_1 - 6\ {}^3P_1 - 6\ {}^1S_0$	46	114	–	.001**
AGR,ADR	Ca	$4p^2\ {}^1S_0 - 4s4p\ {}^1P_1 - 4s^2\ {}^1S_0$	89	5	6	.324
PDBK	D	$2\ S_{1/2} - 1\ S_{1/2}$	$.125 \cdot 10^9$	0	–	0

Table 1. Atomic cascade experiments of the Bell inequality [8]. The authors' initials are given in the first column.
  * calculated with $\tau_1 = 0$
  ** calculated with L=.5m

lifetime of the intermediate level of the cascade (5 ns), is small compared to 2L/c (40 ns), a detection event on one side and the corresponding change of orientation on the other side are separated by a spacelike interval". And also: "The settings are changed during the flight of the particles". It is clear, from these two statements, that the first photon's lifetime τ_1 is not even considered. The coincidence window, the signal-to-noise ratio, the switching time in the last Orsay's experiment are all tuned on the second photon's lifetime τ_2. As pointed out in [13], the first photon plays only the role of trigger in the couple detection mechanism. And since the probability that the second photon be "clean" is

$$R_2 = 1 - e^{2L/c\tau_2}$$

(i.e. practically 1 in the Orsay experiments), the problem proposed in this section has been simply ignored.

5. FINAL REMARKS

This paper was aimed at showing that many arguments can be raised against the alleged generality and indisputability of the actual experimental results. The literature is full of statements of the type "Einstein was wrong" or "we are compelled to change our very conception of physical reality" and so forth. Phrases like these are so common today that they are not worth mentioning. Physics is the fruit of mathematical and experimental proofs and of daring hypothesis ; sometimes even of bold words. Never the product of facile and superficial sentences.

Maybe Nature is enhancement-free and maybe no information-exchange exists, even if there is time enough for it to take place. But this should be proven either by compelling theoretical arguments or by indisputable experimental results. Alas, these are still to come.

This work has been supported by the EEC twinning contract n° ST2J-0089-2-B.

REFERENCES

[1] C.S. Wu and I. Shaknov, Phys. Rev. 77 (1950) 136; L.R. Kasday, J.D. Ullman and C.S. Wu, Bull. Am. Phys. Soc. 15 (1970) 586; Nuovo Cimento 25B (1975) 633; M. Lamehi-Rachti and W. Mittig, Phys. Rev. 14 (1976) 2543
[2] J.S. Bell, Physics 1 (1965) 195
[3] J.F. Clauser, M.A. Horne, A. Shimony and R.A. Holt, Phys. Rev. Lett. 23 (1969) 880
[4] J.F. Clauser and M.A. Horne, Phys. Rev. 10D (1974) 526
[5] T.W. Marshall, private communication
[6] S. Pascazio, in "Quantum Mechanics versus Local Realism. The Einstein, Podolsky and Rosen Paradox" (Plenum Publ., 1988) F. Selleri ed.
[7] E. Santos, in "Microphysical Reality and Quantum Formalism" (Reidel, Dordrecht, 1988) G. Tarozzi and A. van der Merwe eds.; F. Selleri, in "Symposium on the Foundation of Modern Physics" (World Scientific, Singapore, 1986) P. Lahti and P. Mittelstaedt eds.
[8] S.J. Freedman, J.F. Clauser, Phys. Rev. Lett. 28 (1972) 938; R.A. Holt, F.M. Pipkin, Harvard University preprint (1974); J.F. Clauser, Phys. Rev. Lett. 36 (1976) 1223; E.S. Fry, R.C. Thompson, Phys. Rev. Lett. 37 (1976) 465 A. Aspect, P. Grangier, G. Rogers, Phys. Rev. Lett. 47 (1981) 460 ; Phys. Rev. Lett. 49 (1982) 91; A. Aspect, J. Dalibard, G. Roger, Phys. Rev. Lett. 49 (1982) 1804; W. Perrie, A.J. Duncan, H.J. Beyer, and H. Kleinpoppen, Phys. Rev. Lett. 54 (1985) 1790
[9] A. Garuccio, F. Selleri, Phys. Lett. 103A (1984) 99 and refs. therein; M. Ferrero, T.W. Marshall and E. Santos, in "Quantum Mechanics versus Local Realism. The Einstein, Podolsky and Rosen paradox". (Plenum Publ., 1988) F. Selleri ed. and refs. therein.
[10] J.F. Clauser, A. Shimony, Rep. Progr. Phys. 41 (1978) 1681
[11] A. Garuccio, private communication
[12] S. Pascazio, Phys. Lett. 111A (1985) 339 ; 118A (1986) 47
[13] S. Pascazio, J. Reignier, Phys. Lett. 126A (1987) 163

ON BELL-TYPE INEQUALITIES IN QUANTUM LOGICS

Jarosław Pykacz
Institute of Mathematics, University of Gdańsk
ul. Wita Stwosza 57, 80-952 Gdańsk
Poland

ABSTRACT. Bell-type inequalities are studied within the framework of quantum logic approach. It is shown that violation of Bell's inequalities indicates that pure states are not dispersion-free, whenever they are Jauch-Piron states which is true both in classical and in quantum mechanics. This result completes the results obtained by Santos [4] who has shown that violation of Bell's inequalities implies that a lattice of propositions for a physical system is not distributive. Connections between Jauch-Piron properties of states, non-dispersive character of pure states and distributivity of a logic are studied and it is shown that if a logic is finite the former two properties imply the latter.

The famous inequalities of Bell [1] have been originally introduced for spin measurements. The experimentally proved [2] violation of those inequalities shows the impossibility of constructing any local hidden-variables model for quantum mechanics. As it was pointed out already by Bell his inequalities should in principle hold for all yes-no observables. For this reason the framework of quantum logic in which yes-no observables appear as basic notions seems to be particularly well-suited to study both origin and consequences of those inequalities. The advantage of quantum logic approach is also such that it allows to study the problem in a purified mathematical manner.

The Bell-type inequalities were studied within the Piron version of quantum logic approach by Aerts in his Ph.D. thesis and subsequent papers [3]. However, this version of quantum logic approach makes no use of probabilistic concepts. For example, states of a physical system are not represented by probability measures defined on a lattice of propositions but by atoms of the lattice. Since there are exactly probabilistic concepts which are involved in Bell-type inequalities, this version of quantum logic approach does not seem to be very well suited to study Bell-type inequalities.

The considerable step forward in studying Bell-type inequalities within the quantum logic approach was made by Santos [4]. He has shown that Bell-type inequalities follow in fact from more simple inequalities which involve only three elements and which can be geometrically interpreted as triangle inequalities. Before I quote Santos results let me remind definitions of basic quantum logic notions which will be used in the sequel.

Def.1. A <u>quantum logic</u> (or simply a <u>logic</u>) L is an orthocomplemented orthomodular lattice i.e. a partially ordered set in which meet $a \wedge b$ and join $a \vee b$ exist for all pairs $a,b \in L$, the orthocomplementation map $':L \to L$ such that : i) $a'' = a$, ii) $a \vee a' = I$ (= maximal element in L) and iii) $a \leq b$ implies $b' \leq a'$ is admitted and the orthomodular identity $a \leq b \Rightarrow b = a \vee (a' \wedge b)$ holds.

Def.2. A <u>probability measure</u> (<u>state</u>) p on a logic L is a map $p:L \to [0,1]$ for which the following conditions are satisfied :
i) $p(\emptyset) = 0$ ($\emptyset$ = minimal element in L)
ii) $p(a \vee b) = p(a) + p(b)$ whenever $a \leq b'$ (such elements are called <u>orthogonal</u> and denoted $a \perp b$).

Elements of a logic, called <u>propositions</u>, represent elementary (yes-no) experiments i.e. experiments with only two possible outcomes. States of a physical system are represented by probability measures on a logic and a number $p(a)$ is interpreted as a probability of obtaining positive i.e. "yes" result for the experiment represented by "a" when the physical system is in the state represented by "p". Several systems of (more or less) physically plausible postulates were proposed in order to achieve the desired structure of a logic *).

The standard examples of logics of physical systems are the following :
1. Lattice L(H) of closed linear subspaces of a Hilbert space in quantum mechanics.
2. Boolean algebra B(Γ) of subsets of a phase space in classical mechanics.
Of course conditions used in the definition of a quantum logic were chosen to extract basic lattice-theoretical properties of L(H). The second example shows that we should not put too much stress on the word "quantum" in the definition 1 since it covers also the case of classical mechanics. Actually, any boolean algebra is an orthocomplemented lattice in which the orthomodular identity is fulfilled because a boolean algebra is a distributive lattice, i.e.

$$a \wedge (b \vee c) = (a \wedge b) \vee (a \wedge c) \quad \text{for all } a,b,c \in L. \tag{1}$$

Santos put his results in a slightly more general framework, assuming only that the logic of a physical system is an orthocomplemented lattice. He introduced in his paper [4] the following, very useful notion :

Def.3. The <u>separation</u> between two propositions $a,b \in L$ in the state p is the real number

$$S(a,b) = p(a) + p(b) - 2p(a \wedge b). \tag{2}$$

---
*) An excellent reviev of quantum logic approach can be find in [5].

Using this notion he proved the following theorem :

Theorem 1. (Santos [4]) In a boolean (i.e.distributive) orthocomplemented lattice the separation fulfils the triangle inequality

$$S(a,b) + S(b,c) \geq S(a,c). \tag{3}$$

It is easy to show, by writting triangle inequalities for suitable triples of propositions and combining them together that if in a state p triangle inequalities are fulfilled, then the following Bell-type inequalities are fulfilled as well :

$$S(a_1,a_2) + S(b_1,a_2) + S(b_1,b_2) \geq S(a_1,b_2) \tag{4}$$

$$p(a_1 \wedge a_2) + p(b_1 \wedge a_2) + p(b_1 \wedge b_2) - p(a_1 \wedge b_2) \leq p(b_1) + p(a_2). \tag{5}$$

According to the results of Santos, violation of Bell's inequalities proves that the logic of a physical system is non-distributive. However, this interesting conclusion is not the only one which can be drawn from the violation of Bell's inequalities. It can be shown [6] that triangle inequalities (3), and therefore also Bell-type inequalities (4) and (5), are satisfied as well whenever all pure states of a physical system are so called dispersion-free Jauch-Piron states.

Def.4. A state p on a logic L is called pure if it cannot be represented in the form of a convex combination of other states.

Def.5. A state p on a logic L is called dispersion-free if for any $a \in L$ either $p(a)=0$ or $p(a)=1$.

It is easy to notice that only pure states can be dispersion-free. This is caused by the fact that if $p = \mu r + (1-\mu)s$, $\mu \in (0,1)$, $p \neq r,s$ then $p(a)=0$ iff $r(a)=s(a)=0$, resp. $p(a)=1$ iff $r(a)=s(a)=1$. Of course if $r \neq s$ then there exists at least one a in L such that $r(a) \neq s(a)$.

Def.6. A state p on a logic L is called Jauch-Piron state if $p(a)=p(b)=1$ implies $p(a \wedge b)=1$.

Let us notice that from the orthomodular identity and additivity of a probability measure (state) on orthogonal propositions it follows that any state is an order-preserving map, i.e.

$$a \leq b \text{ implies } p(a) \leq p(b) \tag{6}$$

Since $a \wedge b \leq a,b$, for any dispersion-free Jauch-Piron state p the value $p(a \wedge b)$ is uniquely determined by values $p(a)$ and $p(b)$, namely $p(a \wedge b)=1$ iff $p(a)=p(b)=1$, otherwise $p(a \wedge b)=0$. The following table contains all eight possibilities for $p(a)$, $p(b)$ and $p(c)$ when p is a dispersion-free Jauch-Piron state and showes that in all eight cases the triangle inequality is satisfied.

Table 1.

p(a)	p(b)	p(c)	p(a∧b)	p(b∧c)	p(a∧c)	S(a,b) + S(b,c) ? S(a,c)
1	1	1	1	1	1	0 + 0 = 0
1	1	0	1	0	0	0 + 1 = 1
1	0	1	0	0	1	1 + 1 > 0
1	0	0	0	0	0	1 + 0 = 1
0	1	1	0	1	0	1 + 0 = 1
0	1	0	0	0	0	1 + 1 > 0
0	0	1	0	0	0	0 + 1 = 1
0	0	0	0	0	0	0 + 0 = 0

Finally, let us notice that if the triangle inequality holds for all pure states, it necessarily holds for all states since mixed states are represented by convex combinations of pure states. Therefore, we can write down inequalities for pure components of any mixed state, multiply them by suitable coefficients $\mu \in (0,1)$ and add separately left hand sides and right hand sides to obtain desired inequality for any mixed state.

The results of Santos [4] and my, mentioned above, results [6] can be combined together in the following way :

Theorem 2. If
(i) the logic of a physical system is boolean, or
(ii) all pure states on the logic are dispersion-free and Jauch-Piron,
then the separation fulfils the triangle inequality (3) which implies that Bell-type inequalities (4) and (5) are fulfilled as well.

Let us check if the assumptions (i) and (ii) of this theorem are fulfilled in classical and in quantum mechanics. It is a well-known fact (see, for example [5]) that the assumption (i) in classical mechanics is fulfilled but in quantum mechanics is not. However, this fact follows from the theory only since it is rather hard to check experimentally whether the logic of propositions for a given physical system is distributive or not. On the other hand the first half of the assumption (ii) is easily verifiable experimentally. Actually, it is one of the most basic experimental facts that distinguishes classical mechanics from quantum mechanics. It is mentioned usually in the very beginning in textbooks on quantum mechanics as a statement that even when a physical system is in a pure state, there is only a definite probability, usually different from 0 and 1, to obtain a specific result of a single experiment. Now I shall show that the second part of the assumption (ii) is fulfilled both in classical mechanics and in quantum mechanics.

1. Classical mechanics. Propositions are represented by elements of a boolean algebra $B(\Gamma)$ of Borel subsets of a phase space Γ and states are represented by probability measures on $B(\Gamma)$, pure states being measures concentrated on one-point subsets of Γ (Dirac measures). Since for any proposition "a" represented by $A \in B(\Gamma)$ and any state "p" represented by

a probability measure m_p

$$p(a) = m_p(A), \qquad (7)$$

we obtain for a pure state p represented by a measure concentrated on a one-point set P

$$p(a) = 1 \text{ iff } P \in A, \quad p(a) = 0 \text{ iff } P \notin A. \qquad (8)$$

The equalities (8) show that all pure states in classical mechanics are dispersion-free. The equality $p(a)=p(b)=1$ means that the point P belongs both to A and to B, i.e. it belongs to the set-theoretic intersection of A and B. This set-theoretic intersection represents in turn the meet $a \wedge b$, therefore we see that if $p(a)=p(b)=1$, then $p(a \wedge b)=m_p(A \cap B)$ i.e. any pure state is a Jauch-Piron state.

2. Quantum mechanics. Propositions are represented by elements of an orthocomplemented orthomodular lattice L(H) of closed linear subspaces of a Hilbert space H. Let A be a subspace which represents a proposition "a" and let P_A denote the projector onto the subspace A. According to Gleason's theorem, pure states are generated by unit vectors of H and are of the form

$$p_\psi(a) = (P_A \psi, \psi) \text{ for } \psi \in S^1(H) \qquad (9)$$

while all other states can be obtained by taking convex combinations of such expressions (cf, for example, [5], [7]). Now let p_ψ be a pure state and let $p_\psi(a)=p_\psi(b)=1$. According to (9) this means that ψ belongs both to the subspace A and subspace B of H, i.e. it belongs to the set-theoretic intersection of A and B. Again, as it was in the case of classical mechanics, the meet $a \wedge b$ is represented by $A \cap B$ (but join a v b and orthocomplement a´ are not represented by set-theoretic union and complement in this case, cf. [5]). Therefore, if $p_\psi(a)=p_\psi(b)=1$, then according to (9) $p_\psi(a \wedge b) = (P_{A \cap B} \psi, \psi) = 1$ i.e. any pure state is a Jauch-Piron state.

Bell-type inequalities mark the borderline between classical and quantum mechanics. The results of Santos [4] indicate that the difference between these two theories consists in the non-distributivity of logics of quantum systems while my, mentioned above, results [6] indicate that since in these both theories pure states are Jauch-Piron, the difference is as well caused by the fact that in classical mechanics pure states are dispersion-free while in quantum mechanics they are not. Therefore, the natural question arises if distributivity of a logic is somehow connected with dispersion-free character of pure states (if they are Jauch-Piron states, of course). The affirmative answer to this question can be given in the case of a finite orthomodular lattice which set of states is <u>separating</u> i.e. is such that $p(a)=p(b)$ for all states p implies a = b.

Theorem 3. If L is a finite orthomodular lattice with separating set of states and if every pure state on L is a dispersion-free Jauch-Piron state, then L is a boolean lattice.

Proof. Let us first notice that if the set of all states on L is separating and if all pure states are dispersion-free then the set of all states is <u>unital</u> i.e. for every proposition a $\neq \emptyset$ there exists a state p such that p(a)=1. Indeed, if it were not so then we would obtain p(a)=0 for all pure states. Since mixed states are convex combinations of pure states this would imply that p(a)=0 for any state, which, since the set of all states was assumed to be separating, would imply that a = $\emptyset$.

If all pure states are Jauch-Piron, then all states are Jauch-Piron because if p(a) = 1 for a mixed state p, then the same necessarily holds true for all its pure components and vice versa.

To finish the proof it is enough now to apply the result obtained by Ruttimann ([8], theorem 4.3) : the set of states of a finite orthomodular lattice is unital and every state is Jauch-Piron if and only if the lattice is boolean.

The problem of connections between Jauch-Piron properties of states, dispersion of pure states and distributivity of a logic in the case of infinite logics will be studied in the forthcoming paper.

REFERENCES

1. Bell, J.S. 'On the Einstein Podolsky Rosen paradox', Physics, 1, 195 (1964).

2. Aspect, A., Grangier, P. and Roger, G., Phys.Rev.Lett. 47, 460 (1981); 49, 91 (1982).
   Aspect, A., Dalibard, J. and Roger, G., Phys.Rev.Lett. 49, 1804 (1982).

3. Aerts, D. The One and the Many, Ph.D. Thesis, Vrije Universiteit Brussel, 1980; 'The physical origin of the EPR paradox' in Open Questions in Quantum Physics, G. Tarozzi and A. van der Merve, eds., 33-50, D. Reidel, Dordrecht, 1985.

4. Santos, E. 'The Bell inequalities as tests of classical logic', Physics Letters A, 115, No. 8, 363 (1986).

5. Beltrametti, E. and Cassinelli, G. The Logic of Quantum Mechanics, Addison-Wesley, Reading, 1981.

6. Pykacz, J. 'On the geometrical origin of Bell´s inequalities', to be published in Problems in Quantum Physics, World Scientific, Singapore.

7. Mackey, G.W. The Mathematical Foundations of Quantum Mechanics, Benjamin, New York, 1963.

8. Ruttimann, G.T. 'Jauch-Piron states', Journal of Mathematical Physics, 18, No. 2, 189 (1977).

PHYSICAL MEANING OF THE PERFORMED EXPERIMENTS CONCERNING THE EPR PARADOX

F. Selleri
Dipartimento di Fisica, Università di Bari
INFN, Sezione di Bari, Italy

ABSTRACT

Two types of inequalities have been proposed for the experimental study of the Einstein, Podolsky and Rosen (EPR) paradox. Those of the first type ("weak inequalities"), like the original one due to Bell, can be deduced from local realism alone. Those of the second type ("strong inequalities") can be deduced only thanks to the introduction of some assumption additional to local realism. All such "additional assumptions" are reviewed and discussed. They are essentally arbitrary assumptions and no evidence exists supporting their validity. The performed experiments with atomic photon pairs violate only strong inequalities. Therefore local realism is still compatible with the existing empirical evidence.

1. INTRODUCTION

All experiments performed with atomic photon pairs in order to test Bell's type inequalities[1] have been analyzed with the help of some additional hypothesis. An assumption of this type was, for instance, proposed by Clauser and Horne[2] and was formulated as follows: *For every atomic emission the probability of a count with a polarizer in place is not larger than the probability with the polarizer removed.*
Although this and other similar hypotheses (a fuller discussion will be given later) had an important historical role in allowing the performance of meaningful experiments, there is no logical reason for believing that they are true in nature. Once one accepts the idea that the present-day quantum theory predicts only averages and that a more detailed theoretical description should be possible by taking into account additional physical properties, one can describe the *detection* process as being different for different objects. This is enough for violating in a natural way all the additional assumptions.
A research program in which the negation of the additional assumptions is adopted is being developed. The first paper of this type was by Marshall, Santos and the present writer[3] who showed that the results

of the performed EPR-type experiments with only detectors and polarizers (one per photon) can be fitted within errors with reasonable local realistic models. Other important results were obtained by Marshall[4], who investigated "the distance separating quantum theory from reality", Marshall and Santos[5], who investigated the EPR correlations of generally elliptical polarization states, and Ferrero and Santos[6], who produced a peculiar local realistic model that agrees completely with the quantum theoretical predictions for not too high detector efficiencies. A complete review of these and other developments will soon be available[7].

The present paper is devoted to a detailed reconstruction of the role of the additional assumptions in the conceptual and experimental study of the EPR paradox. The conclusion will be obtained that in spite of the good agreement of the experimental results with the quantum-theoretical predictions the EPR paradox is still fundamentally unsolved.

2. THE WEAK INEQUALITIES

Consider an ensemble formed by a very large number of atomic cascades, each of them leading to the emission of two photons, here called α and β. Suppose that the dichotomic observable $A(a)$ is measured on α, while in a distant region of space another dichotomic observable $B(b)$ is measured on β.

The observables $A(a)$ and $B(b)$ have been taken to depend on the arguments a and b, respectively, which are the directions of the axes of the polarizers set on the trajectories of the two photons. When measurements of these observables are made on all the N cascades of the given ensemble, the first apparatus will obtain a set of results $\{A_1, A_2, \ldots A_N\}$, while the second apparatus will collect a similar set of results $\{B_1, B_2, \ldots B_N\}$, all relative to fixed values of the parameters a and b. The results of the two sets are correlated in the sense that A_1 and B_1 pertain to the photons α and β, respectively, arising from the first cascade; A_2 and B_2 are similarly associated with the second cascade, and so on. By definition, these results in every case equal ± 1.

The correlation function $P(a,b)$ of the results A_i and B_i is defined as the average product of the results obtained by the two apparatuses from the same cascades. Since every product $A_i B_i = \pm 1$, it follows that

$$-1 \leq P(a,b) \leq +1. \qquad (1)$$

If one defines

$$\Delta \equiv |P(a,b) - P(a,b')| + |P(a',b) + P(a',b')| \qquad (2)$$

it is possible to deduce from <u>local realism</u>

$$|\Delta| \leq 2, \qquad (3)$$

that is Bell's inequality. The latter is only the simplest in an infinite set of independent inequalities that can be deduced from local realism as shown by Garuccio and the present author[8]. More recently Lepore could show[9] that the widest set of inequalities is the one holding for linear combinations of joint probabilities of specified results of measurements made on the two photons of an EPR pair.

The inequality (3) has never been tested experimentally and can be considered a "weak" inequality if compared with much stronger inequalities that can be deduced from local realism if some suitable "additional assumptions" are made. To these matters the following sections are devoted: Here we will merely notice that what breaks down in practical experiments is the idea that a dichotomic observable can really be measured, in the sense that the choice between +1 and -1 can be made in every individual act of measurement. It happens almost invariably that there is a third possibility, that the photon goes undetected. And since the efficiency of photodetectors is around 10-20%, the third possibity turns out to be the dominant one!

3. THE ADDITIONAL ASSUMPTIONS

The problem of the low efficiency of photodetectors has traditionally been "solved" by means of <u>ad hoc</u> assumptions concerning the nature of the transmission/detection process. The additional assumption made in 1969 by Clauser, Horne, Shimony and Holt[10] was the following:

<u>Given that a pair of photons emerge from two regions of space where two polarizers can be located, the probability of their joint detection from two photomultipliers is independent of the presence and of the orientation of the polarizers.</u>
(CHSH assumption)

This assumption allowed one to deduce a new inequality, much stronger than Bell's, that was violated by the quantum mechanical predictions in concrete experimental conditions. We will see the additional assumptions at work in the following sections, while here we will only review them. In 1974 Clauser and Horne[11] proposed a slightly different additional assumption which they formulated as follows:

<u>For every photon in the state λ the probability of a detection with a polarizer in place on its trajectory is less than or equal to the detection probability with the polarizer removed.</u>
(CH assumption)

The two previous additional assumptions refer to EPR experiments in which a one-way polarizer is put on the path of each photon. This is however not a very convenient configuration, since the dichotomic choice is between photon transmission and lack of transmission (absorption, reflection). Now, of course, an absorption cannot be detected and a considerable amount of information is therefore lost inside the polarizer where the photon is absorbed. A better experiment is one in which a truly binary choice is made, and where the two alternatives are both detected. In 1981 Garuccio and Rapisarda studied an experiment in which a piece of calcite, monitored by two detectors put on the ordinary and on the extraordinary ray, was used as analyzer for each of the two photons[12]. They noticed that also for this type of experiments an additional assumption was needed, which they formulated as follows:

> For every photon in the state λ the sum of the detection probabilities in the "ordinary" and in the "extraordinary" beams emerging from a two-way polarizer does not depend on the polarizer's orientation.

(GR assumption)

A different additional assumption that can be applied to polarizers of all types was made by Aspect in 1983[13]:

> The set of pairs actually detected for given polarizers' orientations is an unbiased representative sample of the set of pairs emitted from the source.

(A assumption)

All these assumptions are today completely beyond experimental control. Consider the A assumption for example. To state that the set of detected pairs is a faithful sample of all pairs means, from an empirical point of view, that one should compare experimentally the set of detected pairs with the set of all pairs and check that they give the same results. But this means that one should be able to detect also the undetected pairs, which is impossible by definition! Therefore a strict positivist should declare the A assumption completely meaningless. Similar considerations can easily be made for the other three additional assumptions.

It is better, however, not to apply strict positivistic criteria, which have fortunately become unfashionable. The previous additional assumptions should therefore not be declared meaningless, also because the consequences that can be deduced from them (and jointly from local realism) can be (and have been) falsified. It must however be insisted that no direct evidence of any type exists about the validity of the additional assumptions, and that the solution of such a fundamental issue of modern physics as the EPR paradox should not be based on the presupposition that they are certainly true.

4. THE STRONG INEQUALITIES

In this section we will compare the weak inequalities with those deduced with the help of the additional assumptions (strong inequalities). The following definitions are used:

$p(a)$: measured probability that photon α crosses the polarizer with parameter a and that it is subsequently detected.

$q(b)$: measured probability that photon β crosses the polarizer with parameter b and that it is subsequently detected.

$D(a,b)$: measured joint probability that both α and β cross their respective analyzers with parameters a and b, and that they are both detected.

$D(a,\infty)$: measured joint probability that α crosses the polarizer with axis a and that it is subsequently detected, and that β is detected when no polarizer is put on its path.

$D(\infty,b)$: measured joint probability that α is detected if no polarizer is put on its path, and that β crosses the polarizer

PHYSICAL MEANING OF EPR PARADOX EXPERIMENTS   125

with axis b and that it is subsequently detected.
As it is well known two types of inequalities can be deduced[14]:
(i) <u>Weak inequalities (no additional assumptions)</u>

$$-1 + p(a') + q(b) \leq \Gamma \leq p(a') + q(b) ; \qquad (4)$$

(ii) <u>Strong inequalities (additional assumptions needed)</u>

$$-D_0 + D(a',\infty) + D(\infty,b) \leq \Gamma \leq D(a',\infty) + D(\infty,b) . \qquad (5)$$

These different inequalities are expressed in terms of the same quantity Γ, defined by

$$\Gamma = D(a,b) - D(a,b') + D(a',b) + D(a',b') . \qquad (6)$$

We should next compare the inequalities (4) and (5) with the quantum mechanical predictions. The existing theory predicts:

$$D(a,b) = \tfrac{1}{4}\{\epsilon^1_+ \epsilon^2_+ + \epsilon^1_- \epsilon^2_- \cos 2(a-b)\}\eta_1 \eta_2 \qquad (7)$$

$$D(a,\infty) = \tfrac{1}{2} \epsilon^1_+ \eta_1 \eta_2 \qquad (8)$$

$$D(\infty,b) = \tfrac{1}{2} \epsilon^2_+ \eta_1 \eta_2 \qquad (9)$$

$$p(a') = \tfrac{1}{2} \epsilon^1_+ \eta_1 \qquad (10)$$

$$q(b) = \tfrac{1}{2} \epsilon^2_+ \eta_2 \qquad (11)$$

$$D_0 = \eta_1 \eta_2 \qquad (12)$$

By substituting (7) in (6) it is easy to show that

$$(\Gamma)_{\substack{max \\ min}} = \tfrac{1}{2}(\epsilon^1_+ \epsilon^2_+ \pm \epsilon^1_- \epsilon^2_- \sqrt{2})\eta_1 \eta_2 \qquad (13)$$

Typical numerical values of the experimental parameters are

$$\epsilon^1_+ = \epsilon^2_+ = 1 \quad ; \quad \epsilon^1_- = \epsilon^2_- = 0.95 \quad ;$$

$$\eta_1 = \eta_2 = 0.1$$

If these values are substituted in all the relations (4) - (13) one gets

$$-0.9 \leq \Gamma \leq 0.1 \qquad \textbf{(weak inequality)} \qquad (14)$$

$$0 \leq \Gamma \leq 0.01 \qquad \textbf{(strong inequality)} \qquad (15)$$

and

$$(\Gamma)_{min} = -0.00138 \quad ; \quad (\Gamma)_{max} = 0.01138 \qquad (16)$$

One can thus see that <u>while the strong inequality is violated by the</u>

quantum theoretical predictions, the weak ones are fully compatible with them. The very good agreement of the experimental results with the theoretical predictions then shows that all empirical evidence is in agreement with the weak inequalities, but disagrees with the strong ones. One of the assumptions on which the strong inequalities are based must therefore be false. The choice is between dropping local realism or declaring the additional assumptions false. That the second choice is more natural will be shown in the following sections.

5. DETERMINISTIC LOCAL MODELS

A class of <u>local</u> and <u>deterministic</u> models which violate the strong inequalities has recently been studied by A. Zeilinger and the present writer[15]. These models were obtained by generalizing the deterministic approach discussed by Wigner[16], by including additional variables which determine whether a specific photon will trigger the detector or not. In these models each individual photon pair is described by the set of ten variables

$$(s, s', \sigma, \sigma', \delta; t, t', \tau, \tau', \epsilon)$$

with the first five variables pertaining to photon α and the second five variables to photon β. Each of the ten variables can only be zero or unity. In detail:

$s = 1$ ($s' = 1$)   determines that photon α will traverse its polarizer oriented along direction a (a');

$\sigma = 1$ ($\sigma' = 1$)   determines that photon α will be registered by its detector after having crossed the polarizer oriented along direction a (a');

$\delta = 1$ determines that photon α will be registered by its detector if no polarizer is on its path;

$t = 1$ ($t' = 1$)   determines that photon β will traverse its polarizer oriented along direction b (b');

$\tau = 1$ ($\tau' = 1$)   determines that photon β will be registered by its detector after having crossed the polarizer oriented along direction b (b');

$\epsilon = 1$ determines that photon β will be registered by its detector if no polarizer is on its path;

$s, s', \sigma, \sigma', \delta = 0$ ($t, t', \tau, \tau', \epsilon = 0$) determines that photon α (photon β) will not pass its polarizer or will not be registered by the detector.

We note that in this way the fate of each individual photon, whether it will traverse its polarizer or not and whether it will be registered by its detector or not, is strictly determined by the whole set of variables the photon carries. Also, it is assumed in this model that, if the photon does not encounter a polarizer on its flight from the source to the detector, the variable $\delta(\epsilon)$ is active at the detector, but it is switched off – and the variable $\sigma(\tau)$ is switched on – upon interaction with the polarizer. We could equally well have assumed only one variable describing whether a photon is detected or not, which could then change upon interaction with the polarizer. Thus our deterministic scheme is the most general one.

All N_o photon pairs emitted by the source may therefore be grouped into subsets according to the specific values of the variables they carry. If we define $n(s,s',\sigma,\sigma',\delta;t,t',\tau,\tau',\epsilon)$ the population of that subset of photon pairs that carry the specific set $(s,s',\sigma,\sigma',\delta;t,t',\tau,\tau',\epsilon)$ of variables, we obtain the normalization condition

$$\sum n(s,s',\sigma,\sigma',\delta;t,t',\tau,\tau',\epsilon) = N_o . \tag{17}$$

Due to the ten dichotomic variables present in this model, it might at first sight appear that one has to deal with $2^{10} = 1024$ different subsets. Yet it is easy to see that subsets for which, say, $s = 0$ and, simultaneously, $\sigma = 1$ are not physically meaningful since they would imply detection with certainty of a photon which has not traversed the polarizer. The same consideration applies to the other three polarizer orientations, and this reduces the number of different subsets by a factor $(3/4)^4$ to 324.

In order to define the experimental observables in terms of sums over the populations of the different subsets we abbreviate

$$n(\ldots) = n(s,s',\sigma,\sigma',\delta;t,t',\tau,\tau',\epsilon) \tag{18}$$

whenever it is understood that the sum is to be taken over the whole ensemble. The various joint probabilities can therefore be written as

$$\begin{aligned}
D_o &= (1/N_o)\sum n(\ldots)\, \delta\epsilon \\
D(a',\infty) &= (1/N_o)\sum n(\ldots)\, s'\sigma'\epsilon \\
D(\infty,b) &= (1/N_o)\sum n(\ldots)\, \delta t\tau \\
D(a,b) &= (1/N_o)\sum n(\ldots)\, s\sigma t\tau \\
D(a,b') &= (1/N_o)\sum n(\ldots)\, s\sigma t'\tau' \\
D(a',b) &= (1/N_o)\sum n(\ldots)\, s'\sigma' t\tau \\
D(a',b') &= (1/N_o)\sum n(\ldots)\, s'\sigma' t'\tau',
\end{aligned} \tag{19}$$

where, as before, $D(a,b)$ is the joint probability of detecting photon α with its polarizer oriented along a and photon β with its polarizer oriented along b, and so on. Similarly, the individual detection probabilities are

$$\begin{aligned}
p(a') &= (1/N_o)\sum n(\ldots)\, s'\sigma' , \\
q(b) &= (1/N_o)\sum n(\ldots)\, t\tau .
\end{aligned} \tag{20}$$

The quantity Γ defined in (6) can be written

$$\Gamma = (1/N_o)\sum n(\ldots) \{s\sigma(t\tau - t'\tau') + s'\sigma'(t\tau + t'\tau')\} \ . \tag{21}$$

The relations (17) - (21) will be useful in the next section for deducing weak and strong inequalities from our deterministic and local models.

6. PROOF OF INEQUALITIES

In order to demonstrate that our model satisfies the weak inequalities we use the relationship[11]

$$-XY \leq xy - xy' + x'y + x'y' - x'Y - Xy \leq 0 \tag{22}$$

which is tautologically valid if $0 \leq x, x' \leq X$ and $0 \leq y, y' \leq Y$. Setting now

$$x = s\sigma \ , \quad x' = s'\sigma' \ , \quad y = t\tau \ , \quad y' = t'\tau' \tag{23}$$

and thus $X = Y = 1$ we obtain

$$-1 \leq s\sigma(t\tau - t'\tau') + s'\sigma'(t\tau + t'\tau') - s'\sigma' - t\tau \leq 0 \tag{24}$$

which inequality is valid for each individual photon pair. Multiplication with the population density $n(\ldots)$, summation over the 324 different populations, and division by N_o yields

$$-1 \leq \Gamma - p(a') - q(b) \leq 0 \tag{25}$$

This is exactly the weak inequality. therefore, as expected, our model is in agreement with Bell's inequality and thus necessarily in disagreement with quantum mechanics.

Also the strong inequalities can be deduced from our model <u>if one of the additional assumptions is explicitly made</u>. Which one to choose is unessential since they lead to the same result. We will therefore focus attention on the CH assumption, for simplicity reasons. This assumption states that <u>for every photon</u> the probability of a detection with a polarizer in place on its trajectory (e.g., $s\sigma$) is less than or equal to the detection probability with the polarizer removed (e.g., δ). By applying this assumption to the two polarizer orientations of each photon we get

$$0 \leq s\sigma, s'\sigma' \leq \delta \ , \tag{26}$$

$$0 \leq t\tau, t'\tau' \leq \varepsilon \ .$$

Since all our variables are dichotomic, these latter relations imply that a photon which is registered by the detector with the polarizer in place would always be detected if the polarizer were removed, but not necessarily <u>vice versa</u>. Using again (22) with the definitions (23) and hence because of (26), with $X = \delta$ and $Y = \epsilon$, we obtain

$$-\delta\epsilon \leqslant s\sigma(t\tau - t'\tau') + s'\sigma'(t\tau + t'\tau') - s'\sigma'\epsilon - \delta t\tau \leqslant 0 \quad (27)$$

which inequality is again valid for each individual photon pair. Multiplication with the population $n(...)$, summation over all populations and division by N_0 now gives

$$-D_0 \leqslant \Gamma - D(a',\infty) - D(\infty,b) \leqslant 0 \quad (28)$$

Here we have obtained exactly the strong inequalities which, again as expected, are not violated by our model, <u>provided the CH assumption is taken to be true</u>. Of course the inequalities (28) could have been obtained by the other additional assumptions as well, with the only difference that the reasoning would have been longer.

7. DETERMINISTIC LOCAL MODELS VIOLATING THE STRONG INEQUALITIES

We start by giving a very simple example of a local deterministic model that violates the strong inequalities. For that purpose we assume that for all photon pairs $s\sigma = s'\sigma' = t\tau = t'\tau' = 1$. in such a case the four joint probabilities $D(a,b)$, $D(a,b')$, $D(a',b)$, $D(a',b')$ all equal unity and it follows from (6) that $\Gamma = 2$. The second inequality (28) thus becomes

$$\Gamma \leqslant (1/N_0)\sum n(...)(\delta + \epsilon) . \quad (29)$$

Clearly, this inequality is violated if δ or ϵ vanish even for a single photon only, because of (17). Therefore all deterministic local models having all variables $s = s' = t = t' = 1$ and $\sigma = \sigma' = \tau = \tau' = 1$, but with $\delta = \epsilon = 0$ at least part of the times, violate the strong inequalities (28).

Admittedly the previous models are not very physical ones both because of the assumptions made and because they exhibit "average enhancement", i.e., the coincidence counting rate with polarizers present is larger than that without polarizers, as a consequence of

$$D(a,b) = 1 \quad ; \quad D_0 < 1.$$

Yet our general model, having a large number of adjustable parameters which are the populations of the individual subensembles is certainly rich enough to yield physically more reasonable cases which still violate the strong inequality. This we will show now.

We next explicitly construct a specific example which agrees with quantum theory exactly but violates the strong inequalities. In order to simplify the notation we note that in the inequalities (27) the variables determining photon behaviour with polarizer present appear only in products with the corresponding detector variables. We therefore define the new variables

$$S \equiv s\sigma \quad ; \quad S' \equiv s'\sigma' \quad ; \quad T \equiv t\tau \quad ; \quad T' \equiv t'\tau' \quad (30)$$

This implies that, for example, $S = 1$ determines that photon α will cross its polarizer oriented at direction a and that it will be registered by its detector. Instead $S = 0$ determines either that the same photon is absorbed in its polarizer, or that it crosses it but is not registered by the detector. The populations are now functions of 6 dichotomic variables only and can be written

$$N(S,S',\delta;T,T',\epsilon) \tag{31}$$

Clearly there are only 64 different populations N: Some of them are sums of different populations n discussed above. In order to show that the populations (31) can reproduce the quantum mechanical predictions exactly we assume, as it is certainly possible, the following structure

$$N(S,S',\delta;T,T',\epsilon) = N_0 L(S,S';T,T') M(\delta;\epsilon) \tag{32}$$

where all 16 quantities L and all 4 quantities M are nonnegative and fulfil the normalization conditions

$$\sum_{S,S';T,T'} L(S,S';T,T') = 1 \quad ; \quad \sum_{\delta,\epsilon} M(\delta;\epsilon) = 1 \tag{33}$$

For our explicit example it is sufficient to assume that only the following 9 quantities L are positive and have the values

$$\begin{aligned}
L_1 &= L(1,0;1,0) = \tfrac{1}{4}(1 + \tfrac{1}{\sqrt{2}})\eta_1 \eta_2 \ , \\
L_2 &= L(1,0;0,1) = \tfrac{1}{4}(1 - \tfrac{1}{\sqrt{2}})\eta_1 \eta_2 \ , \\
L_3 &= L(0,1;1,0) = \tfrac{1}{4}(1 + \tfrac{1}{\sqrt{2}})\eta_1 \eta_2 \ , \\
L_4 &= L(0,1;0,1) = \tfrac{1}{4}(1 + \tfrac{1}{\sqrt{2}})\eta_1 \eta_2 \ , \\
L_5 &= L(1,0;0,0) = \tfrac{1}{2}\eta_1 - \tfrac{1}{2}\eta_1 \eta_2 \ , \\
L_6 &= L(0,1;0,0) = \tfrac{1}{2}\eta_1 - \tfrac{1}{2}(1 + \tfrac{1}{\sqrt{2}})\eta_1 \eta_2 \ , \\
L_7 &= L(0,0;1,0) = \tfrac{1}{2}\eta_2 - \tfrac{1}{2}(1 + \tfrac{1}{\sqrt{2}})\eta_1 \eta_2 \ , \\
L_8 &= L(0,0;0,1) = \tfrac{1}{2}\eta_2 - \tfrac{1}{2}\eta_1 \eta_2 \ , \\
L_9 &= L(0,0;0,0) = (1 - \eta_1)(1 - \eta_2) + \eta_1 \eta_2 / 2\sqrt{2} \ .
\end{aligned} \tag{34}$$

$L_1 \ldots L_4$ can give rise to double and single detections, $L_5 \ldots L_8$ can give rise to single detections only, and L_9 can never give rise to any detection. Likewise we assume for the positive quantities M:

$$\begin{aligned}
M_1 &= M(1;1) = \eta_1 \eta_2 \ , \\
M_2 &= M(1;0) = M_3 = M(0;1) = \eta_1 - \eta_1 \eta_2 \ , \\
M_4 &= M(0;0) = (1 - \eta_1)(1 - \eta_2) \ .
\end{aligned} \tag{35}$$

It is now easy to check that the populations (31) – (35) reproduce the quantum mechanical predictions exactly for the case of the maximum violation of the strong inequality. One can easily see, for example, that

$$p(a) = (L_1 + L_2 + L_5)(\, M_i\,) = \tfrac{1}{2} n_1 \, ,$$

$$D_0 = (\, L_i\,) M_1 = n_1 n_2 \, ,$$

$$D(a',\infty) = (L_3 + L_4 + L_6)(M_1 + M_3) = \tfrac{1}{2} n_1 n_2 \, , \tag{36}$$

$$D(a,b) = L_1(\, M_i\,) = \tfrac{1}{4}(1 + \tfrac{1}{\sqrt{2}}) n_1 n_2 \, ,$$

$$D(a,b') = L_2(\, M_i\,) = \tfrac{1}{4}(1 - \tfrac{1}{\sqrt{2}}) n_1 n_2 \, ,$$

and so on. These predictions obviously refer to the case of ideal polarizers, but it is a very simple matter to modify (34) and (35) in such a way as to obtain the fully realistic case. It is also evident that the populations can be chosen such as to violate the strong inequalities in cases of less than maximum violation, thus agreeing with experiment for all angular orientations of the polarizers.

Obviously the results (36) which reproduce the quantum mechanical predictions exactly do not exhibit any form of <u>average</u> enhancement of probabilities. In fact one has, for example

$$D_0 < D(a,b)$$

and so on. Inequalities such as the previous one imply the validity <u>on the average</u> of the CH additional assumption. The same assumption is however violated in individual cases, as we have seen.

8. CONCLUSIONS

Deterministic local models are certainly less general (and less appealing) than truly probabilistic local models. However these two classes of local model are probably generally equivalent from the practical point of view, since no Bell-type theorem exists that could allow one to discriminate, in practice, determinism from probabilism. Formulated in the broadest possible way local models are thus seen to agree with the quantum theoretical predictions and therefore to violate the strong inequalities checked experimentally in EPR-type experiments.

The generally accepted conclusion that locality has been falsified in these experiments is thus seen to be totally unfounded and can be judged to be more an expression of wishful thinking than a sharp scientific statement. New and better experiments are needed with better detectors than hitherto used. The efficiency of the detectors is in fact the crucial issue: It can for example be verified that L_6 and/or L_7 become negative if n_1 and/or n_2 exceed $(1 + 1/\sqrt{2})^{-1} = 0.586$. For high detector efficiency the distinction between weak and strong inequalities tends to vanish, Bell's theorem plays a much more concrete role and no explanation of the quantum mechanical predictions in terms od deterministic local models is possible.

REFERENCES

1. S.J. Freedman and J.F. Clauser, *Phys. Rev. Lett. 28*, 938 (1972);
    R.A. Holt and F.M. Pipkin, Harvard University preprint (1974);
    J.F. Clauser, *Phys. Rev. Lett. 37*, 1223 (1976); *Nuovo Cimento 33B*, 740 (1976);
    E.S. Fry and R.C. Thompson, *Phys. Rev. Lett. 37*, 465 (1976);
    A. Aspect, P. Grangier and G. Roger, *Phys. Rev. Lett. 47*, 460 (1981);
    A. Aspect, P. Grangier and G. Roger, *Phys. Rev. Lett. 49*, 91 (1982);
    A. Aspect, J. Dalibard and G. Roger, *Phys. Rev. Lett. 49*,1804(1982);
    W. Perrie, A.J. Duncan, H.J. Beyer and H. Kleinpoppen, *Phys. Rev. Lett. 54*, 1790 (1985).
2. J.F. Clauser and M.A. Horne, *Phys. Rev. D10*, 526 (1974).
3. T.W. Marshall, E. Santos and F. Selleri, *Phys. Lett. A98*, 5 (1983).
4. T.W. Marshall, *Phys. Lett. 99A*, 163 (1983);
    T.W. Marshall, *Phys. Lett. 100A*, 225 (1984).
5. T.W. Marshall and E. Santos, *Phys. Lett. 107A*, 164 (1985).
6. M. Ferrero and E. Santos, *Phys. Lett. 116A*, 356 (1986).
7. F. Selleri, ed., *Quantum Mechanics versus Local Realism: The Einstein,Podolsky, and Rosen Paradox* (Plenum, London - New York, 1988).
8. A. Garuccio and F. Selleri, *Found. Phys. 10*, 209 (1980).
9. G. Lepore, *New Inequalities from Local Realism*, Univ. of Bari, preprint, (1987).
10. J.F. Clauser, M.A. Horne, A. Shimony and R.A. Holt, *Phys. Rev. Lett. 23*, 880 (1969).
11. See ref. 2.
12. A. Garuccio and V. Rapisarda, *Nuovo Cimento A65* , 269 (1981).
13. A. Aspect, *Thése*, Université de Paris-Sud (1983), page 121.
14. F. Selleri, *Is Bell's Inequality Violated in Nature?* in *Problems in Quantum Physics* (World Scientific, Singapore, 1988).
15. A. Zeilinger and F. Selleri, *A Deterministic Local Model for Einstein-Podolsky-Rosen Experiments*, Univ. of Bari preprint , (1988).
16. E.P. Wigner, *Am. J. Phys. 38*, 1005 (1970).

COMMENTS ON THE "UNCONTROLLABLE" CHARACTER OF NON-LOCALITY

J.P. VIGIER
Institut Henri Poincaré
Laboratoire de Physique Théorique
11, rue P. et M. Curie
75231 PARIS CEDEX 05 (France)

ABSTRACT. The point is made that the proposed Non-Local Hidden Variable Theories also yield, like Quantum Mechanics, uncontrollable non-locality which cannot be utilized to propagate superluminal signals.

1. INTRODUCTION

In a very interesting paper on controllable and uncontrollable non-locality[1] Shimony, developing recent results of Jarrett[2], argues that the non-locality of Quantum Mechanics (now strongly suggested by Aspect et al.'s results in EPR-type experiments[3]) can "peacefully coexist" with Relativity Theory because of its uncontrollable character. He points that one should distinguish between "controllable non-locality" (utilizable for signalling faster than light) and "uncontrollable non-locality" (which implies non local correlations but cannot be used to transmit superluminal informations) and shows that Q.M. leads to "uncontrollable non-locality". Shimony then asks three specific questions to the supporters of non-local hidden variables (i.e. Bohm et al.[4] and Vigier et al.[5], i.e. :
1) "Even though we can acknowledge the general proposition that any statistical physical theory is able to account for the emergence of some stability out of disorder... we nevertheless want to know in detail how the polarization correlations are stably maintained in E.P.R. experiments".
2) "Can it be shown rigourously that the postulated non-locality of the hidden variable theories is "uncontrollable" rather than "controllable"?"
3) "What are the predictions of the proposed non local hidden variables theories about a modification of the Aspect[3] experiment consisting in a strong burst of laser light that intersects the paths of photons after the commutator of that experiment have been actuated, but before the polarization analysis of the photon has been completed ?"

2. THE NON-LOCAL HIDDEN-VARIABLE APPROACH TO QUANTUM MECHANICS

The first question means that, if one accepts the models of random covariant subquantal aether (of the Dirac type[6]), one should explain the stability of the experimentally precise correlation of the polarization of well separated photons (or other type of particles) in an EPR-type experiment[4]. Strictly speaking we have, at present, no complete theory of aether, but only preliminary attempts[7] to build it on the basis of the relativistic hydrodynamics, developed, for example, by Halbwachs[8]. Following this line of research we have explicitly interpreted in a causal way the results of Aspect's experiments and calculated[9] the evolution equation for the physical quantities involved in it. For the proper time evolution of the spin vector $S_{1\mu}$ of the particle 1 we found :

$$S_{1\mu} = -\frac{i}{2}\mu\nu\alpha\beta \ S_1^{\alpha\beta}(U_1^\lambda \partial_{1\lambda} U_1^\rho) \tag{1}$$

where $S_{1\mu\nu}$ is the spin tensor and $U_{1\mu}$ the four velocity of the particle 1. Of course this derivative depends in a non local way on the $A_{2\mu}(x_1,x_2)$, i.e. the field of the particle 2 contained in the definition of $S_{1\mu\rho}$. From this results that the non-local interaction is carried by a quantum torque that, (like the well-known quantum potential[10]) directly reflects the behaviour of the wave function. It is strictly non local if the wave function is non factorizable. The concept of non-local quantum potential and quantum torque, which are the starting points for the causal interpretation of Quantum Mechanics[10], can, in a very natural way, be connected with the Stochastic Interpretation of Quantum Mechahics (S.I.Q.M.)[11] in the sense that quantum potentials and quantum torques can be considered as the global reaction of the subquantic aether to the motion of a particle imbedded into it.

3. STABILITY and STOCHASTICS

This exposition shows that non-locality emerges as a necessity both in the frame of QM and in the HV-attempt in the frame of the Stochastic Interpretation of QM. But there is still another point to be clarified : since non-locality is mediated by the Quantum Potential and expresses the fundamentally chaotic, random character of the covariant subquantal medium (Dirac-aether), the question raised by Shimony on the stability of quantum HV-predictions is completely justified.
However we can show[12] :
(1) that in the SIQM the Markov processes appropriate to reproduce the Klein-Gordon statistics[13] are not a simple generalization of the Einstein-Smoluchowski theory of Brownian motion, but imply non-locality as a basic feature. Indeed these processes include particle-antiparticle-transitions that establish non-local correlations of space-like separated regions ; a fact absent in the relativistic extension of classical Brownian motion
(2) that these processes satisfy an H-theorem which is derived as a relativistic generalization of the Bohm-Vigier[14] H-Theorem for the

Guerra-Ruggiero-Vigier relativistic Markov processes that reproduce the Klein-Gordon equation. This H-theorem establishes the relativistic quantum predictions as a mean equilibrium solution and implies that every deviation from this equilibrium (e.g. due to random chaotic aether-motions) immediately initiates a process that restores equilibrium. This answers Shimony's first question.

4. UNCONTROLLABLE NON-LOCALITY

To this proposed line of research Shimony asks, in his second question, what is the precise character of the non local correlations carried by the physical random aether, i.e. : are they controllable or uncontrollable ? The answer has to be made on two levels :

- on the same level which classical Quantum Mechanics adresses, the statistical predictions of the two theories are the same. In fact the crucial differences in the two theories only arise in their predictions of the outcome of specific experiments on isolated particles specially devised to test the physical reality of the de Broglie's waves, i.e. to disprove the "wave packet collapse" of the Copenhagen Interpretation of QM (CIQM)[15]. Thus at the quantum level non-local H-V imply uncontrollablelable non-locality.

- A more embarrassing way to put the question would be to ask what happens at a subquantum level in a non-local aether ? In this non-locality controllable in principle or not ? Neglecting the problem of their physical measurability, we can attribute to our random covariant form of the aether some general properties even in the absence of a complete theory. For example we can choose between two possibilities : to build a non-local theory completely deterministic or still containing some irreductible indeterministic character at that level. If our theory is a completely deterministic, covariant, non-local theory we claim that no causal anomalies can arise. In fact with "completely deterministic theory" we mean that we can uniquely predict, starting with initial conditions given on a space-like surface, all the world lines of all aether elements by solving a Cauchy problem (with or without action-at-a-distance). It is an advance of the last ten years that this program has been shown to be compatible with a covariant formalism by means of suitable constraints imposed on actions-at-a-distance[16] and a satisfactory result is the fact that the quantum potential satisfies the said compatibility conditions[9]. In such a world model the word "signals" is largely meaningless[17] : in fact in ordinary language a signal suggests the intervention of a <u>free will</u> external to the physical world which, at a given time, decides to modify a regular behaviour in order to send a message. From this standpoint, if subquantal physics is completely deterministic, the world is a unique configuration of events evolving in space-time and a problem will not be how to modificate events but to discover general laws that express the space-time disposition of events. If, on the other hand, we keep some degree of real randomness at the subquantum level, that is if we admit that at this level, some events cannot be predicted by an aether theory we must, of course, check whether the non local interactions are still uncontrollable : but at present this only represents a logical possibility and we have no such theory to test.

The last question assumes some physical intervention occurring in Aspect's experimentation after the photon has been switched (but before the polarization measurement) and asks what does the non-local hidden variable program predict concerning its effects. That problem allows us to make some remarks on the concept of a measuring process. In the Stochastic Interpretation of the Quantum Mechanics a measurement is a particular type of interaction and nothing more : No place is there for an "observer" role. However a measurement is a "particular" interaction in the sense that it determines an evolution of the state of our system which passes through non equilibrium states so that it can not be described by a Schrödinger equation. Some ideas on the precise mechanism of that evolution have been already discussed[10,21]. In that sense the difference between a measurement and a simple interaction is not in the presence of an external consciousness which "looks" at the results of a measurement, but in the effects that they produce on the system.

Shimony proposes now a modification of the Aspect's experiment which consists in a physical intervention on the path of one of the two photons and asks what does the non-local hidden variable program predict concerning its effects, id est would the postulated non-local quantum potential correlating the two photons (instantaneous in their rest frame) be perturbed strongly enough to suppress the connection.

The answer is that the effect would be extremely small, otherwise due to the ambiant electromagnetic noise permeating the Dirac vacuum, no correlation would be mediated at all. The quantum potential does not per se decrease with the wave amplitude R (since $Q \simeq \Box R/R$) i.e. with distance, nevertheless, due to the vacuum temperature induced by de Broglie waves, its effect might vanish at some distance (as suggested by Baracca et al.[20], i.e. the field $\psi(x_1, x_2)$ of the correlated particles might go to $\Psi_1(x_1) \Psi_2(x_2)$ beyond a certain distance λ_c (so we have a macroscopic local world which satisfies Bell's inequalities at our level).

If so the proposed strong laser burst would tend to reduce λ_c. The reduction would most probably be function of the ratio of the beam power density integrated on the paths to the ambiant noise. Stronger (resonant) effects are probable if the laser wavelength is tuned to the photon energy. As no experiment has fixed a value for λ_c (a lower limit being the photon distances involved in Aspect's experiments) no qualitative prediction can be made.

In the frame of the CIQM no effect is expected at all due to the vanishingly small cross-section of photon-photon interactions. This suggests an extension of Shimony's proposal to other EPR experiments with proton pairs instead of photons : Let us call a spin - flipper a device which inverts the spin of protons of known spin. If we place such a spin - flipper along one of the branches of the EPR experiment, what effect would it have on the correlations ? In the frame of the CIQM the protons going through it would have no definite spin, i.e. they are in a symmetrical mixture of spin states. Therefore no modification is expected. In the frame of the SIQM, spin values really exist, even if not measured ; the proton will change spin when passing through the spin-flipper, but so will the correlated proton do, due to the quantum potential connection. Therefore the coincidence rates will not be changed at all.

5. THE IMPLICATION OF NON-LOCAL CAUSALITY

We proceed with a supplementary remark : it concerns the motivation of Einstein to defend locality in Relativity Theory. Evidently his main motivation was his belief in determinism[18], i.e. in the facts :
a) that any causal sequence of events should be described along a time-like path... so that its order in time should be observer independent ;
b) that only positive energy can propagate forward in time within the forward light cone so that there is no retroaction in time ;
c) that the laws of Mechanics should be able to predict the mechanical evolution of any given system from a given set of initial conditions defined on a space-like surface... so that one can solve the Cauchy problem in the forward (or backward) time direction.

Now, the main new fact is the discovery[16] that the conditions a), b) and c) are also compatible with non-local interactions between N particles provided supplementary conditions are satisfied. Indeed with relativistic predictive Mechanics one can establish the equivalence between a free Klein-Gordon system and a relativistic Hamiltonian system of directly interacting classical particles, for which the interaction is described by non-local potentials. Then one can show[22] that the predictivity condition implies the vanishing of the Poisson brackets of the Hamiltonians :

$$\{H_i, H_j\} = 0 \quad \text{with} \quad H_e = H_{oe} + V_e \quad (2)$$

where H_{oe} represents the free particle Hamiltonians and V_e the non-local potentials. This condition is shown to be satisfied identically in the whole of phase space. Hence causal non-local HV-Theories explain from first principles the postulated quantum features, in the form of the Klein-Gordon equation.

6. THE QUANTUM "INDISTINGUISHABILITY"

But also the reverse can be proven. Indeed recent work[23] has shown that the - classically incomprehensible - principle of indistinguishability of particles, which is the basic assumption for the derivation of Bose-Einstein statistics, is not needed, provided one assumes a random stochastic disturbance of a set of distinguishable particles correlated by action-at-a-distance. In this frame, an assumption of random stochastic weights for the probability of the states in configuration space immediately forces a set of distinguishable particles away from Maxwell-Boltzmann and into the Bose-Einstein statistics. Moreover it has been shown[24] that this stochastic probability assumption of [23] is a straightforward consequence of the non-local correlation between particles mediated by the action-at-a-distance of the Quantum Potential. Consequently the fundamental property of quantum entities results in the absence of "bare" elements of the B-E statistics, i.e. in the (non-locally) correlated character of the particles in the ensemble, in agreement with a qualitative suggestion of Einstein[25] that the differences between the Boltzmann and the B-E counting "expresses indirectly a certain hypothesis on a mutual influence of the molecules which for the time being is of a quite mysterious nature". If one further notes[26] that any correlation of the particles with local potentials inevitably leads to Maxwell-Boltzmann statistics, then one conclusively deduces that quantum statistic necessarily implies non-locality.

7. CONCLUSION

We have thus, in the present stage of evolution of non-local HV-theories in the frame of SIQM, reached a point of relative selfconsistency of the scheme. At the price of introducing a causal non-locality, which is a straightforward consequence of the theory, QM is thus now interpreted in causal stochastic terms. Furthermore, causal non-locality is shown to be equivalent to the usually postulated quantum features, namely the quantum statistics. Finally the non-local correlations are of such a nature that they guarantee the stability of (equilibrium) quantum predictions. This we claim is very encouraging despite the absence of detailed aether theory. But evidently we do not claim that this replaces the need for an aether theory, an indispensable, vast program entirely open in front of us.

Just to conclude, we wish to stress a point that is absent in Shimony's reasoning. In fact, in the effort to outline the equivalence of the non-local, causal HV-approach with the basic quantum features, another point often remains underestimated. The two approaches, SIQM and the Copenhagen Interpretation of QM are not equivalent or identical. In all cases SIQM is not opposed to the quantum formalism of the Quanta theory as a whole, but only to the Copenhagen School and especially to Bohr's wave packet collapse concept[27] in agreement with Cini's new measurement theory without wave packet collapse[21]. Furthermore a major point of difference concerns the physical existence of de Broglie waves, currently tested by Rauch et al.[28] and Gozzini[29]. For SIQM they represent collective excitations on the top of Dirac's aether and for CIQM only probability - potentiality waves. These differences in predictions[27] are items that can now be submitted to experiment so that the Bohr-Einstein controversy can be settled on the table of the laboratory.

REFERENCES

1. A. Shimony : "Controllable and uncontrollable Non-Locality" (Boston University, preprint, 1983) published in the Proceedings of the "International Symposium on the Foundations of Quantum Mechanics" Tokyo, Aug. 29-31, 1983, p. 225-230.

2. J. Jarrett : "Bell's theorem, Quantum Mechanics and Local realism" (Ph.D. Thesis, Chicago University, 1983).

3. A. Einstein, B. Podolsky and N. Rosen : Phys.Rev. $\underline{47}$, 777 (1935).
   A. Aspect, G. Grangier and G. Roger : Phys.Rev.Lett. $\underline{47}$, 460 (1981).
   " " " : Phys.Rev.Lett. $\underline{49}$, 91 (1982).
   A. Aspect, J. Dalibard and G. Roger : Phys.Rev.Lett. $\underline{49}$, 1804 (1982).

4. D. Bohm and B. Hiley : Found.Phys. $\underline{5}$, 93 (1975).
   D. Bohm : "Wholeness and Implicate Order" (Rowtledge and Kegan Paul, London, 1980).

5. Ph. Droz-Vincent : Phys.Rev. $\underline{D19}$, 702 (1979).
   N. Cufaro-Petroni and J.P. Vigier : Phys.Lett. $\underline{A81}$, 12 (1981).
   A. Garuccio, V.A. Rapisarda and J.P. Vigier : Lett.N.Cim. $\underline{32}$, 451 (1981).
   J.P. Vigier : Astrón.Nachr. $\underline{303}$, 55 (1982).

6. P.A.M. Dirac : Nature $\underline{168}$, 906 (1951).
   " : Nature $\underline{169}$, 702 (1952).
   K.P. Sinha, C. Sivaram and E.C.G. Sudarshan : Found.Phys. $\underline{6}$, 62 (1976).
   K.P. Sinha and E.C.G. Sudarshan : Found.Phys. $\underline{8}$, 823 (1978).

7. J.P. Vigier : Lett.N.Cim. $\underline{29}$, 467 (1980).
   N. Cufaro-Petroni and J.P. Vigier : Found.Phys. $\underline{13}$, 253 (1983) ;
   " " : Nuov.Cim. $\underline{81B}$, 243 (1984).

8. F. Halbwachs : "Théorie Relativiste des fluides à spin" (Gauthier-Villars, 1960).

9. N. Cufaro-Petroni, Ph. Droz-Vincent and J.P. Vigier : Lett.N.Cim. $\underline{31}$, 415 (1981).
   N. Cufaro-Petroni and J.P. Vigier : Phys.Lett. $\underline{A88}$, 272 (1982).
   " " : Phys.Lett. $\underline{A93}$, 383 (1983).

10. D. Bohm : Phys.Rev. $\underline{85}$, 166, 180 (1952).

11. E. Nelson : Phys.Rev. $\underline{85}$, 150, 1079 (1966).
    L. de la Pena-Auerbach : J.Math.Phys. $\underline{12}$, 453 (1971).
    W. Lehr and J. Park : J.Math.Phys. $\underline{18}$, 1235 (1977).
    F. Guerra and P. Ruggiero : Lett.Nuov.Cim. $\underline{23}$, 529 (1978).
    J.P. Vigier : Lett.N.Cim. $\underline{24}$, 265 (1979).
    N. Cufaro-Petroni and J.P. Vigier : "Random motions at the velocity of light and Relativistic Quantum Mechanics" (Bari University, preprint, 1983).

12. A. Kyprianidis, D. Sardelis, Lett.Nuov.Cim. $\underline{39}$, 337 (1984).

13. F. Guerra, P. Ruggiero, Lett.Nuov.Cim. $\underline{23}$, 529 (1978) ; J.P. Vigier, Lett.Nuov.Cim. $\underline{24}$, 258, 265 (1979).

14. D. Bohm, J.P. Vigier, Phys.Rev. $\underline{96}$, 208 (1954).

15. A. Garuccio, K. Popper and J.P. Vigier : Phys.Lett. $\underline{A86}$, 397 (1981). A. Gozzini : "Proceedings of the Symposium on Wave particle dualism" (Reidel, Dordrecht, 1983).

16. Ph. Droz-Vincent : Phys.Scrip. $\underline{2}$, 129 (1970).
L. Bel : Ann.Inst.Henri Poincaré, $\underline{3}$, 307 (1970).
Ph. Droz-Vincent : Ann.Inst.Henri Poincaré $\underline{27}$, 407 (1977).
A. Komar : Phys.Rev. $\underline{D18}$, 1881, 1887, 3617 (1978).
L. Bel : Phys.Rev. $\underline{D18}$, 4770 (1978).
Ph. Droz-Vincent : Ann.Inst.Henri Poincaré $\underline{32}$, 377 (1980).
R.P. Gaida : Sov.J.Pact.Nucl. $\underline{13}$, 179 (1982).

17. N. Cufaro-Petroni : "Proceedings of the symposium Open Questions in Quantum Physics" (Bari, 1983), in print Reidel, 1984.

18. A. Naess : in "Old and New Questions in Physics, Cosmology, Philosophy and Theoretical Biology" (Plenum, New York, 1983).

19. A. Einstein : "Investigation on the theory of the Brownian movement" (Methuen London, 1926).

20. A. Baracca, D. Bohm, B. Hiley and A.E.G. Stuart, N.Cimento $\underline{B28}$, 453 (1975).

21. M. Cini, Nuov.Cim. $\underline{73B}$, 27 (1983).

22. A. Garuccio, A. Kyprianidis, J.P. Vigier : Relativistic Quantum Potential ; The N-Body case, to be published in Nuovo Cimento B.

23. J. Tersoff, D. Bayer, Phys.Rev.Lett. $\underline{50}$, 553 (1983).

24. A. Kyprianidis, D. Sardelis, J.P. Vigier, Phys.Lett. 100I, 228 (1984).

25. A. Einstein, Sitzungsber, Preuss, Akad.Wiss.Phys.Math. Kl. p. 3 (1925).

26. Z. Maric, M. Bozic, D. Davidovic, Randomness and Determinism in the Kinetic equations of Clausius and Boltzmann, Proceedings of the Boltzmann meeting, Vienna, 1981.

27. A. Garuccio, A. Kyprianidis, D. Sardelis, J.P. Vigier, Lett.Nuov. Cim. $\underline{39}$, 225 (1984).

28. H. Rauch, Proceedings of the Bari Conference "Open Questions in Quantum Physics", Reidel 1984 and references quoted herein.

29. A. Gozzini : Communication at the Symposium on Wave Particle Dualism, Perugia, April 1982.

Part 4

Real or gedanken experiments and their interpretation

SOME BASIC DIFFERENCES BETWEEN THE COPENHAGEN AND DE BROGLIE INTERPRETATION OF QUANTUM MECHANICS LEADING TO PRACTICAL EXPERIMENTS

J. R. Croca
Departamento de Física
Faculdade de Ciências - Universidade de Lisboa
Campo Grande, Ed. C1, Piso 4 - 1700 Lisboa
PORTUGAL

ABSTRACT. Some important differences between the Copenhagen and de Broglie interpretation of quantum formalism are seen in order to provide a theoretical background for practical crucial experiments.

1. Introduction

The interpretation of quantum formalism has not been an easy problem. Polemics have ravaged the welfare of the scientific community from the very beginning[1]. On the one hand the quantum formalism can be seen as a powerful mathematical tool, very useful to predict the outcome of the experiments, but having nothing to do with an hidden hypothetical reality; on the other hand it can be interpreted as a reasonable representation of some physical deep reality. As it is well known the first position represents the usual interpretation of the quantum formalism, and it is the one generally accepted. The second way of understanding the quantum formalism is the one proposed by the causal theories. These theories are now able to explain almost every phenomenon of the microphysical domain. One of such theories, perhaps the most developed of them, and with which we shall be concerned, is the Causal theory of de Broglie also called the double solution theory.

For a long time this problem of the existence or non existence of a more deep reality beyond the quantum formalism could not be cleared up! As long as no experimental evidence can, without doubt, confirm the causal interpretation, its acceptance shall remain only a matter of faith. In the best case the causal theory shall be no more than an alternative, unnecessary and much more complicated way of arriving at the same final experimental data. Thus it is not a surprise that much effort has been recently centered on the study of possible experimental situations where the opposite theories predict different results. One of the possible ways of solving this problem could in principle be based on the different meanings given by the Copenhagen school and the causal interpretation to the wave function ψ. As it is well known[2] for the usual theory ψ_u (where u stands for the usual theory) is a probability wave function carrying all available information on the microphysical system, therefore it is not a real entity, but only a previsional tool. For the causal theory of de Broglie[3] instead ψ_c (c means causal) is a real wave which carries and guides the singularity. In this interpretation a quantum system is at the same time a wave, with no appreciable energy, plus a high energetic singularity always localized in spacetime. In order to shed a better light on this problem it is useful to consider an experimental situation.

2 - Double slit experiment

Let us consider a wave packet, a neutron for instance, coming from a source, emitting one at a time, impinging on an impenetrable barrier with two slits (1) and (2). In order to predict a fringe pattern, at the detection zone, one must make the assumption that the neutron wave packet passes the two slits, otherwise no interference would be seen. In a way the neutron must be divided into two smaller pieces, each crossing one slit. In such conditions if one places two detectors, one in front of the slit (1) the other in front of the slit (2), both would click at the same time, reacting to a fractional energy of the incident neutron. All experimental evidence is against this last hypothesis. No one has ever detected half of a neutron! What really happens is that some times the detector placed in front of the slit (1) clicks, other times is the one placed in front of the slit (2), in any case they never react simultaneously.

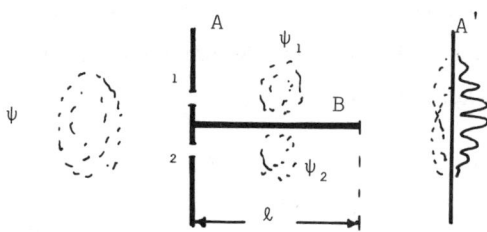

Fig.1 - Double slit experiment seen by the usual interpretation of quantum mechanics. The probability wave packet ψ crosses the two slits, and the two resulting wave packets ψ_1 and ψ_2 overlap at A' giving origin to a fringe pattern. A barrier B, with a length of ℓ, for separation of the two wave packets ψ_1 and ψ_2 is shown.

The Copenhagen school as it is well known interprets this phenomenon by denying the reality of the neutron, before measure. The only important element is the probability wave $\psi_c \equiv \psi$. The initial neutron probability wave packet represented by ψ when arriving at the barrier A crosses the slits originating two probability wave packets ψ_1 and ψ_2 which interfere in the interference region.

According to de Broglie's causal theory the neutron is a singularity plus a wave. So upon arriving at the barrier the wave crosses the two slits while the neutron-singularity passes through one or the other slit. That is, through one slit passes one wave plus the singularity, through the other passes a wave without singularity, an **empty wave**. In order to fix the notation, the real empty wave shall be represented by the symbol θ, and the wave with singularity by ϕ. The reasons of this choice will soon be clear. At the interference region these two waves mix and interfere guiding the singularity to the points of higher total wave intensity.

As the double slit experiment is explained in a satisfactory way both by the causal and usual theories, it can not therefore be used as a deciding criteria. Nevertheless if there happens to be some difference in the physical properties of these two real waves, one with singularity ϕ, and the other without singularity θ, the empty wave, it should be possible, in principle, to modify slightly the experimental setup so that that the the two theories predict different effects. The usual probability waves ψ_1 and ψ_2 share the same properties. So the results of the double slit experiment shall be the same regardless the free independent path length ℓ of the two wave packets.

Let us now consider the double slit experiment as seen by the causal theory with the supplementary hypothesis that there is a difference in the physical behaviour of the wave with singularity ϕ and the wave without singularity (the empty wave) θ.

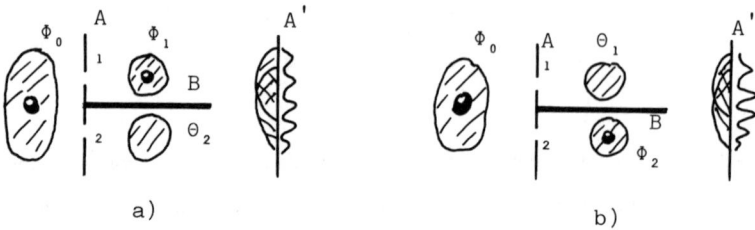

Fig. 2 - Double slit experiment seen by the causal theory of the Broglie. a) The singularity crossed the slit (1). b) The singularity crossed the slit (2). The wave always crosses the two slits and the resulting waves overlap in the interference region guiding the singularity preferentially to the points of higher total wave intensity.

In Fig.2 there are depicted the two possible interesting cases of the double slit experiment as seen by the causal theory of de Broglie.

As it is clearly seen the causal real wave and the usual probabilistic wave have generally the same probabilistic scheme, in the sense that both, the real and the probabilistic waves, cross the slit *one* and the slit *two*, and scheme. The particle-singularity, on the contrary, as it crosses *one* or the *other* slit has a probabilistic or scheme. However, if it is possible to find experimental situations where the two real waves, one with singularity ϕ and the other devoid of singularity θ, exhibit different properties the probabilistic scheme of these waves will change accordingly and no longer shall remain equal to the one of the usual wave.

In order to treat the problem with a certain degree of generality we consider that the waves arriving at the detection region A' have a certain transmittance factor, so that

$$\theta_1 = b_1\phi_0; \qquad \theta_2 = b_2\phi_0; \qquad \phi_1 = b_{s1}\phi_0; \qquad \phi_2 = b_{s2}\phi_0. \tag{1}$$

the total wave at the interference zone shall be

$$\phi = \phi_1 + \theta_2 \quad or \quad \phi = \theta_1 + \phi_2. \tag{2}$$

In the first case the singularity passed the slit (1), the second represents the other possibility of the singularity crossing the slit (2). A general formula including the two possible cases can be written

$$\phi = (1-\varepsilon)(\phi_1 + \theta_2) + \varepsilon(\theta_1 + \phi_2) \qquad (3)$$

where ε is a numeric factor having the two possible values

$$\varepsilon = 0; \qquad \varepsilon = 1, \qquad (4)$$

with certain probabilities. Formula (3) can be expressed in terms of the incident wave ϕ_0 by substituting the waves θ_i and ϕ_i by their values given by (1)

$$\phi = (1-\varepsilon)(b_{s1} + b_2 e^{i\chi})\phi_0 + \varepsilon(b_1 + b_{s2} e^{i\chi})\phi_0, \qquad (5)$$

where χ is the phase factor between the waves (ϕ_1, θ_2) and (θ_1, ϕ_2) which, obviously, is the same. The probability intensity distribution of finding one singularity at A' is given by

$$I' = k \mid \phi \mid^2, \qquad (6)$$

k is a proportionality factor, substituting in this formula the value of ϕ given by (5) one gets

$$I' = k \mid \phi_0 \mid^2 [(1-\varepsilon)^2(b_{s1}^2 + b_2^2 + 2b_{s1}b_2 \cos\chi) + \varepsilon^2(b_1^2 + b_{s2}^2 + 2b_1 b_{s2} \cos\chi)]. \qquad (7)$$

After N particle arrivals, some times the singularity passing through slit (1) or other times by the slit (2) the total intensity distribution shall be

$$I = NI'. \qquad (8)$$

Supposing that the singularity has equal probability of passing through one or the other slit

$$P(\varepsilon = 0) = P(\varepsilon = 1) = 1/2,$$

expression (8) will assume the form

$$I = \frac{1}{2}kN \mid \phi_0 \mid^2 [(b_{s1}^2 + b_2^2 + 2b_{s1}b_2 \cos\chi) + (b_1^2 + b_{s2}^2 + 2b_1 b_{s2} \cos\chi)]. \qquad (9)$$

furthermore, if we suppose that the transmission wave factors depend only whether the wave carries the singularity or not and are independent of the crossed slit

$$b_{s1} = b_{s2} = b_s; \qquad b_1 = b_2 = b, \tag{10}$$

which by substitution in (9) gives

$$I = kN \mid \phi_0 \mid^2 (b_s^2 + b^2 + 2b_s b \cos\chi). \tag{11}$$

2.1 - The empty wave is absorbed by a physical device

As was stated, in the conceptual framework of the causal theory of de Broglie, the probabilistic scheme is richer than that of the usual interpretation. In the Copenhagen scheme, the probability wave crosses the slit (1) and the slit (2); in de Broglie picture, the wave keeps the same probabilistic and scheme, but the singularity, as it crosses the slit (1) or the slit (2), gets a different probabilistic scheme. Nevertheless as the interference pattern results from the wave and scheme, and not from the singularity scheme, this experiment, as it stands, is not conclusive. The predictions of the two theories are precisely the same. Although it is possible to modify slightly the experiment in such conditions to break the and scheme of de Broglie's model, while keeping the usual one unchanged. In this new probabilistic scheme the singularity and the wave have the same or scheme, therefore the modified double slit experiment could be conclusive.

Looking at formula (11) one sees two transmission coefficients b_s the one of the wave with singularity ϕ, and b, that of the empty wave. It seems reasonable to admit that the full wave ϕ remains essentially unchanged as long as the singularity it carries lives. So the mean life of the full wave is just the mean life of the particle. So under the usual experimental conditions the transmission coefficient remains approximately constant

$$b_s = c^{te}. \tag{12}$$

But the wave θ, devoid of singularity, could in principle behave differently. For instance, it is possible to think of an hypothetical physical device with the property of stopping the empty wave

θ, but allowing the wave with the singularity to pass. That is a kind of physical barrier transparent to the singularity, but opaque to waves without singularity. In this case, after the empty wave stopper (EWS) the transmission factor for the θ waves becomes zero

$$b = 0. \qquad (12')$$

The probabilistic scheme of the waves changes from and to or, therefore the interference disappears. This conclusion can be checked placing (12') into formula (11)

$$I = kN \mid \phi_0 \mid^2 b_a^2. \qquad (13)$$

This formula shows that by interposing the empty wave stopper the interference pattern disappears. According to the usual interpretation empty waves do not exist so nothing is changed. The empty wave stopper is completly transparent to full waves therefore the scheme and remains unaltered and the intensity shall be

$$I \propto \mid \psi_1 + \psi_2 \mid^2, \qquad (14)$$

that is an interference pattern.

2.2 - The empty wave progressively looses its properties

Another possibility that remains open is the one developed by Koh[4] which he named "in-flight wave function reduction". Basically the idea consists in supposing that the mean life of the empty wave is shorter than that of the full wave. When propagating through the subquantum medium[5] the empty wave progressively looses its properties. According to Koh the greater the mass of the particle the shorter the path traveled by the free empty wave.

For photons, the Janossi experiment[6], done with an interferometer with the arms more than 33m long, seems to prove that at least the free photonic empty wave can travel this distance without apparent change of its properties.

In the interference experiments with electrons, done by Möllenstedt- Bayh[7] with a biprism, fringes were reported to a maximum length of 52cm, which may be a symptom that after that distance the electron empty wave lost its properties. Therefore assuming that the empty wave progressively vanishes in its free way the transmission coefficient b becomes a function of the length ℓ

$$b = b_0 e^{-\mu \ell}, \tag{15}$$

where b_0 is the initial empty wave factor of transmission and μ is the attenuation coefficient, ℓ being the free traveled distance. Substituting (15) into (11) one gets

$$I = kN \mid \phi_0 \mid^2 (b_s^2 + b_0^2 e^{-2\mu \ell} + 2b_s b_0 e^{-\mu \ell} \cos \chi), \tag{16}$$

which gives a visibility parameter

$$V = \frac{I_M - I_m}{I_M + I_m} \tag{18'}$$

of

$$V = \frac{2b_s b_0 e^{-\mu \ell}}{b_s^2 + b_0^2 e^{-2\mu \ell}}. \tag{18}$$

If the exponential term is very small

$$e^{-2\mu \ell} \ll 1, \tag{19}$$

and for the case of transmission coefficients b_s and b_0 of the same order (18) becomes

$$V = 2 \frac{b_0}{b_s} \exp^{-\mu \ell}. \tag{20}$$

Thus, if the assumption of the progressive loss of properties of the empty wave, as it travels free, is valid interferometric experiments done with different free path lengths shall present an increasing loss of the fringe visibility given by formula (18), which is in clear contradiction with the usual interpretation of quantum mechanics. After all the experiment remains conceptually the same, only the free independent wave path is made longer, consequently, in this interpretation, as one must consider that the probability wave crosses one slit and the other, no alteration in the visibility fringe should be expected.

3 - Experiments based on the collapse of the wave function

The other possible alternative crucial experiment, considered here, is based on the famous collapse of the wave function. The advantage of this last method lies in the fact that it is not founded upon some hypothetical unproven different properties of the waves ϕ and θ, but only on sounded accepted facts.

The essential features of the method are:

a) - Production of empty waves, taking advantage of the reduction of the wave train[8]

b) - Mixing, in an interferometer, incoherently the θ waves with the usual waves coming from a different source.

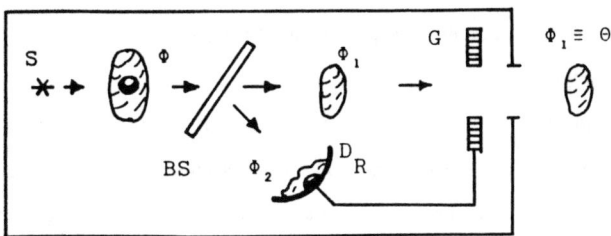

Fig.3 - Empty wave generator (EWG)

To produce an empty wave generator (EWG) it is necessary to dispose of a source of microparticles (photons, electrons, neutrons, etc.) able to emit one at a time. This quantum system a neutron, for instance, emited by the source S and represented by ψ strikes on a beamsplitttr aned is divided into two parts ψ_1 and ψ_2, each following well separated trajectories. If a detector D_R placed in the path of the wave train ψ_2 is triggered it means that the particle chose that way, therefore the probability of being at the other path turns instantly to zero, $\psi_1 = 0$. This phenomenon is known in the Copenhagen interpretation as the collapse or reduction, of the wave train. For the causal interpretation of de Broglie this same very phenomenon is seen in a completely different way. The initial real wave with a particle-singularity, upon arriving at the beamsplitter, is divided into two waves: one carrying the singularity ϕ and other without the the singularity θ. When the detector D_R placed at the path (2) is triggered it means that along the path traveled the wave ϕ carrying the singularity with sufficient energy to start the detector. Naturally at the other path (1) travels the empty wave, which is unaffected by the interaction of the particle-singularity with the detector D_R. This detector is electronically connected with a gate G which opens for a short time, just sufficient to allow the wave packet θ to pass. In the alternative cases when the particle singularity chooses the path (1) at the detector D_R arrive only the empty wave, which has no energy to trigger the detector D_R, so the gate G remains closed, see Fig.3, and no singularity leaves the apparatus.

This theoretical conclusion can perfectly be tested experimentally by placing in front of the (EWG) a detector, which if the source S is really able to deliver one particle at a time should count nothing besides the usual noise. In this circumstances according to the causal school of de Broglie the apparatus (EWG) emits empty waves θ which are not seen directly because they have not enough energy to activate a common detector. Only the singularity gets energy-momentum to do it. For the Copenhagen interpretation, the empty wave generator is no more than a fiction because empty waves do not exist.

In order to prove the existence of such waves it is necessary to build a special kind of detector for them. There are several possible ways in which these hypothetical empty waves can be detected[1,9]. Here it will be discussed only the incoherent wave mixing, method which has the great advantage of being a Yes-No type experiment valid, in principle, for any kind of microparticle.

In the following only the photon case shall be considered, nevertheless the whole process remains essentially the same for neutrons or electrons.

Consider Fig.4 where is schematically represented the case of the superposition of three waves. Two empty waves θ_1 and θ_2 coming from the empty wave generator plus a common wave ϕ.

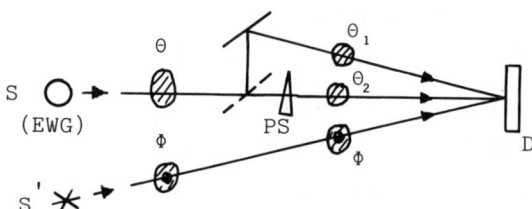

Fig.4 - Schematic representation of the superposition of three waves. The empty waves θ_1 and θ_2 are coherent between themselves but incoherent relative to the full wave ϕ

Accordingly at the detector D the three waves superimpose, and the total wave $\phi_T = (\theta_1 + \theta_2 + \phi)$ shall guide the photon-singularity to the points of higher wave intensity. In the interference region the photon shall be guided not by the wave that initially had carried it, but by the conjugated effect of the three present waves. Therefore the expected intensity predicted by the causal theory shall be

$$I_c \propto |\theta_1 + \theta_2 + \phi|^2 \tag{21}$$

this expression can be written

$$I_c \propto |\theta_1 + \theta_2|^2 + |\phi|^2, \tag{22}$$

because the θ waves are coherent between themselves, they came from the same source S, but incoherent relative to the wave ϕ coming from a different source S'. Considering for simplicity the unnecessary hypothesis of equal wave amplitude

$$|\theta_1|^2 = |\theta_2|^2 = |\phi|^2 \tag{23}$$

one gets, after some calculation

$$I_c \propto \left(1 + \frac{2}{3}\cos\delta\right), \tag{24}$$

where δ is the relative phase shift of the θ waves. In the Copenhagen interpretation the empty waves do not exist therefore formula (21) shall be written in a simple way

$$I_u \propto |\phi|^2. \tag{25}$$

The difference in the previsions of the two interpretations of quantum formalism is striking as shown in expressions (24) and (25). These values are plotted in Fig.5

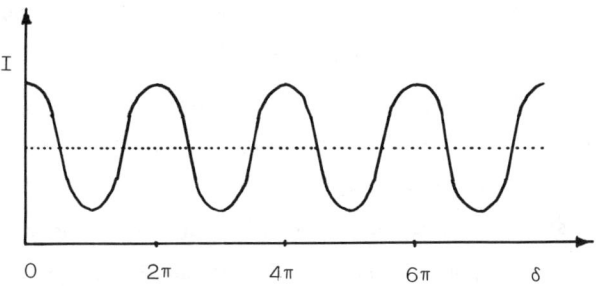

Fig.5 - Predicted results: Usual interpretation dotted line; Causal interpretation continuous line.

The Copenhagen interpretation predicts an intensity distribution constant, while the causal school of de Broglie predicts an interference pattern modulated by the difference of the phase δ between the waves θ_1 and θ_2 which is controlled by the phase shifting device Ps.

A practical way, slightly different from the one discussed in reference [11] is depicted in Fig.6

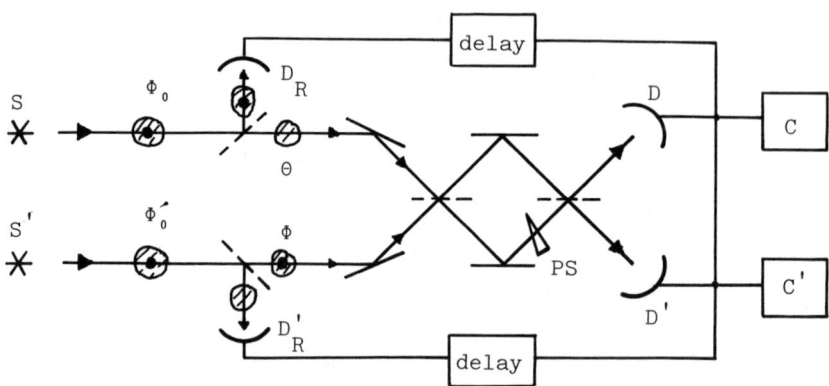

Fig.6 - Proposed experiment for detection of photonic empty waves. S and S' are photon sources emitting at the same time single photon quantum states and Ps is the phase shifting device. The events registered at C and C' are the ones in which there are coincidence counting between $[(D_R, C), (D_R, C')]$ and $[(D'_R, C), (D'_R, C')]$. Only the case $[(D_R, C), (D_R, C')]$ is represented.

The two identical photon sources S and S' are built in the same way (same intensity, frequency, etc.) and emit at the same time photons one by one[10]. Here, in this proposed experiment, the role of the empty wave generator is in a certain way fulfilled by the electronic setup. So, in the case represented in Fig.6, the empty wave entering the Mach-Zehnder interferometer comes from the source S. When the detector D_R reacts it sends a triggering signal which feeds a delay line, and a little later opens the detectors D and D' at the time of arrival of the empty wave packet θ. If the photomultipliers D or D' detect a photon-singularity it could be emitted only by the source S'. Therefore, in this case of detection at D_R, coincidences are made between detectors (D_R,D) and (D_R,D'). The symmetric case not represented in Fig.6 is the one in which D'_R is activated and consequently the empty wave comes from S' with coincidences made between (D'_R,D) and (D'_R,D'). In either case at D or D', there are present four waves, two from S and other two from S'. The superposition of these four waves guide the particle-singularity if present. To be experimentally secure that, when D_R detects the singularity, an empty wave from S is injected in the interferometer coincidences between (D_R,D) and (D_R,D') without the source S' must be made. Symmetrically coincidences with (D'_R,D) and (D'_R,D') with the source S canceled must be carried out. If S and S' are good sources fulfilling the requirement of emitting one photon at a time, no other counting besides the usual noise should be expected. In order to do the calculation one considers only the case depicted in Fig.6, bearing in mind that in the symmetric case when D'_R and not D_R is activated the roles of the θ and ϕ change but calculations are just the same. To help visualize the wave superimposition the Mach-Zehnder interferometer and the respective waves are represented in Fig.7

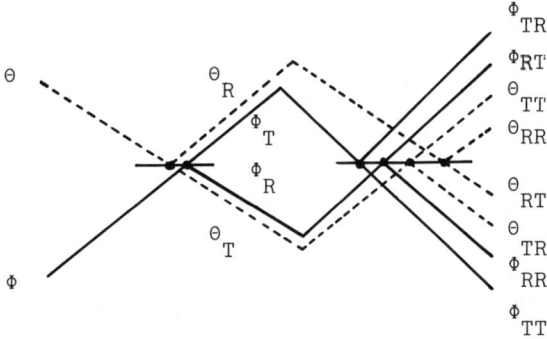

Fig.7 Schematic diagram of wave superimposition in the Mach-Zehnder interferometer.

The counting rate expected at D and D' shall be proportional to the total present waves

$$I_c \propto |\phi_{TR} + \phi_{RT} + \theta_{TT} + \theta_{RR}|^2 \qquad (25)$$

$$I'_c \propto |\phi_{TT} + \phi_{RR} + \theta_{TR} + \theta_{RT}|^2 . \qquad (25')$$

But S and S' are independent sources, therefore incoherent, so one could write

$$I_c \propto |\phi_{TR} + \phi_{RT}|^2 + |\theta_{TT} + \theta_{RR}|^2 \qquad (26)$$

$$I'_c \propto |\phi_{TT} + \phi_{RR}|^2 + |\theta_{TR} + \theta_{RT}|^2, \tag{26'}$$

and by developing the first relation (26) one gets

$$I_c \propto |\phi_{TR}|^2 + |\phi_{RT}|^2 + |\theta_{TT}|^2 + |\theta_{RR}|^2 + 2|\phi_{TR}||\phi_{RT}|\cos\delta_\phi - 2|\theta_{TT}||\theta_{RR}|\cos\delta_\theta, \tag{27}$$

where δ_ϕ and δ_θ are the relative phase shifts of the waves ϕ and θ. Considering the ideal case of equal splitting and no absorption it is reasonable to assume that

$$|\phi_{TR}|^2 = |\phi_{RT}|^2 = |\theta_{TT}|^2 = |\theta_{RR}|^2 = \frac{1}{2}|\phi_0|^2 \tag{28}$$

substituting these relations into (27) one has

$$I_c = \frac{1}{2}I_0\left(1 + \frac{1}{2}\cos\delta_\phi - \frac{1}{2}\cos\delta_\theta\right), \tag{29}$$

also the relation (26') gives

$$I'_c = \frac{1}{2}I_0\left(1 - \frac{1}{2}\cos\delta_\phi + \frac{1}{2}\cos\delta_\theta\right), \tag{29'}$$

because the relative phase shift of the two branches is $\delta_i \pm \pi$, the total intensity counted in C and C' is given by

$$I_0 = 4k|\phi_0|^2,$$

where k is a proportionality constant.

Looking at those two expressions (29) and (29') one sees two constant phase shifts which can be related by

$$\delta_\theta = \delta_\phi + \alpha, \qquad 0 \leq \alpha \leq \pi \tag{30}$$

where the value of the α parameter depends on the experimental conditions.

A different set of predictions is given by the Copenhagen interpretation. In this theory empty waves do not exist, therefore the expressions (25) and (25') must be written without the θ waves

ON THE COPENHAGEN AND DE BROGLIE INTERPRETATIONS

$$I_u \propto |\phi_{TR} + \phi_{RT}|^2, \tag{31}$$

$$I'_u \propto |\phi_{TT} + \phi_{RR}|^2, \tag{31'}$$

taking in account the anterior calculation one is allowed to write

$$I_u = \frac{1}{2}I_0(1 + \cos\delta_\phi), \tag{32}$$

$$I'_u = \frac{1}{2}I_0(1 - \cos\delta_\phi). \tag{32'}$$

The most striking difference between (29) and (32) occurs when the two phase shifts are equal $\delta_\theta = \delta_\phi$ ($\alpha = 0$). In this case (29) and (29') turn into

$$I_c = \frac{1}{2}I_0, \tag{33}$$

$$I'_0 = \frac{1}{2}I_0. \tag{33'}$$

That is the conjugated effect of the two empty waves completely swallows the interference pattern, as shown in Fig.8

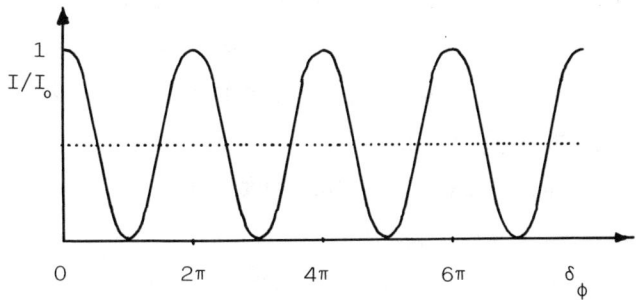

Fig.8 - Predictions for the case $\alpha = 0$, $\delta_\theta = \delta_\phi$: continuous line usual theory; dotted line causal theory.

If the experimental conditions are such that $\alpha = \pi$, or $\delta_\theta = \delta_\phi + \pi$, expressions (39) and (29') shall become

$$I_c = \frac{1}{2}I_0(1 + cos\delta_\phi),$$

$$I'_c = \frac{1}{2}I_0(1 - cos\delta_\phi),$$

and the predictions of the two theories shall be equal. All intermediate gradation of fringe blurring are obtained for other different values of the α parameter.

This work was supported by the "Instituto Nacional de Investigação Cientifica - Portugal"

The author also wants to thank the "Fundação Calouste Gulbenkian" for a grant allowing him to be at the Delphi Conference.

References

1 - F. Selleri,*Le Grand Débat de la Theorie Quantique*, (Flamarion, Paris, 1986)

2 - W. Heisenberg,*The Physical Priciples of Quantum Theory* (Dover, N.Y.,1930)

3 - L. de Broglie, *The Current Interpretation of the Wave Mechanics,*

A Critical Study, (Elsevier, Amsterdam, 1964)

4 - Y.Koh and T. Sasaki, in *Microphysical Reality and Quantum Formalism,*

eds. A. van der Merwe and al.(Reidel, Dordrecht, in print)

5 - D. Bohm and J.P. Vigier, Phys. Rev. 96(1954)208

6 - L. Janossy and Zs.Naray, Suppl. Nuovo Cim. 9(1958)588

7 - G. Möllenstedt and W. Bayh, Naturwiss, 48(1961)400

8 - J. R. Croca, Found. Phys. 17(1987)971

9 - F. Selleri, in *Wave-Particle Dualism*, eds. S.Diner et al. (Reidel, Dordrecht, 1984)

10 - M. Dagenais and L. Mandel, Phys. Rev.A, 18(1978)2217

- S. Friber, C. Hong, and L. Mandel, Phys.Rev. Lett.,54(1985)2011

- P. Grangier, G.Roger and A. Aspect, Europhys. Lett. 1(1986)173

11 - J. R. Croca, Phys. Lett. 124(1987)22

SEARCH FOR NEW TESTS OF THE EPR PARADOX FROM ELEMENTARY PARTICLE PHYSICS.

DIPANKAR HOME
Physics Department
Bose Institute
Calcutta 700009, India.

ABSTRACT. A significant recent trend has been the quest for examples of the Einstein-Podolsky-Rosen (EPR) paradox from hitherto unexplored domains in order to subject the incompatibility between quantum mechanics and local realism to a critical scrutiny. This motivates the present talk concerning the investigations related to an EPR-type example involving the decay of a $J^{PC} = 1^{--}$ vector meson into a pair of correlated neutral pseudo-scalar mesons; a typical example is the decay of the spin-1 Φ resonance, by strong interaction, into a pair of neutral kaons $K° - \bar{K}°$ (1-4). In this context, we discuss a new twist to the EPR paradox by taking into account CP noninvariance.

1. EPR-TYPE EXAMPLE FOR $\Phi \to K° - \bar{K}°$

Invoking charge conjugation invariance of the strong interaction, the wave function of the $K° - \bar{K}°$ pair at the time of production (t=0) from the decay of the $J^{PC}=1^{--}$ state is given by

$$|\Psi_o> = [|K°>_L|\bar{K}°>_R - |\bar{K}°>_L|K°>_R]/\sqrt{2} \qquad \ldots (1)$$

where L(R) refers to the left (right) hemisphere. The subsequent time development of the $K° - \bar{K}°$ pair is described in terms of the eigenstates of the effective Hamiltonian which includes weak interactions. In the situation under consideration, the weak interactions induce decays of both $K°, \bar{K}°$ and also give rise to $K° - \bar{K}°$ transitions. The effective Hamiltonian is written as $H = M - i\Gamma/2$ where M and Γ are the hermitian mass and decay matrices respectively. The eigenstates of this H are $|K_L>$ and $|K_S>$ with the eigenvalues $\lambda_L = m_L - i\gamma_L/2$ and $\lambda_S = m_S - i\gamma_S/2$ respectively, where m_L (m_S) and γ_L (γ_S) are, respectively, the mass and the decay width of $|K_L>$ ($|K_S>$). We assume CP invariance; the implications of CP violation in this type of example will be discussed later (Section 3). Note that

$$|K_L> = [|K°> + |\bar{K}°>]/\sqrt{2}, \text{ and } |K_S> = [|K°> - |\bar{K}°>]/\sqrt{2} \qquad \ldots (2)$$

They time evolve as $U(t,0)|K_L> = |K_L> \exp(-i\lambda_L t) + |\phi_L(t)>$, with a corresponding equation for $|K_S>$. Here $|\phi_L(t)>$ ($|\phi_S(t)>$) represents the decay products from $|K_L>$ ($|K_S>$); $|\phi_L>$ ($|\phi_S>$) is orthogonal to the state $|K_L>$ ($|K_S>$). CP invariance requires $<K_L|K_S> = 0$. In terms of the states $|K_L>$ and $|K_S>$, the wave function $|\Psi_o>$ given by Eq. (1) can be written as

$$|\Psi_o> = [|K_S>_L |K_L>_R - |K_L>_L |K_S>_R]/\sqrt{2} \quad \ldots \quad (3)$$

The time evolution of the non-separable form of the two-particle wave function $|\Psi_o>$ correlates the oscillations between the $|K°>$ and $|\bar{K}°>$ states such that it carries the essence of EPR-type non-local correlation. If the left (right) kaon is observed to be a K° (Strangeness S = + 1) at a particular instant then the right (left) kaon can be predicted with certainty to be observed as a $\bar{K}°$ (S = -1) at that same instant. Alternatively, if the left (right) kaon decays in the K_S mode (CP = + 1), then the right (left) kaon is bound to decay as a K_L (CP = -1) at some future instant. It is to be noted that there is a subtle distinction between the $K°-\bar{K}°$ and K_L-K_S correlations; while the former holds only for equal proper times, the latter is a time-independent consequence of the non-separable form of the wave function. This aspect was pointed out by Selleri (4).

Six (3) suggested an experimental test of this EPR-type situation through measurement of the joint probability $P_{--}(t_1,t_2)$ of a double $\bar{K}°$ observation (i.e., on two sides), at times t_1 and t_2 on the left and right respectively. The quantum mechanical prediction for $P_{--}(t_1,t_2)$ is given by

$$P_{--}(t_1,t_2) = |<\bar{K}°_L \bar{K}°_R|\Psi(t_1,t_2)>|^2$$

where $|\Psi(t_1,t_2)>$ is the state evolved from $|\Psi_o>$ at t = 0:

$$|\Psi(t_1,t_2)> = (1/\sqrt{2})\{|K_S>_L|K_L>_R \exp-i(\lambda_S t_1 + \lambda_L t_2)$$
$$- |K_L>_L|K_S>_R \exp-i(\lambda_L t_1 + \lambda_S t_2)\} \quad \ldots \quad (4)$$

whence one obtains

$$P_{--}(t_1,t_2) = (1/8) \{\exp-(\gamma_S t_1 + \gamma_L t_2) + \exp-(\gamma_L t_1 + \gamma_S t_2)$$
$$-2 \exp(-\gamma(t_1+t_2)) \cos \Delta m(t_1-t_2)\} \quad \ldots \quad (5)$$

where $\gamma = (\gamma_L + \gamma_S)/2$ and $\Delta m = m_L - m_S$.

Selleri derived an upper bound on $P_{--}(t_1,t_2)$ for the $K°-\bar{K}°$ system ($P_{--}^u (t_1,t_2)$) using a general argument based on the notion of local realism:

$$P_{--}^u(t_1,t_2) = (1/8) \{\exp-(\gamma_S t_1 + \gamma_L t_2) + \exp-(\gamma_L t_1 + \gamma_S t_2)\} \quad (6)$$

This local realistic upper bound differs from the quantum mechanical prediction (5) by the absence of the interference term. Quantum mechanics, therefore, leads to a prediction that violates Eq.(6) whenever the interference term is positive, that is, whenever $\cos \Delta m(t_1-t_2) < 0$.

It is important to note that the experimental study envisaged in the context of the Eqs.(5) and (6) has an intrinsic handicap: for meaningful results, t_1 and t_2 must be shorter than the life-times of K_L and K_S, i.e., one requires t_1, $t_2 \lesssim 10^{-10}$ s. The uncertainties involved in ensuring that the observations are at the specified instants t_1 and t_2 would be quite appreciable within such a small time interval. This difficulty may, however, be circumvented by considering the time-integrated joint probabilities. This aspect has been recently examined by Datta and Home (5) for the case of the $B^\circ - \bar{B}^\circ$ system. This system is almost identical to the $K^\circ - \bar{K}^\circ$ system, the only difference being that $\gamma_L = \gamma_S (= \gamma \sim 10^{12} \hbar s^{-1})$ for the eigenstates of the $B^\circ - \bar{B}^\circ$ system which are analogous to the $|K_L\rangle, |K_S\rangle$ states. They are denoted by $|B_H\rangle$ and $|B_L\rangle$ with masses m_H and m_L respectively ($m_H > m_L$).

2. EPR-TYPE TEST USING THE $B^\circ - \bar{B}^\circ$ SYSTEM

Current experiments on the decay of the spin - $1\Upsilon(4s)$ vector meson into a pair of neutral pseudoscalar mesons $B^\circ - \bar{B}^\circ$ have attracted considerable attention in view of the search for evidence of the $B^\circ - \bar{B}^\circ$ mixing. Datta and Home (5) have analysed the possibility of investigating experimentally the EPR-type quantum non-local correlations within the framework of these experiments. Here one considers the time-integrated joint probabilities, remembering that B° and $\bar{B}^\circ$ can be identified by their characteristic semi-leptonic mode of decays: $B^\circ \to l^+ \nu \chi$; $\bar{B}^\circ \to l^- \nu \chi$ where l and χ denote lepton and hadron respectively. From the decay kinematics of $\Upsilon(4s) \to B^\circ \bar{B}^\circ$ it can be shown that the spatial separation between $B^\circ - \bar{B}^\circ$ is of the order of 0.1 mm (>> the de Broglie wavelength of the particles involved) after a time-interval of the order of the life-time of the decaying particles.

The experimental arrangement currently in use to study $\Upsilon(4s) \to B^\circ \bar{B}^\circ$ is designed to measure the parameter R defined as follows:

$$R = N_{++} + N_{--}/N_{+-} + N_{-+} \quad \quad \quad (7)$$

where N_{++} = Total number of double $\bar{B}^\circ$ decays (corresponding to the observation of double l^+ decay products on both sides); N_{--} = Total number of double B° decays (corresponding to the observation of double l^- decay products on both sides); N_{+-} = Total number of $\bar{B}^\circ$ decays on the left associated with B° decays on the right (corresponding to the observation of l^+ decay products on the left associated with l^- decay products on the right); N_{-+} = Total number of B° decays on the left associated with $\bar{B}^\circ$ decays on the right (corresponding to the observation of l^- decay products on the left associated with l^+ decay products on the right). The quantum mechanical prediction for R is given by

$$R_{QM} = \chi^2/(2 + \chi^2) \quad \quad \quad (8)$$

where $X = \Delta m/\gamma$. The result given by Eq. (8) hinges on the quantum non-separability which is built into the wave function (1) and is assumed to be maintained even after the particles get separated in space. The experimental verification of Eq.(8) will, therefore, constitute an interesting test for quantum non-separability in this EPR-type situation.

It should be noted that R_{QM} is model dependent. There are two types of B° mesons: B°_d ($b\bar{d}$ quark-antiquark bound state) and B°_s ($b\bar{s}$ quark-antiquark bound state). $\Upsilon(4s)$ decays into the $B^\circ_d \bar{B}^\circ_d$ system only (The $B^\circ_s \bar{B}^\circ_s$ channel is forbidden by the kinematic considerations). In this case, the precise prediction of the standard model of electro-weak interactions as regards $\Delta m/\gamma$ has certain inherent uncertainties which are now being debated (see, for example, Datta et al.(6)). The most recent experimental study indicates strong evidence for substantial B°_d-$\bar{B}^\circ_d$ mixing and the value of R is claimed to be 0.21 ±0.08 (7). Datta and Home (5) have shown that in this case Furry's hypothesis leads to the prediction R = 1, which is evidently ruled out by the experimental result.

To suggest further work along these lines we may point out that, apart from calculating the parameter R using the various local realist models (analogous to the types invoked for analysing the EPR atomic-cascade experiments), it seems important to derive general bounds on R from local realism, independent of the details of any particular model. This would enable us to make decisive use of the current experimental studies on R in order to constitute a valuable complement to the other EPR experiments.

3. EPR PARADOX AND CP NONCONSERVATION

In this section we discuss the controversial work by Datta, Home, and Raychaudhuri (DHR) who have pointed out a new facet of the EPR paradox using CP noninvariance (8). The example (similar to the ones discussed earlier) involves a pair of correlated neutral pseudo-scalar mesons (M°-$\bar{M}^\circ$) originating from the decay of a $J^{PC} = 1^{--}$ vector meson. The exponentially decaying states, with the associated masses and lifetimes, are $|M_L\rangle$ and $|M_S\rangle$. In the presence of CP nonconservation, $|M_L\rangle$ and $|M_S\rangle$ are non-orthogonal. This non-orthogonality of the physically relevant states (unique characteristic of the quantum mechanical treatment of CP nonconservation) gives a new twist to the EPR paradox.

The two-particle wave function at the time of production (t = 0) of the M°-$\bar{M}^\circ$ pair is given by

$$|\Psi_0\rangle = [|M_S M_L\rangle - |M_L M_S\rangle]/N \quad \quad \ldots (9)$$

where N is a normalization factor and the first (second) member of each pair refers to the left (right) hemisphere.

The time evolved wave function can be written in the form

$$|\Psi(t)\rangle = c_1|M_L\phi_S\rangle + c_2|M_S\phi_L\rangle + c_3|\chi\rangle \qquad \ldots \qquad (10)$$

where c_1, c_2, c_3 are time-dependent constants, and $|\chi\rangle \sim |M_S M_L\rangle - |M_L M_S\rangle$ represents the undecayed piece with $\langle\chi|\chi\rangle = 1$. $|\phi_L\rangle (|\phi_S\rangle)$ corresponds to the decay products on the right from $|M_L\rangle (|M_S\rangle)$. The important difference between Eq.(10) and the EPR-type correlations in other standard examples lies in the fact that $|M_L\rangle$ and $|M_S\rangle$ are non-orthogonal eigenstates of the effective Hamiltonian $H = M - i\Gamma$ where M and Γ are non-commuting due to CP noninvariance. In writing Eq.(10), we have not considered those components of the wave function which contain decay products on the left, as they are not relevant for our subsequent discussion which is focussed on the flux of, say, undecayed $|\tilde{M}^°\rangle$ on the left.

It can be easily seen from Eq.(10) that the above flux on the left would involve a contribution due to the overlap between the decay product states $|\phi_L\rangle$ and $|\phi_S\rangle$ on the right, which is non-vanishing in the presence of CP noninvariance. Note that $\langle \phi_L(t)|\phi_S(t)\rangle$ is proportional to $\langle M_L|M_S\rangle$; contribution to this non-orthogonality comes essentially from the common decay products of M_L and M_S (in the presence of CP violation). That the statistical property of the particles on one side has some formal dependence pertaining to the overlap between the physical states of the particles on the other side is the key feature of the example suggested by DHR. Whether this overlap can be physically tampered (even in principle) by suitable 'measurements' is the intriguing issue raised by this example.

In the absence of CP violation, the mutually orthogonal $|\phi_L\rangle$ and $|\phi_S\rangle$ states can be unambiguously distinguished. However, in the presence of a small but non-vanishing CP violating interaction, if one can partially discriminate between $|\phi_L\rangle$ and $|\phi_S\rangle$ on one side by exploiting the differences in their physical attributes, there arises a possibility to affect the above overlap, thereby leading to a curious non-local effect concerning the flux of undecayed kaons on the other side. Such a scheme, of course, envisages measurements partially destroying the coherence of the original pure state and leading to statistical mixtures of non-orthogonal states. Concept of this type of measurement ('partial collapse'), though unconventional, is not prima-facie inadmissible and can be dealt with, in principle, by appropriate generalisation of the standard quantum theory of measurement, as has been shown by various authors (9). Parenthetically it may be noted that there are various examples of realistic measurements such as approximate measurements or non-ideal measurements which cannot be described by the standard scheme based on orthogonal projections only. Ivanovic (10) has recently analysed the viability of possible non-standard schemes to differentiate between non-orthogonal states. A simple example of a non-orthodox measurement in which the final states of the measuring apparatus are not orthogonal is a Stern-Gerlach experiment for spin-1/2 atoms where the magnetic field is very weak and the counters are placed so close together that each of the two separated beams has a finite probability of being registered in both the counters. M. Namiki has pointed out (private discussion) that the many Hilbert-spaces formulation of the quantum measurement theory (see, for example, Machida and Namiki (11)) appears to provide a suitable framework to deal with such non-ideal measurements.

In their original treatment, DHR had first assumed the collapse to a mixed state comprising of non-orthogonal components to be 'total' and then the error involved (due to overlap between the probability distributions of the invariant masses of the decay products corresponding to the non-orthogonal states) was estimated. It was argued that the error could be, in principle, made small compared to a suitably defined measure of the non-local effect. However, the scheme for estimating this error has certain ambiguity, depending upon the choice of the parameter used as a measure of the error. This 'ambiguity' can be avoided by directly incorporating the notion of 'partial collapse' (in terms of a specific ansatz) and by properly taking the probability conservation into account through a formal density matrix treatment, as shown by DHR in their subsequent paper (12). Of course, if one chooses to confine one's attention only to standard quantum measurements involving, for example, the invariant masses of the individual and mutually orthogonal decay product components of $|\phi_L\rangle$ and $|\phi_S\rangle$, then there will be no non-local effect at the statistical level (13-14). However, the essential new feature in this case is whether one can envisage non-orthodox or generalised measurements to tinker with the overlap between $|\phi_L\rangle$ and $|\phi_S\rangle$ which contributes to the flux of undecayed kaons on the other side. One such possibility, albeit at the gedanken level, is to exploit the difference in the life-times of the states $|M_L\rangle$ and $|M_S\rangle$ to select out partially the decay products corresponding to, say, the $|\phi_L\rangle$ state. This may be done, for example, by using an apparatus one side which registers only the decay products originating within a specified time interval (say, ΔT) around the time of the order of the life-time of $|M_L\rangle$. These decay products will predominantly come from $|M_L\rangle$. There will also be a small but non-vanishing probability (which can be made as small as we like at the level of a thought experiment by assuming the life-times of $|M_L\rangle$ and $|M_S\rangle$ to be widely different) of the decay products from $|M_S\rangle$ being registered during ΔT. But the relevant point is whether the overlap between the common decay products from $|M_S\rangle$ (which originated well before ΔT) and those from $|M_L\rangle$ during ΔT can be affected by this process which can be regarded as a non-orthodox measurement of the decay times within a specific time interval; a simple model of the measuring apparatus which can register the decay products along with their time of origin has been discussed by Sudbery (15).

Hall (16) and Ghirardi et al. (17) have argued on the basis of the operation-effect formalism (using the First Representation Theorem (18)) that the type of wave function collapse envisaged in the DHR example necessarily corresponds to a 'measurement' which simultaneously affects the two separated subsystems; this would of-course ensure that no action at a distance is involved here. However, generality of this type of abstract argument (particularly the assumptions underlying the First Representation Theorem) to cover all possible models of non-orthodox measurements needs to be carefully examined before drawing any firm inference.

In conclusion, we summarise the key questions raised by the DHR example :

(a) Is the peculiarity of the M°-$\bar{M}^\circ$ type example essentially due to the incompleteness of the conventional quantum mechanical formalism (with its inherent approximations) used to describe the decaying systems and CP nonconservation ?

(b) Since CP noninvariance implies time-reversal asymmetry, does this example suggest that time-irreversible interaction, in general, introduces a new element in the quantum mechanical treatment of the EPR-type examples ? Or, does the use of the notion of non-orthodox measurement, in itself, lead to a new testable feature of the EPR paradox ? This is presently under investigation.

ACKNOWLEDGEMENTS

I gratefully acknowledge warm hospitality and inspiring discussions at this historic conference commemorating the birth centenary of one of the creators of quantum mechanics. I wish to thank the Indian Council of Philosophical Research and the Department of Science and Technology (Govt. of India) for their generous financial support.

REFERENCES

1. T.D.Lee and C.N.Yang, unpublished; D.R.Inglis, Rev. Mod. Phys. **33**, 1 (1961).

2. B.d'Espagnat, Conceptual Foundations of Quantum Mechanics (W.A. Benjamin, London, 1976) pp. 85-86.

3. J.Six, Phy. Lett. B **114**, 200 (1982).

4. F.Selleri, Lett. Nuovo Cimento **36**, 521 (1983).

5. A Datta and D.Home, Phys. Lett. A **119**, 3 (1986).

6. A Datta, E.A.Paschos and U.Turke, Phy. Lett. B **196**, 376 (1987).

7. H.Albrecht et al., Phys. Lett. B **192**, 245 (1987).

8. A.Datta, D.Home and A.Raychaudhuri, Phys. Lett. A **123**, 4 (1987); D.Home, in: Quantum Mechanics versus Local Realism, ed. F. Selleri (Plenum, New York, 1988).

9. I. Bloch and D.A.Burba, Phys. Rev. D **10**, 3206 (1974); K. Kraus, in: Foundations of Quantum Mechanics and Ordered Linear Spaces, eds. A.Hartkamper and H.Neumann (Springer-Verlag, Berlin, 1974); E.B.Davies, Quantum Theory of Open Quantum Systems (Academic Press, 1976); H.P. Yuen, Phys. Lett. A **91**, 101 (1982).

10. I.D.Ivanovic, Phys. Lett. A **123**, 257 (1987).

11. S.Machida and M.Namiki, in: Proc. 1st. Int. Symp. Foundations of Quantum Mechanics, eds. S.Kamefuchi et al. (Phys. Soc. of Japan, Tokyo, 1984) pp. 127-135, 136-139.

12. A.Datta, D.Home and A.Raychaudhuri, to be published.

13. E.Squires and D.Siegwart, Phys. Lett. A **126**, 73 (1987).

14. J.Finkelstein and H.P.Stapp, Phys. Lett. A **126**, 159 (1987).

15. T.Sudbery, in: Quantum Concepts in Space and Time, eds. R.Penrose and C.J.Isham (Clarendon Press, Oxford, 1986) pp. 65-83.

16. M.J.W.Hall, Phys. Lett. A **125**, 89 (1987).

17. G.C.Ghirardi, R.Grassi, A.Rimini and T.Weber, Europhys. Lett. (in press).

18. K.Kraus, States, Effects and Operations (Springer-Verlag, Berlin, 1983).

ON THE ROLE OF CONSCIOUSNESS IN RANDOM PHYSICAL PROCESSES

Robert G. Jahn and Brenda J. Dunne
School of Engineering and Applied Science
Princeton University
Princeton, NJ 08544 U.S.A.

ABSTRACT. Extensive data from a variety of man/machine experiments indicate that human operators can influence random device output distributions in accordance with pre-stated intentions. Deviations of these output distribution mean values from theoretical expectations or calibration data accumulate in individually characteristic and replicable patterns of achievement, many of which achieve high statistical significance. Despite wide variations in individual performance, the composite data base for 33 operators also departs substantially from chance expectation. Histograms of the deviations achieved in 87 separate experimental series confirm the significant mean shifts in the intended directions, and in addition distribute with greater than chance variance. In contrast, the histogram of baseline deviations, obtained under null intentions of the operators, centers on the appropriate chance mean, but with a variance considerably smaller than chance. When all such baseline and directional mean deviations are combined in an appropriately balanced mixture, a normal chance histogram is reconstituted. Such behavior raises the possibility that the basic combinatorial processes undergirding classical or quantum statistical mechanics may reflect, at least to a marginal degree, some processes of human consciousness. This hypothesis, along with the empirical data on which it is based, is consistent with a quantum wave mechanical model of the interactions of consciousness with its physical environment that predicts experiential eigenfunctions indexed by appropriate quantum numbers of both the consciousness and the physical system, and influenced by the degree of resonance between them, in much the same fashion as the Heitler-London treatment of molecular bonds.

Over the past decade, an experimental program to assess the role of operator intention or volition on the performance of a variety of physical devices embodying some form of random or pseudorandom process has been in progress at Princeton University's School of Engineering and Applied Science. Experiments have been chosen for their immediate and longer term relevance to modern engineering practice, and for their amenability to controlled laboratory study. Physical and technical parameters of these man/machine interactions have been the primary concern; investigation of psychological or physiological correlates has been held secondary to the accumulation of very large data bases by a substantial number of human operators, all of whom have been anonymous volunteers, with no claims of extraordinary abilities.

One typical experiment belongs to a genre of random event generators (REGs) widely used in contemporary studies of this class of phenomena as well as in diverse engineering applications. The specific device employed is based on a commercial microelectronic noise source (Elgenco #3602A15124), whose output is transcribed by appropriate circuitry into a random train of positive and negative pulses, suitable for sampling and counting.$^{(1)}$ For all experiments described here, this device is set to generate "trials" of 200 pulses each at a rate of 1000 per second, and to count and display the number of those pulses that conform to the regular alternation: +, -, +, -, +, -,.... Various display and recording units show the operator the results of the counting and insert them on-line into a digital data base and computational system.

The operator, seated a few feet from the device, initiates each run by pressing a remote switch and then attempts to influence the process to produce a higher number of counts (PK^+) or a lower number of counts (PK^-), or to generate a baseline (BL), in accordance with pre-recorded intentions or instructions. Data are generated in "runs" of 50 trials and accumulated in "sessions" comprising a minimum of fifteen runs. While session lengths are left to the preference of the operator, a complete experimental "series" -- which is the major unit for all subsequent statistical analysis -- requires a full 7500 trials, or 50 runs in each of the three directions of intention.

Device qualification tests include built-in pulse sequences to confirm that sample generation and counting circuits are unbiased, oscilloscope test points for accessing design voltages and signal waveforms, and independent external counters that display the actual number of positive and negative pulses processed in sequences of 2×10^6 binaries. The output of the device, both in calibration and experimental operation, is regularly subjected to numerous analytical tests. Repeated 1000-trial (200,000 sample) calibrations entail t- and z-score calculations, using both empirical and theoretical variances, chi-square comparisons against the expected normal distributions, and comparisons with arcsine probability predictions, all of which invariably show only chance deviations from the appropriate theoretical expectations. Autocorrelation tests with lags up to 50 also display only the expected variations, and Fourier analyses on single and concatenated data sets indicate only white noise. Integrated periodgrams with Kolmogoroff-Smirnoff limits show no significant deviations in Fourier spectrum amplitudes. In sum, all ongoing calibration data are fully random and normally distributed, and are in no way significantly distorted by the finite sequence lengths employed in the experiments.

Beyond these analytical calibrations, the machine has built-in technical failsafes which shut it down in the event of any voltage variations or total count errors. The temperature of the diode is constantly monitored to maintain it well within specification. Contingency protocols specify actions to be taken in the event of any form of malfunction, and complete records of any such aberrations are maintained. Environmental conditions, such as temperature, humidity, and barometric pressure are routinely recorded and examined for correlations with the experimental data. None such have been found.

Ultimately, however, the tri-polar protocol -- wherein data are collected in interspersed sequences of PK^+, PK^-, and BL, with all other variables and parameters held constant save the operator's intention -- is the best protection against artifact, since any spurious influence would itself need to be strongly correlated with those intentions to influence the data in the observed form.

An example of the type of results obtained in this experiment is shown in Fig. 1(a) as a distribution of scores for some 5000 baseline trials (one million samples) generated by one operator, superimposed on the theoretical Gaussian approximation to the appropriate binomial combinatorial statistics. With reference to the same theoretical distribution, Fig. 1(b) displays the results when this same operator attempted to shift the distribution to higher (PK^+) or lower (PK^-)

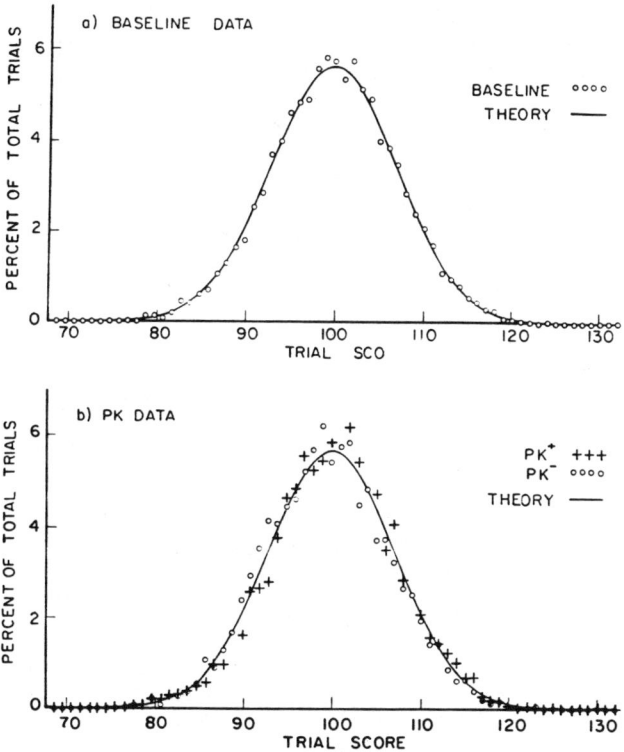

Figures 1a,b. REG Baseline and Experimental Distributions on Theory

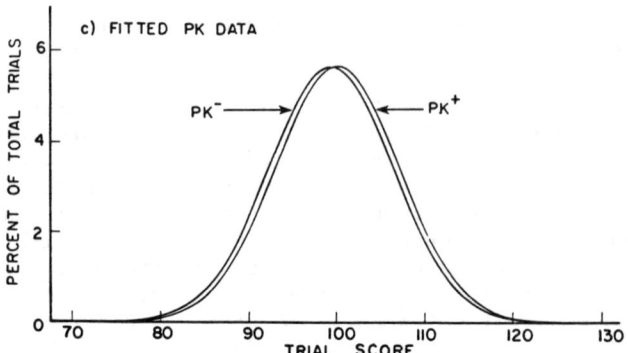

Figure 1c. REG Gaussian Fits to Experimental Data

counts over a comparable number of trials. Figure 1(c) plots the best Gaussian fits to these PK^+ and PK^- data, illustrating the scale and character of the effect that is typically observed in such experiments.

To track the regularity of these small shifts of the mean and to display their statistical significance as a function of data base size, the accumulated deviation of the trial counts from the theoretical chance mean for each of the three operator intentions are graphed in Fig. 2 as a function of the number of trials processed. While all three traces display the stochastic variations to be expected in a random process of this kind, the PK^+ and PK^- data also show systematic, almost linear, deviations from chance that compound to progressively larger values as the number of trials increases, while the baseline data remains close to the theoretical expectation. On this figure, the dashed parabolas are the loci of the .05 chance expecta-

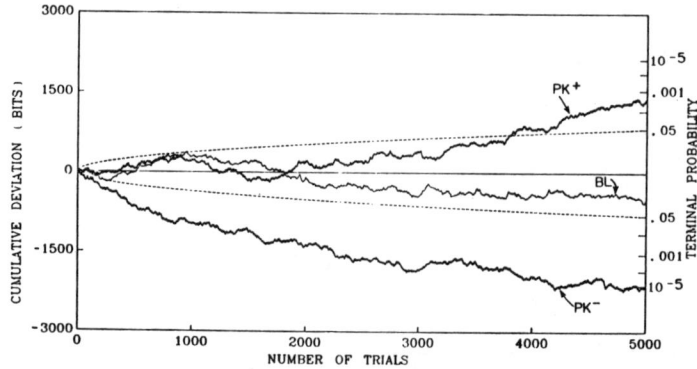

Figure 2. REG Cumulative Deviations from Theoretical Mean: Opr. 010

tion of reaching that accumulated deviation at that number of trials, and the scale on the right ordinate indicates the range of terminal chance probabilities. The terminal means of these PK+ and PK- data, 100.264 and 99.509 respectively, differ from chance expectation by several standard error units, with the composite achievement unlikely by chance to the order of 10^{-6}.

Such cumulative deviation graphs are found to be quite operator-specific, to the extent that we refer to them as "signatures." Some operators achieve in only one direction of effort, some in both, some in neither, and some consistently achieve effects opposite to their intentions. The PK^+ and PK^- traces are usually not symmetrical, and for many operators results are found to be dependent upon certain secondary conditions of the experiment, such as whether each of the 50 trials in the run is generated manually whenever the operator feels "ready," or automatically at some regular repetition rate, or whether the operator chooses or is randomly assigned the direction of intention for each run. One operator's sensitivity to this latter "volitional/instructed" parameter is illustrated in Figs. 3a and b.

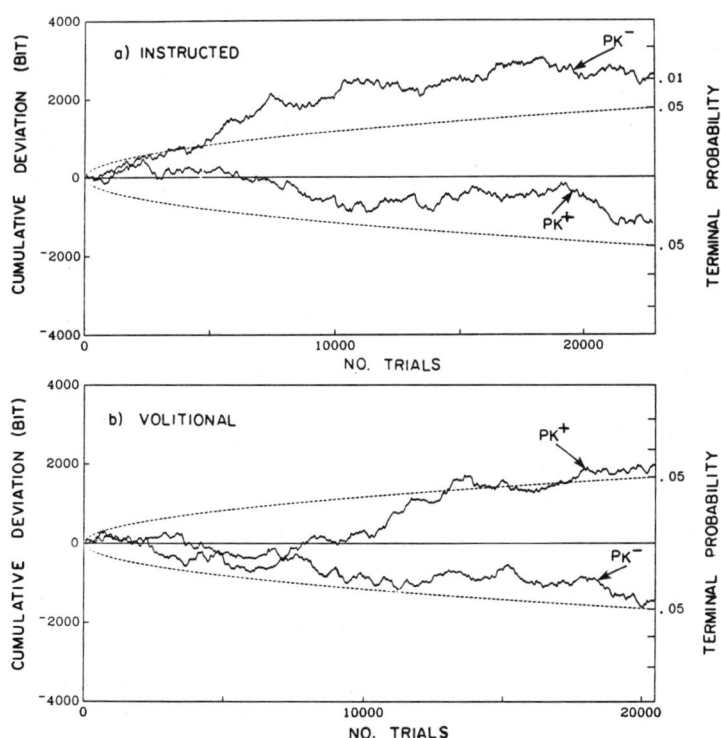

Figures 3. REG Cumulative Deviations from Theoretical Mean: Opr. 055
a) Instructed b) Volitional

The individual replicability of signature characteristics is illustrated in Fig. 4, as one operator's cumulative deviation graph concatenated over 15 full series that were performed over an eight-year period.

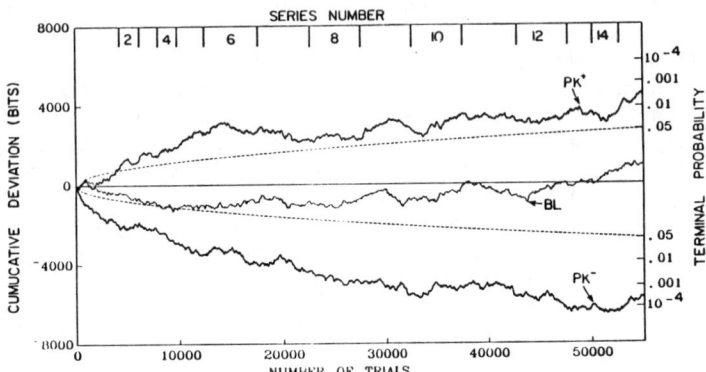

Figure 4. REG Cumulative Deviations: Opr. 010, All Data

The total data base accumulated with this REG experiment since its inception nine years ago consists of 87 full series or over 750,000 trials of 200 samples each, performed by 33 different operators.[2] If all of these data are combined, the composite results deviate from theoretical expectation with a probability against chance of .004 for the PK^+ data, .007 for the PK^- data, and 2×10^{-4} for the combined efforts. The composite baseline data remain well within chance expectation at a probability of .389 (Fig. 5a). Since the different operators represented in this concatenation have performed widely different numbers of complete series, it is also instructive to display a balanced composite graph including only the first series performed by each of the 33 operators [Fig. (5b)].

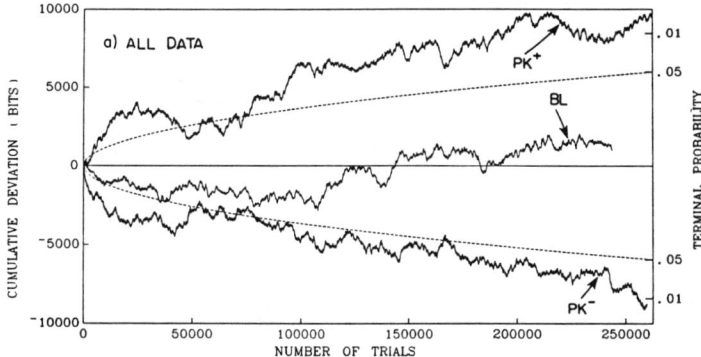

Figure 5a. REG Cumulative Deviations: All Data, 33 Operators, 87 Series

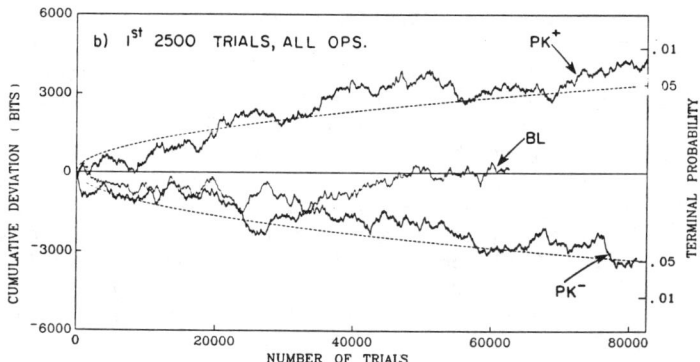

Figure 5b. REG Cumulative Deviations: All Operators, First Series Only

Given the robust and straightforward character of the effect, namely a small shift of the distribution mean with little distortion of the variance or higher moments, simple z-score assessments are legitimate. Nonetheless, the significance of the results has been confirmed by a number of more sophisticated statistical tests, including multi-factor analysis of variance and various graphical representations such as fits to the integrated arcsine probability curves. Again, the possible effects of finite sequence lengths have been assessed and dismissed, and even application of an ultra-conservative "log log N law" to the overall results confirms a statistically significant yield with a probability against chance of a few parts per thousand.

Obviously, effects such as these raise a hierarchy of questions regarding the basic nature of the phenomenon and its broader ramifications. For example, one may speculate whether the physical behavior of the noise source itself is affected during the experimental efforts. An obvious first strategy is to perform a Fourier analysis on the source output to determine whether it deviates from the random behavior of the calibration tests; no such deviation has been found. Another is to replace the source with other units and compare results. Several similar noise boards have been used, with no discernible differences in the overall pattern of the data. A more aggressive step is to substitute a deterministic, pseudorandom source for the thermal noise unit. Such a device, consisting of 31 microelectronic shift registers that produce a determinate repeating sequence of 2×10^9 bits at a set clock frequency has so been employed, in a mode wherein the pseudorandom sequence cycles continuously at a repetition period of about 60 hours (a full experimental series requires about 5 hours of machine operation). Thus, the only remaining non-deterministic aspect of the experiment is the time of incursion into the bit sequence initiated by the operator. Once again, the results of 29 experimental series generated by 10 operators are found to be quite comparable with the original REG data, and statistically significant with a probability against chance of .003. Most interestingly, many of the individual operator signatures on this pseudo REG bear strong similarities to those obtained with the physical random source.

To explore the device-specificity of the effect yet further, a totally different category of physical machine has been employed. This macroscopic apparatus, termed a random mechanical cascade (RMC), is some 6 ft. x 10 ft. in size and allows 9000 3/4 inch polystyrene spheres to trickle downward through a quincunx array of 330 3/4 inch diameter nylon pegs whereby they are scattered into 19 collecting bins across the bottom, filling them in close approximation to a Gaussian distribution.[2,3] The growing population of each bin is tracked photoelectrically and displayed via LED counters and is simultaneously recorded on-line in an appropriately coded computer file. The operator, seated on a sofa about eight feet from the machine, attempts to shift the distribution to the right (PK$^+$), to the left (PK$^-$), or to generate a baseline, following a tri-polar protocol similar to that employed in the REG and pseudo-REG experiments. Once again, operator-specific signatures of achievement are observed, in many cases bearing strong similarities to those developed on the microelectronic devices (Figs. 6a-c).

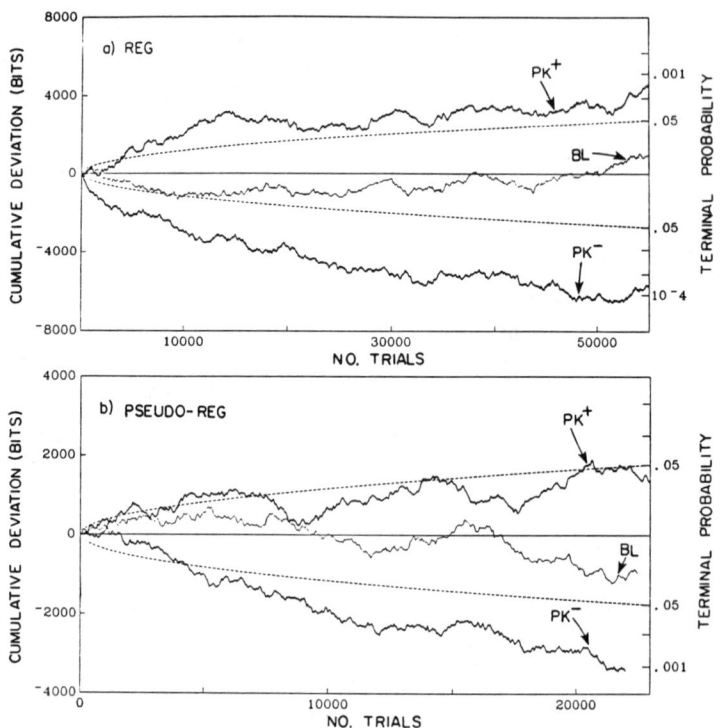

Figures 6a,b. Cumulative Deviations, Operator 010
                a) REG   b) Pseudo-REG

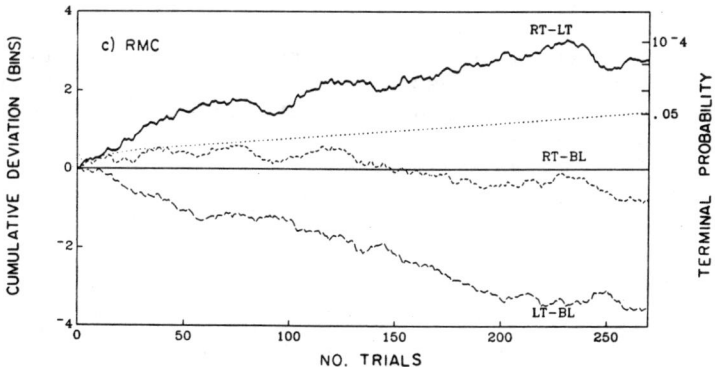

Figure 6c. Cumulative Deviations, Operator 010: c) RMC

The composite data base of 87 series, or 3393 runs by 25 operators who have completed at least one series of 10 or 20 runs per intention, again displays a statistically significant aberration from chance correlated with operation intention, at a probability level of the order of 3×10^{-4} (Fig. 7).

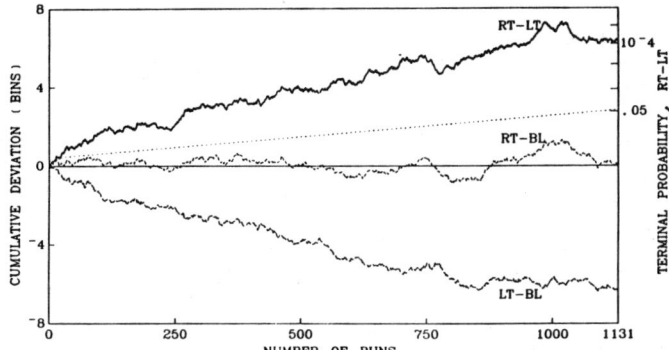

Figure 7. RMC Cumulative Deviations from Empirical Baseline: All Data, 25 Operators, 87 Series

One ancillary finding of all of these experiments that may bear on the ultimate interpretation of the data can be demonstrated by forming histograms of the frequency of terminal z-scores of complete series. Figures 8a-c present such displays for all of the 87 REG experimental series. While the mean values of the fitted Gaussians are consistent with the terminal values of the cumulative deviation traces of Fig. 7a, the PK^+ and PK^- series score distributions clearly exhibit significantly larger than expected variances. Conversely, the distribution of the baseline scores is significantly constricted around the theoretical mean and is totally devoid of any scores outside the one-tailed significance criterion, $z > \pm 1.645$. Since the baseline data are generated under conditions identical to the PK data, save for the absence of a stated directional intention on the part of the operator, one is led to hypothesize that the intention or desire to generate a "good" baseline may also bear on the output distributions. Another curious, and possibly important indication from this format is that recombination of all the data for equal numbers of baseline and intentional efforts, essentially reconstitutes the theoretical chance Gaussian (Figs. 8a-d).[4]

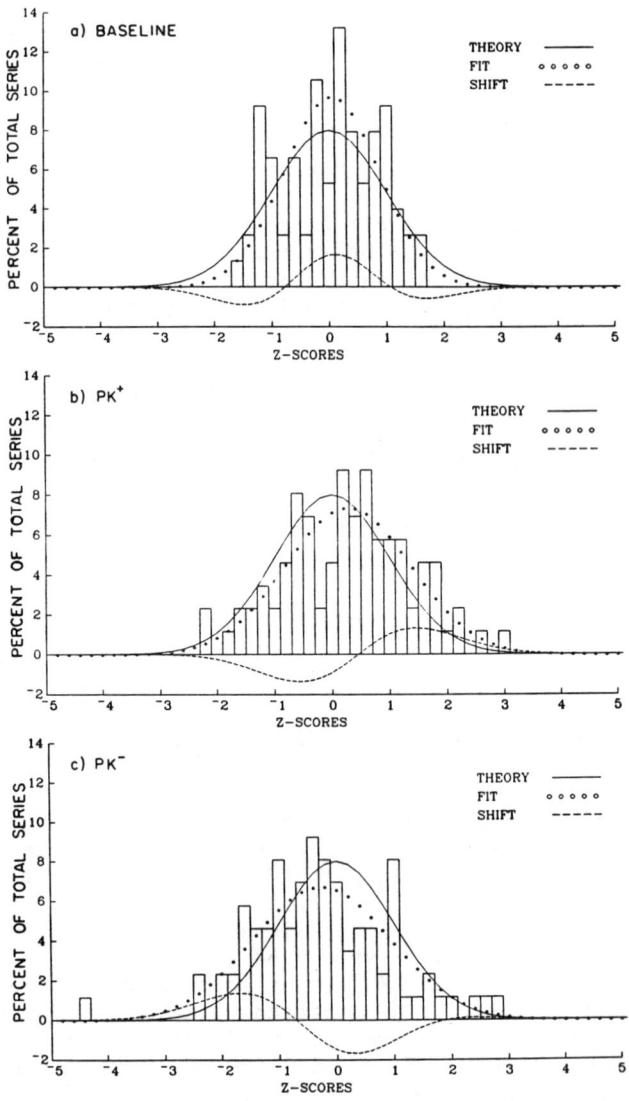

Figures 8a-c. REG Series z-Scores: a) BL b) PK$^+$ c) PK$^-$

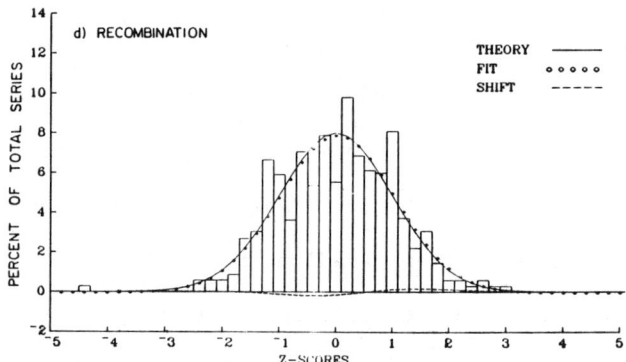

Figure 8d. REG Series z-Scores: Recombined BL, PK^+, and PK^- Data

From data such as those presented above and much more not here included, the following hierarchy of general results emerges:

1) Marginal shifts of the means of the output distributions of such devices show statistically significant correlations with the pre-stated intentions of the operators.

2) Individual operator "signatures," depicted as graphs of the cumulative deviations from the theoretical chance mean, are keenly operator-specific, yet individually quite replicable.

3) For some operators, performance is sharply affected by specific secondary parameters; for others, it is not.

4) For most operators, the signatures are not device-specific, that is, their cumulative deviation graphs for widely different devices appear quite similar.

5) Even ostensible baseline data are not immune to the effect; full series baseline scores are distributed with excessively small variance.

Such empirical evidence seems to argue against phenomenological interpretations involving direct physical intervention in the various random processes, in favor of models that address features generic to all of the systems, e.g. the information content of the output distributions. Space does not allow development here of a theoretical wave-mechanical approach presented elsewhere[5-6] that has proven useful in correlating such data, designing new experiments, and offering some basic insight into these phenomena. Very briefly, the model invokes Schrödinger formalism in a quantum mechanical metaphor that represents consciousness via wave functions, the environment of the consciousness by appropriate potential profiles, and experience, or reality, as the eigenfunctions established by the interaction of the two. Interpretation of these eigenfunctions requires definition of a consciousness metric, a consciousness coordinate system, and consciousness quantum numbers, but with these in place, a number of wave mechanical and quantum mechanical effects avail to represent the experimental anomalies. Perhaps most powerful of these is the concept of a resonant bond between the operator and the device, represented in much the same format as the Heitler-London treatment of molecular structure. Also emerging from the model, and consistent with the empirical variance data, is the intriguing possibility that one function of consciousness is to select from the grand chance distribution of possible outcomes of any given random process, subset distributions that support its teleological purpose, e.g. to obtain high, low, or baseline values, without violating the grand distribution, per se.

While these concepts are obviously speculative and somewhat radical in the prevailing scientific context, they are not totally inconsistent with the Copenhagen paradigm, nor with certain metaphysical ideas advanced by that patriarch of quantum probability whose centennial this conference has celebrated, Erwin Schrödinger. As he put it,

"...Mind has erected the objective outside world of the natural philosopher out of its own stuff. ...It is the same elements that go to compose my mind and the world. This situation is the same for every mind and its world, in spite of the unfathomable abundance of 'cross-references' between them. The world is given to me only once, not one existing and one perceived. Subject and object are only one. The barrier between them cannot be said to have broken down as a result of recent experience in the physical sciences, for this barrier does not exist."

"The only possible inference ...is, I think, that I--I in the widest meaning of the word, that is to say, every conscious mind that has ever said or felt 'I'--am the person, if any, who controls the 'motion of the atoms' according to the Laws of Nature."[7]

References

1. R. D. Nelson, B. J. Dunne, R. G. Jahn, 'An REG Experiment with Large Data-Base Capability, III: Operator Related Anomalies,' Technical Note PEAR 84003, September 1984 (159 pages).
2. R. G. Jahn, B. J. Dunne, R. D. Nelson, 'Engineering anomalies research,' Journal of Scientific Exploration, 1, pp. 21-50, October 1987.
3. R. D. Nelson, B. J. Dunne, R. G. Jahn, 'Operator Related Anomalies in a Random Mechanical Cascade Experiment,' Technical Note PEAR 88001, (∿ 175 pages), in preparation.
4. R. G. Jahn, R. D. Nelson, B. J. Dunne, 'Variance Effects in REG Series Score Distributions,' Technical Note PEAR 85001, July 1985 (14 pages).
5. R. G. Jahn and B. J. Dunne, 'On the quantum mechanics of consciousness, with application to anomalous phenomena,' Foundations of Physics, 16, pp. 721-772, August 1986.
6. R. G. Jahn and B. J. Dunne, Margins of Reality: The Role of Consciousness in the Physical World. San Diego: Harcourt Brace Jovanovich, 1987.
7. E. Schrödinger, What Is Life? and Mind and Matter. Cambridge: University Press, 1967, pp. 131, 137, 93.

Recent Experiments in the Foundations of Quantum Mechanics

Arthur G. Zajonc
Physics Department
Amherst College
Amherst, Massachusetts 01002
USA

ABSTRACT. Modern experimental methods have brought thought-experiments from the mind of the theorist into the laboratory. Three specific experiments are discussed: 1) macroscopic quantum phenomena in SQUIDS, 2) single-atom detection of quantum jumps, and 3) delayed-choice experiments in quantum optics. Each experiment intensifies the puzzling character of quantum physics.

1. INTRODUCTION

Over the last several years experiments have been performed that bear rather closely on the foundations of quantum physics. Three areas in particular have produced results of remarkable sensitivity that demonstrate (sometimes rather dramatically) the puzzling nature of quantum phenomena.

Those three areas are:

1. Macroscopic quantum phenomena,
2. Neutron interferometry,
3. Quantum optics.

In order to provide something of an overview, I will speak about a set of experiments complementary to those reported by Aspect and Sumhammer elsewhere in this volume. In particular I would like to illustrate the progress toward laboratory realization of isolated, macroscopic quantum systems by reference to the work of Washburn, Webb and Tesche of IBM, Yorktown. I will then follow this by a review of several so-called single-atom, "quantum-jump" experiments. To close I will briefly treat the two delayed-choice experiments completed at the Max Planck

Institute for Quantum Optics in Garching.

2. MACROSCOPIC QUANTUM EFFECTS

As this year is the centenary of Schroedinger's birth I would like to begin by reminding you of one of Schroedinger's best known quantum paradox's: Schroedinger's cat.[1]
Very briefly, a microscopic quantum superposition state is amplified to create an apparently impossible situation. In Schroedinger's example the end result is that an unobserved cat exists in the superposition state of "cat alive" plus "cat dead." The cat does not actually die in fact (according to Schroedinger's presentation of the quantum formalism) until the wavefunction "collapses" to either "cat alive" or "cat dead."

There are various strategies for avoiding the consistent and apparently compelling argument put forward by Schroedinger and often repeated by Wigner and others, but one of the favorites is to point to the impossibility of isolating any macroscopic object (much less a cat) as is required by the argument. If one includes coupling even through thermal radiation with the surroundings (that are treated classically) one finds that the cat's wavefunction decays to alive or dead essentially instantaneously. Recently rather detailed calculations by Joos and Zeh[1] for example show how sensitive the wavefunction is to coupling to the environment.

Yet while this is certainly true, Leggett[3] has suggested that isolated macroscopic quantum systems are in fact now realizable using modern cryogenic techniques and SQUIDs (Superconducting Quantum Interference Device). He and others point out that in such devices one has a true superposition of $\sim 10^{23}$ particles rather than essentially macroscopic two particle states as in simple superconductivity. The introduction here of the Josephson effect is crucial.

The typical SQUID consists of a superconducting ring broken by a Josephson junction of normal (i.e. non-superconducting) material. The complete dynamics of the system can be given in terms of the collective coordinates $\Delta\theta$ (the phase difference across the junction) and ϕ (the magnetic flux through the ring). The potential function for the system has a characteristic "washboard" form. Two important instances are shown in Fig. 1.

Two macroscopic quantum effects are of special interest. The potential function of Fig. 1a is suited to Macroscopic Quantum Tunneling (MQT), while the potential in Fig. 1b is expected to display coherent macroscopic quantum oscillations.

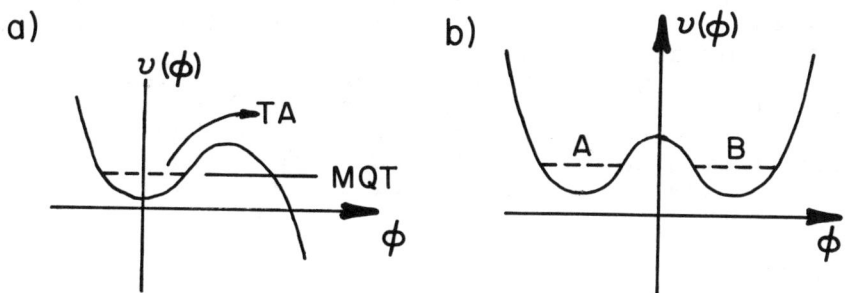

Figure 1. SQUID potentials.

For a potential of the form found in Fig. 1a a metastable state at $\phi = 0$ exists. Two modes exist for the decauy of the metastable state: thermal activation and tunneling. The rates for the two processes are:

$$\Gamma_T = A_T \, e^{-(U_0/kT)}$$

$$\Gamma_Q = A_Q \, e^{-B_Q}$$

where B_Q reflects the barrier characteristics. At sufficiently low temperatures one would expect the rate for thermal activation to give over its normally dominant role to quantum tunneling. Indeed this is what experiments in fact show.

By a careful study of the I-V characteristics of Josephson junctions over a range of low temperatures the IBM group of Washburn, Webb and co-workers[4], for example, have succeeded in clearly demonstrating the existence of macroscopic quantum tunneling.

With the potential of Fig. 1b one moves quite a bit closer to realizing "Schroedinger's cat" at very low temperatures. In this instance two macroscopic quantum states exist, A and B, that correspond to clockwise and counterclockwise current through the ring. There is predicted to be an oscillation from one state to the other at a rate proportional to the tunneling frequency. While not the coherent superposition state originally envisioned by Schroedinger the quantunm tunnel from A to B and back ("alive" to "dead" and back again) would be dramatic additional evidence for the validity of quantum mechanics for macroscopic, many particle systems. The experimental difficulties are formidable and are discussed in detail by Claudia Tesche[5] also of IBM.

In sum, although Schroedinger's room-temperature cat remains well described by classical probabilities, cryogenic devices have now demonstrated that the transition from quantum to classical effects is not simply a function of the number of particles involved.

2. QUANTUM JUMPS

In 1952 two articles appeared in the British Journal for the Philosophy of Science authored by Erwin Schroedinger[6] entitled, "Are There Quantum Jumps?" In Bohr's early treatment of quantum phenomena, quantum jumps between stationary states were invoked to account for such empirical facts as the spectral lines of the elements. Schroedinger points out in these articles that nowhere in the formalism of modern quantum mechanics do quantum jumps appear. In fact "the principle of superposition .... completely does away with the prerogative of the stationary state [Schroedinger's emphasis]. The epithet stationary has become obsolete."[7]

As an example of the logical impossibility of quantum jumps Schroedinger points to experiments using interferometers whose arms are of different lengths. Consider Fig. 2.

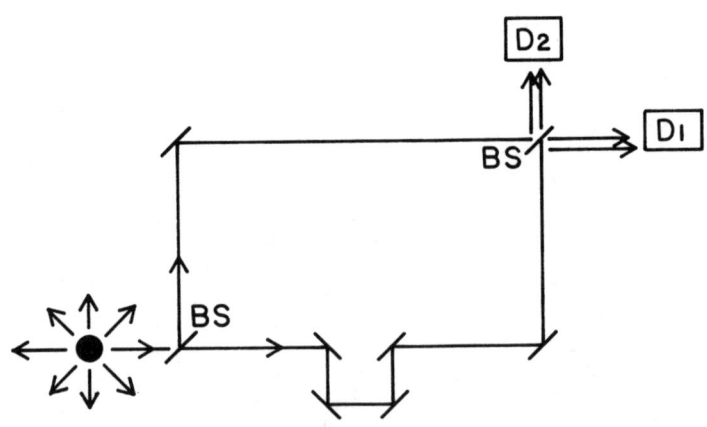

Figure 2. Schroedinger's interferometer.

An atom undergoes a quantum jump giving off a quantum of light at a time t_o. After passing through the two arms of the interferometer one of the detectors fires at a time t_1: $t_o = t_1 - \ell/c$. But in general there may be different path lengths for the two arms. As long as the "coherence length" is greater than the path difference $\Delta\ell$, interference will be seen. Therefore, when did the quantum jump take place, at a time gotten by tracing back along path 1 or path 2? How could it be instantaneous and produce the coherent wavetrain required for the interferometer?

From this argument and from theoretical considerations Schroedinger maintained quantum jumps as merely a heuristic fiction of the same status as Ptolemy's epicycles. With this as background I would like to discuss recent so-called "shelved-atom" experiments.

A typical title indicates the dilemma already: "Observation of Quantum Jumps in a Single Atom."[8] In its essentials the experiments consist of a single ion held in an ion trap and exposed to two laser beams. The ion acts as a single three-level system as shown in Fig. 3a. One transition (the S to P) is a strong transition with ~10^8 photons per second scattered and the second is a much weaker transition to a metastable state with only a few transitions per second. As originally pointed out by Demehlt[9] in 1975 such an arrangement could be used to monitor the presence or absence of the atom in its ground state. If the atom is in its ground S-state, a strong fluorescent signal will be detected. If, however, it makes a transition to the metastable state, the strong fluorescence will switch off. Thus by monitoring the strong fluorescent signal one might hope to see a single quantum jump to the D-state as Bohr's early quantum theory predicts. Fig. 3b shows the results of the experiment of Bergquist et al. One sees a clear quantum jump signature.

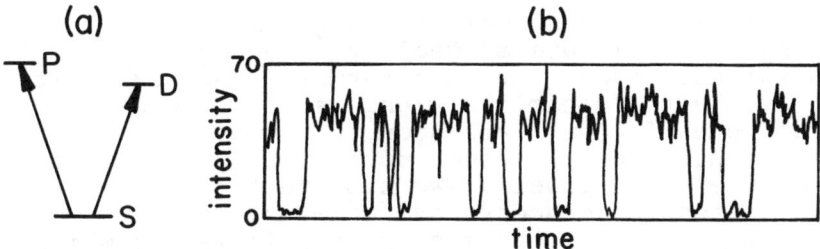

Figure 3. Typical quantum-jump data.

Was Schroedinger mistaken? Has Bohr's heuristic picture been proven correct? It seems that the situation is more subtle than that. Schroedinger was certainly correct in that the equations for the time evolution of probability amplitudes do not show discontinuous behavior except at the instant of detection. Theoretical treatment of these recent experiments[10] never predict quantum jumps as such, only correlation functions and the like. But they do point to sometimes subtle aspects of the measurement process as critical for an understanding of these experiments. Putterman and Porrati[10] in particular point not only to the strong fluorescence signal as a measurement, but also the dark period. Both are important for an understanding of the single-atom, quantum-jump experiments.

3. DELAYED-CHOICE EXPERIMENTS

The struggle between wave and particle theories of light goes back to the struggle between Huygens (and later Young) with the giant Newton. In his Opticks Newton framed as a question what soon became dogma, "Are not the Rays of Light very small bodies emitted from shining substances?"[11] The argument took a new twist at the turn of this century when the wave-particle nature of light and matter became evident. The situation with light or "the photon" became even more confusing when it was realized that one cannot define a position coordinate for the photon in the usual quantum mechanical way.[12]

These and related considerations lead one into the enormous difficulties of comprehending nature in any tradition sense via quantum mechanics. For Erwin Schroedinger this was profoundly disquieting. He possessed a deep faith in the comprehensibility of nature, and was hopeful that the present apparent incomprehensibility of quantum phenomena was a transient aberration. In "Our Conception of Matter" he wrote:

> A widely accepted school of thought maintains that an objective picture of reality - in any traditional meaning of that term - cannot exist at all. Only the optimists among us (and I consider myself one of them) look upon this view as a philosophical extravagance born of despair in the face of a grave crisis. We hope that the fluctuations of concepts and opinions only indicate a violent process of transformation which in the end will lead to something better than the mess of formulas that today surrounds our subject.[13]

Have we come any closer to comprehending nature since Schroedinger wrote these lines? Let us ask this question as we approach the specific example of the delayed-choice experiments[14] performed at The Max Planck Institute for Quantum Optics. Are delayed-choice experiments comprehensible in any traditional meaning of the term?

The history of delayed-choice experiments can be traced to conversations between Bohr and Einstein.[15] In 1931, C.F. von Weizsacker[16] focused the question further, but it had to wait for J.A. Wheeler's[17] independent analysis for the issue to reappear in contemporary discussions of the foundations of quantum mechanics.

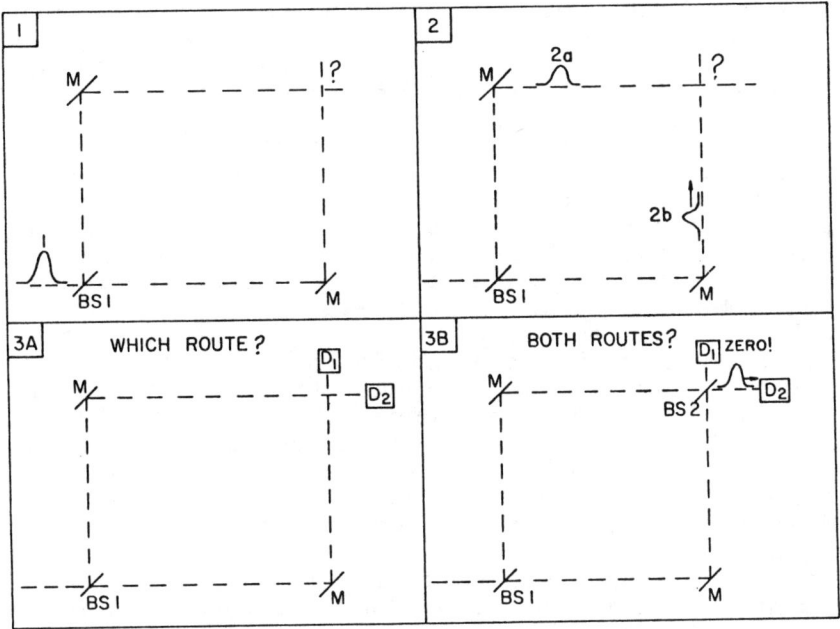

Figure 4. Delayed-choice interferometer.

Fig. 4 shows the basic conceptual arrangement. A single-photon light pulse arrives at beamsplitter one (BS1). Upon passage through the beamsplitter the wavefunction becomes a superposition of relfected and transmitted modes. We can ask two questions which are complementary in Bohr's sense of the term. Formally we may ask for the intensity in each detector D_1 and D_2 by the configuration shown in 3A of Fig. 4, or we may give up this information and focus on the relative phases of the wave amplitudes by inserting a second beamsplitter. Suitable Hermetian operators can be formed for these measurements (see Ref. 14).

Quite simply these correspond to asking which route the "photon" has traveled or, giving up such knowledge, establishing a phase measurement through interference of light that has traveled both ways. The feature of delayed-choice is added by waiting until the quantum of light has already past the first beamsplitter before completing the experimental setup. In Fig. 4 this corresponds to the choice between experiment 3A or 3B. In 3A one determines which path the photon has traveled; in 3B one gives up this information by inserting the second beamsplitter (only after the photon has past BS1) and sees interference which implies the photon has traveled both paths. As Wheeler points out, it therefore appears that after the measurement we have a right to say something about the already past history of the photon.

This experiment has been realized in two forms at The Max Planck Institute for Quantum Optics. The first parallels the above discussion quite closely by using a Mach-Zehnder interferometer with a fiber-optic delay line. The second experiment is a time-analog of The Mach-Zehnder experiment. That is, the interference now appears in the form of single-atom, quantum beats. I refer the reader to reference 14 for details of each experiment.

In both instances delaying the choice of wave vs particle experiment makes no difference to the experimental results. In one sense this is not at all surprising. Yet in another sense it intensifies the discomfort that Schroedinger, and those contemporaries who share his opinions, must feel in the face of quantum phenomena. In these and related instances nature does indeed seem to defy comprehension in the traditional meaning of the term.

Several responses can be formed in the face of such experiments. One is to follow Bohr who takes all such paradoxes as evidence that we must give up our attempts to understand quantum phenomena in terms of classical concepts. Others maintain that to acquiesce in this manner is to give up the ultimate goal of physics, namely the understanding of nature. They propose alternate realistic theories of the way nature is ordered. The protagonists in this debate are well represented by contributions elsewhere in the volume. Each, of course, must account for all the phenomena as determined by the methods of modern experimental physics. If they can do so in all situations and to arbitrary accuracy, how shall we choose between them?

Whatever the outcome I think we may agree with Einstein when in 1951 he remarked:

> All the fifty years of conscious brooding have brought me no closer to the answer to the

question: what are light quanta? Of course today every rascal thinks he knows the answer, but he is deluding himself.

REFERENCES

1. E. Schroedinger, Naturwissenschaften 23, 807, 823, 844 (1935).

2. E. Joos and H.D. Zeh, Z.Phys. B-Condensed Matter 59, 223 (1985).

3. A.J,. Leggett, Contemp. Phys. 25, 583 (1984).

4. S. Washburn, R.A. Webb, R.F. Voss and S.M. Faris, Phys. Rev. Lett. 54, 2712 (1985, and S. Washburn and R.A. Webb, New Techniques and Ideas in Quantum Measurement Theory, Edited by Daniel Greenberger, (Ann. N.Y. Acad. Sci., New York, 1986), Vol. 480, pp. 66-77.

5. C. Tesche, New Techniques and Ideas in Quantum Measurement Theory, pp 36-50.

6. E. Schroedinger, Brit. J. Phil. Sci. 3, 109, 233 (1952), also in Collected Papers, Vol. 4, pp. 478-502.

7. Schroedinger, p. 114.

8. J.C. Bergquist, R.G. Hulet, W.M. Itano and D.J. Wineland, Phys. Rev. Lett. 57, 1699 (1986); Th. Sauter, W. Neuhauser, R. Blatt, P.E. Toschek, Phys. Rev. Lett. 57, 1696 (1986); W. Nagourney, J. Sandberg and H. Dehmelt, Phys. Rev. Lett. 56, 2797 (1986); R.G. Hulet, D.J. Wineland, J.C. Berquist, and W. M. Itano, Phys. Rev A 37, 4544 (1988).

9. H.G. Dehmelt, Bull. Am. Phys. Soc. 20, 60 (1975).

10. D.T. Pegg, R. Loudon, and P.L. Knight, Phys. Rev. A, 33, 4085 (1986); M. Porrati and S. Putterman, Phys. Rev. A, 36, 929 (1987); D.T. Pegg and P.L. Knight, Phys. Rev A, 37, 4303 (1988).

11. I. Newton, Opticks (1704), Query #29 (Dover, New York, 1952).

12. E. Wigner and T.D. Newton, Rev. Mod. Phys. 21, 400 (1949).

13. E. Schroedinger, *What is Life? and Other Scientific Essays,* (Doubleday-Anchor Books, Garden City, NY, 1956) pp 161-62.

14. T. Hellmuth, H. Walther, A. Zajonc, and W. Schleich, *Phys. Rev. A*, 35, 2532 (1987).

15. N. Bohr, in *Quantum Theory and Measurement* edited by J.A. Wheeler and W.H. Zurek, *Princeton Series in Physics* (Princeton University Press, Princeton, NJ, 1983).

16. C.F. von Weizsacker, *Z. Phys.* 70, 114 (1931); 118, 489 (1941).

17. J.A. Wheeler, in *Mathematical Foundations of Quantum Theory,* edited by A.,R. Marlow (Academic Press, New York, 1978), p.9.

NUMERICAL SIMULATIONS OF REDUCTION OF WAVEPACKETS

Y. Murayama
Advanced Research Laboratory
Hitachi, Ltd.
Kokubunji
Tokyo 185
Japan

and

M. Namiki
Department of Physics
Waseda University
Okubo, Shinjuku
Tokyo 160
Japan

ABSTRACT. According to the Machida-Namiki theory a physical information can be acquired from quantum measurement processes by averaging quantum events over many-Hilbert spaces. The averaging process causes an irreversibility. In this paper several numerical simulations were performed to testify how quantum coherence is lost during measuring processes and how the reduction of wavepacket is eventually caused.

1. INTRODUCTION

It is really a long-lasting puzzle that quantum coherence of a single particle wave is never destroyed when the interaction between measuring apparatus and particle wave is specified by a potential analogous to the scattering problem, since it only causes a phase shift[1]. This kind of measurement is describable in terms of a unitary transformation and, hence, the interaction does not cause the reduction of wavepacket, so long as the event is limited within a single Hilbert space.

On the other hand, it is a reasonable notion that an ensemble average is vitally important to deduce a classical deterministic physical quantity on measurement. Quantum mechanics is essentially of a probabilistic character; it is completely meaningless to talk about a single event from the point of view how a single particle wave makes an acausal jump into a localized, specific state. In order to get a macroscopic information at least a macroscopic number of events are necessary. Following this idea the Machida-Namiki theory[2] took into account that, since each event belongs to its own Hilbert space, a macroscopic quantity must be an outcome averaged over many-Hilbert spaces. According to their original papers a critical assumption was that a detector seems to be an open system, where the number of the constituents of detector interacting with the measured object is actually indefinite.

In the many-Hilbert spaces theory the quantum mechanics of each scattering event is ascribed to its own Hilbert space and, since Hilbert spaces are independent of one another, all events can be taken independent. As a matter of course they are exclusive. The quantum mechanics of the whole system is constructed on discrete and continual super-selection rule spaces[3] and, accordingly, any measured quantity is presented by an average over all Hilbert spaces of concern. This concept corresponds to ordinary experiments in particle physics; a measurement is done based on counting rates. Recently Tonomura showed[4] in a 16m/m movie that an interference pattern of electron beams in the Aharonov-Bohm type geometry experiment is nothing other than a pile-up of single events, i.e., spots on a fluorescent screen. Thus an interference pattern is also the ensemble average of numerous events.

Take an Aharonov-Bohm type interferometer or a neutron interferometer using a single crystal Si. A particle wave is decomposed into a couple of paths; if there is no disturbance imposed on the beam paths, then an ideal interference will be seen after superposition. Otherwise, measuring processes would cause a phase shift on the particle wave. If the phase shift on each event varies from one to another, averaging will bring a smeared out interference pattern. This is a model of detection in the Machida-Namiki theory.

This paper deals with numerical simulations of this model detection process. That is, a detection is modeled to be a series of scattering processes. Each scattering causes a phase shift on each particle wave. If all particles experience different environments, then any physical quantities derived from the whole events will become a classical one in the meaning that there is no longer an interference between the two decomposed paths. For example, consider how we can determine the path of particle waves. The paths, say a and b, are exclusive. After the path is determined definitely to be the a channel, then the interference between the a and b channels are lost. When the variation in the environment of each event is modeled appropriately, it is possible to calculate each scattering event by solving the time-dependent Schrödinger equation and to average the probability density of superposed waves over a number of events. This is a job suited for computer simulation.

2. MODEL

In this investigation a model for measurement, or detection, is as follows. The simplest case of measurement is to know whether a particle wave passes channel a or b in the neutron interferometry type experiment. A silicon interferometer is ordinarily used to decompose waves. As is well known, without measurement two states through the two paths make an interference after being superposed and escaping the whole system. There a counter records the intensity of neutrons. Figure 1 shows conceptually the experimental system. We start from a wavepacket from a neutron source. It is hard to deal with plane waves; since they spread over the whole space, initial states themselves must be specified by the scattering potential.

Due to an appropriate measurement the path of particle wave is determined. This means that on measurement the superposed state shows a pattern compatible with classical measurement without interference. ψ_b is a freely flying wavepacket, which may be given a phase shift χ through the phase shifter and becomes ψ'_b. The other state of the wave is ψ_a, which is scattered by the potential in the detector and becomes ψ'_a. Superposed wave is specified by the probability $|\psi'_a+\psi'_b|^2$. The present calculation is limited within one dimension.

As for scattering model, we take a series of Gaussian potentials. As was discussed by Machida-Namiki[2], a detector is modeled to be an open system. The number of scatterers, N, may fluctuate around a mean value. A Gaussian distribution is assumed to describe the fluctuation. This surely causes a reduction of wavepacket, or in other words, the loss of coherence in the limit of an infinite number of scatterers, as was shown in Monte Carlo simulation by Nakazato and others[5].

In the scattering problem of this kind it is known that the phase shift caused by the potential is given in terms of $\bar{k}L$, where $\bar{k}$ is an effective k-vector of the wavepacket. L is the size of the region within which the wavepacket is scattererd. The fluctuation in the phase term is the sum of $\bar{k}\delta L$ and $L\delta\bar{k}$. If this is large enough, it may be clear that averaging over an ensemble must cause an

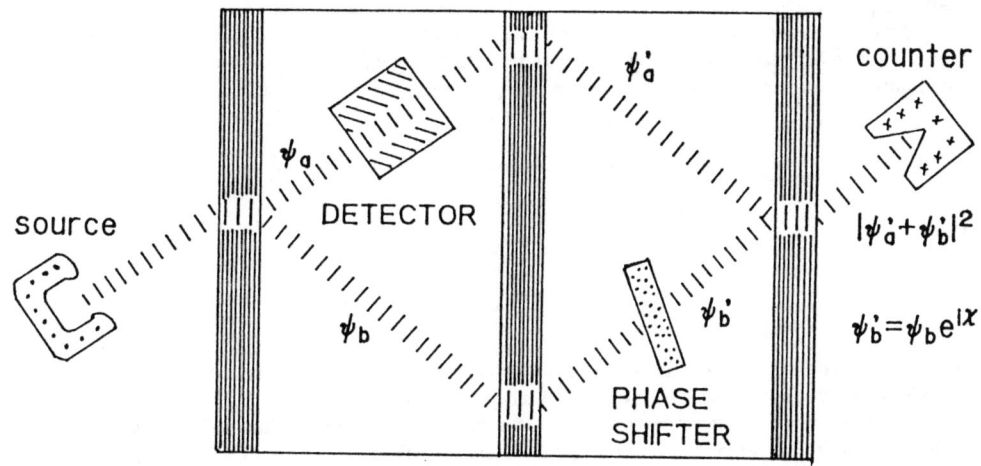

Figure 1. Schematic illustration of neutron interferometer with measurement to determine which path a neutron passes.

interference-free pattern similar to that in classical physics.

The occurrence of fluctuation, δL, is rather easy to understand, so long as the detector can be modeled to be an open system and the positions of scatterers vary from one to another according to thermal vibration in solid detectors or thermal motion in gas chambers. Meanwhile, the fluctuation in $\bar{k}$ seems to occur through, for example, random variation in the potential strength. The fact that its randomness can really cause a reduction of wavepacket was first demonstrated in the film[5]. This is partly because in realistic three dimensions the location of scatterers may deviate from the path of the wave. Their potentials must have effectively different magnitudes, when projected on to the path. It is thus assumed that the potential strength may also fluctuate randomly.

For simulation we consider the following three possibilities. The position of the ith scatterer is denoted by a_i and the average spacings, $\Sigma_i \, (a_{i+1}-a_i)/(N-1)$ by $<a>$, and its fluctuation by δa.

(1) Displacement of scatterers: $<a>=3$ A with
    $\delta a=0$ A (REGULAR lattice),
    $\delta a=0.3$ A (GAUSSIAN distribution),
    or 5 A $\geq |a_i-a_{i+1}| \geq 1$ A (uniformly RANDOM distribution).

(2) Scattering potential: $\eta_i V$ where
    $\eta_i=\{0 \text{ to } 1\}$ (uniformly RANDOM),
    or $\eta_i=1$ (CONSTANT).

[V is assumed to be 1.5% of the incident energy, 2.5 eV in this study.]

(3) Number of scatterers: $<N>=50$ with
    $\delta N=0$ (constant DEFINITE number)
    or $\delta N=3$ (INDEFINITE number specified by a Gaussian distribution).

Scattering potential function is also taken to be Gaussian-shaped with $\delta x=1$ A.

Scattering problem is solved using the time-dependent Schrödinger equation. The counter records the number of superposed wavepackets by integrating calculated probability density over x. Here Fourier transform is utilized to obtain scattered wavepackets, as follows:

$$\psi_0^p(x,t) = \Sigma A_k^{0p} u_k e^{-itE_k^0/\hbar}, \quad u_k = e^{ikx}/\sqrt{S}$$

and

$$\psi^p(x,t) = \Sigma u_i X_{ij} e^{-itE_j/\hbar} (X^{-1})_{jl} A_l^{0p},$$

where a diagonalizing matrix X is defined by $X^{-1}HX=EI$. H is the total Hamiltonian. S is the linear size within which the system is quantized and taken to be 10^5 A here. The Gaussian envelop function, A_k^{0p}, has the central value specified by p[6]. In the succeeding calculation the width of wavepacket in k-space, Δp, is 10^{-3} times p, and accordingly the span in real space, ΔX, is 14.4 A. Throughout this study waves no less than 461 are taken within a wavepacket. ψ_0^p is the well-known solution of a freely flying wavepacket. In this methodology solutions are limited only to the region after scattering, since E is the eingenvalue with full scattering potential included.

3. RESULTS OF SIMULATION

Figure 2 refers to the experimented counting rates in neutron interferometer with a few phase shifters inserted within a path[7]. We are interested in how much the amplitude is as a function of the magnitude of phase shift χ. As is easy to imagine, no shifter is ideal; it causes a slight measurement on neutrons. The degree of measurement was already discussed[8] in the case of neutron interferometry. When two Al shifters and one Ge are inserted (the rightmost data in Fig.2), amplitude is somewhat depressed. In this respect Summhammer noted that, when using more pure, monochromatic neutron flux, the depression in amplitude is recovered[9]. This point is quite interesting if we consider that the degree of measurement is given in terms of $\bar{k}\delta L<$ or >1, where δL is $(<N>-1)\delta a$ and/or $\delta N<a>$, depending whether the fluctuation comes from that in a and/or in N[2]. The case with more monochromatic neutron flux will be discussed in another paper[10].

The first result shown in Fig.3 is for the case with scatterers distributed following the GAUSSIAN law and their number DEFINITE, $N=50$. Potential is assumed CONSTANT. In this figure each small segment denotes a neutron and an ensemble of 50 neutrons is to average. The mean values are denoted by thick, long segments. Ideal variation is a smooth curve: $1+\cos\chi$. In this case only a GAUSSIAN fluctuation with $\delta a=0.3$ A is considered for the displacemet of scatterers, which is insufficient to induce a complete determination of path. That is, $\bar{k}\delta L \ll 1$. Such a case is called "imperfect measurement".

Thus the degree of the perfectness of measurement is given by the amplitude of sinusoidal variation in ensemble averages of a number of neutrons which reach the counter. In the case of Fig.3, the measurement is the slightest to disturb coherence.

As for the phase shift seen in Fig.3 it should be noted that the phase difference between ψ'_a and ψ'_b is not vanishing, even when $\chi=0$. So far we discussed the fluctuation in the phase of $\bar{k}L$. The constant phase term is $<\bar{k}L>$. Fluctuations around the constant value are given by $L\delta\bar{k}$ and/or $\bar{k}\delta L$. For GAUSSIAN or uniformly RANDOM fluctuation, criterion is given in terms of the latter: $L\delta\bar{k} <$ or > 1, whereas for fluctuation in potential strength it should be replaced by $\bar{k}\delta L <$ or > 1. The INDEFINITE-ness in the number of scatterers seems to give more fluctuation in $L=(N-1)a$ than in $\bar{k}$, as will be seen later.

NUMERICAL SIMULATIONS OF REDUCTION OF WAVEPACKETS 193

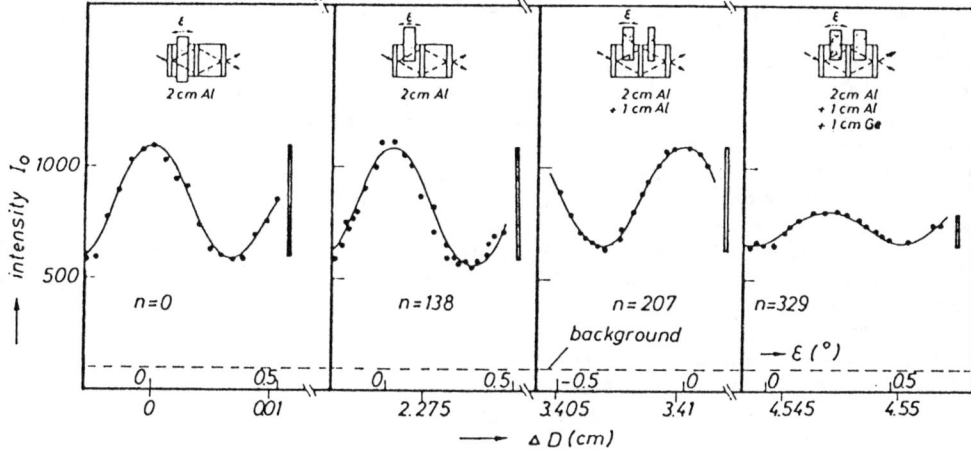

Figure 2. Experimental data of neutron counting after passing a Si interferometer with a few phase shifters inserted. After Rauch[7].

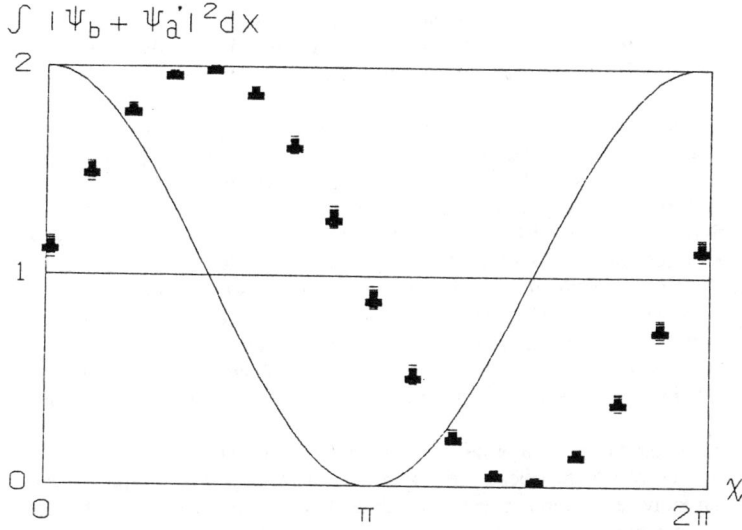

Figure 3. A result of simulation of quantum measurement processes. Displacement of scatterers follows a GAUSSIAN distribution, while their number is DEFINITE and potential is CONSTANT.

First of all the value of $\bar{k}$ should be noted. When we assume a REGULAR lattice with 50 scatterers and vary the spacing between them uniformly around 3 A, it is possible to estimate $\bar{k}$. In this case calculation gives $\bar{k}=1.16\times 10^7$/cm. If $\delta L=(50-1)\delta a$, then $\bar{k}\delta L$ becomes about 1.7.

Figure 3 is contrasted with the case of an INDEFINITE number of scatterers, Fig.4, where according to the Gaussian law the number varies from 41 to 60 ($\delta N=3$). The displacement of scatterers is also assumed to be GAUSSIAN. According to INDEFINITEness in their number the amplitude of variation is considerably depressed.

The next result of simulation is shown in Fig. 5. As was already mentioned before, we take a uniformly RANDOM displacement to be within ± 1 A around 3 A. This kind of randomness is greater than the previously assumed Gaussian distribution with $\delta a=0.3$ A. For this RANDOM case δa should be some value more than 1 A, so that $\bar{k}\delta L$ product becomes more than 5.7. This large value causes much more degree of measurement, or in other words, the reduction of wavepacket.

Figure 6 depicts the case with uniformly RANDOMly distributed potential strengths. Each potential V has a factor η_i specified by a RANDOM number between 0 and 1. Thus fluctuation in phase term is sufficiently large, as is seen in the figure. Displacement is taken to be GAUSSIAN similar to Fig. 3. So the comparison between Figs. 3 and 6 shows exclusively the effect of randomness in the potential strength.

Finally the most interesting case is shown in Fig. 7. This includes every possibility causing the reduction of wavepacket: uniformly RANDOM displacement, INDEFINITE number of scatterers, as well as uniformly RANDOM potential strengths. Here the degree of perfectness of measurement seems to be full, coming from fluctuations in both $\bar{k}$ and L. The total of origins of fluctuation, i.e., $L\delta\bar{k}$ and $\bar{k}\delta L$, must exceed unity largely.

4. DISCUSSION AND CONCLUSION

So far we showed by simulation how the reduction of wavepacket really occurs on quantum measurement processes. The counting rates obtained with a neutron counter give an information about how many neutrons reach it. When superposed, neutron wave gives a full magnitude, disappears, or takes a value in-between, the probability of which is given by $\int |\psi'_a+\psi'_b|^2 dx$. This calculation is limited to one dimension; hence, a disappearing probability means that the neutron wave is repeled on the occasion of superposition. In actual experimental geometry shown in Fig. 1, superposed neutrons come into the directions of a and b[7]. The sum of the intensities in both directions give a full constant magnitude, independent of phase shift.

Through these simulations it is shown that the criterion that $\delta(\bar{k}L) <$ or > 1 seems plausible. From a REGULAR lattice configuration of scatterers the $\bar{k}$ is estimated. In the most *im*perfect measurement we dealt here with (for example, GAUSSIAN displacement, Fig.3), the product $\bar{k}\delta L$ was something like 1.7; whereas for uniformly RANDOM displacement (Fig.5) it becomes at least 3 times as large. With respect to the INDEFINITEness of the number of scatterers (Fig.4), δN is 3 and $\delta N<a>=9$ A, whereas in the less perfect measurement, GAUSSIAN case (Fig.3), $(<N>-1)\delta a=15$ A. This relationship seems to be reversed. By simulation the former gives more effectiveness of measurement process. However, this is not the case, because $\bar{k}$ may also vary depending on N. The dependence is very hard to estimate, since the phase variation as a function of N can not exactly determine the $\bar{k}$ value because of discreteness of N. Thus we rely only on the result of simulation and conclude that the INDEFINITEness in the number of scatterers, i.e., the fact that detector is an open system, really causes the reduction of wavepacket.

In actual experimental condition the span of neutron wavepacket is never extremely small in comparison with Si interferometer, of which typical size is ≈ 15 cm. This large span in space means a small extension in k-space, or equivalently more monochromatic feature of wavepacket. Such a large wavepacket as covers a lot of scatterers within it at a time seems to give a smaller $\bar{k}$ value than that with a spatial span comparable to a few spacings in scatterer configuration. This small $\bar{k}$ value will bring the system closer to *im*perfect detection. This point will be affirmed by simulation and reported

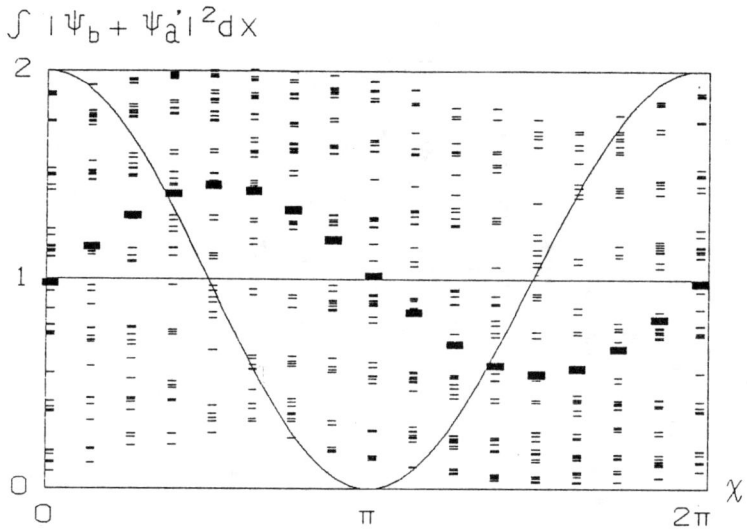

Figure 4. The same as Fig.3. Assumed number of scatterers is INDEFINITE.

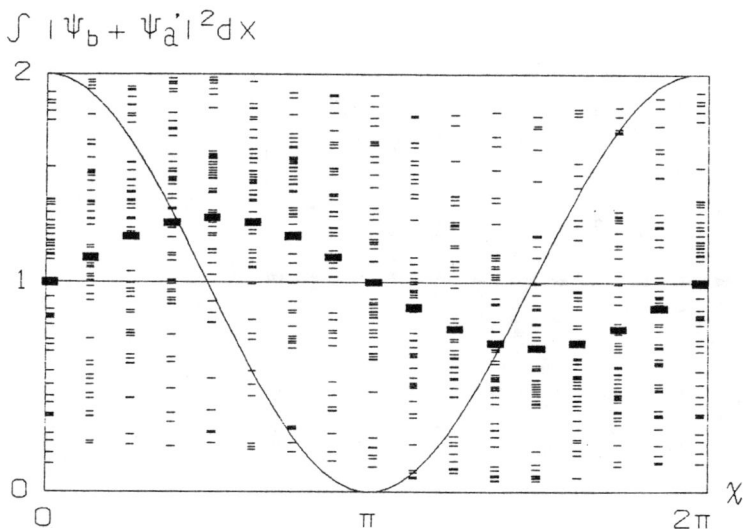

Figure 5. The same as Fig.3. Assumed displacement is uniformly RANDOM instead of GAUSSIAN. Number of scatterers is DEFINITE.

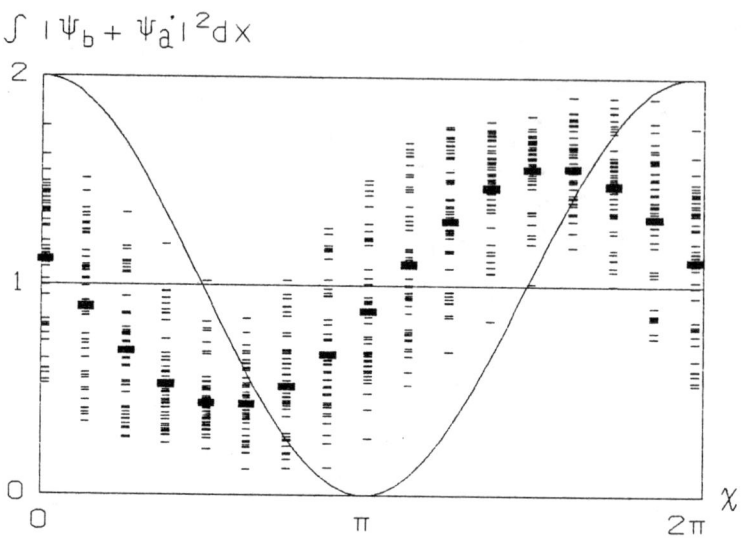

Figure 6. The sme as Fig. 3. Displacement is GAUSSIAN, the number of scatterers is DEFINITE, but potential varies uniormly RANDOMly between 0 and V.

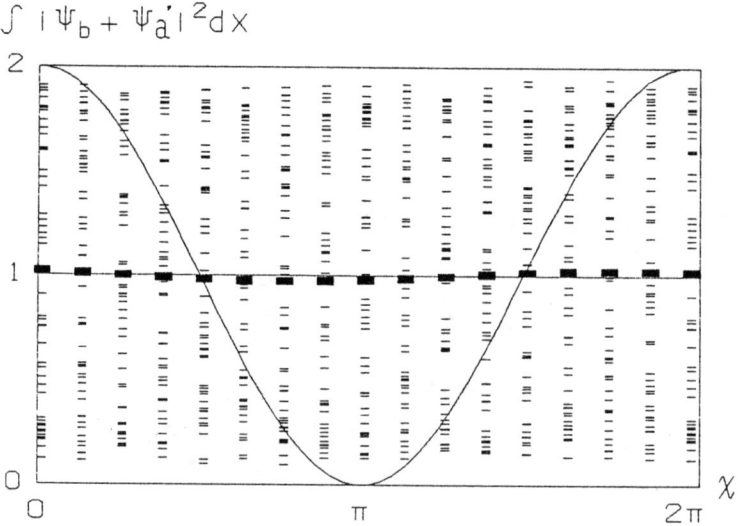

Figure 7. Full origins causing the reduction of wavepackets are included: uniformly RANDOM displacement, INDEFINITE number of scatterers and RANDOM potential strengths.

elsewhere[10].

At last the following point should be stressed. Zurek and Wootters[11] developed a theory of measurement based on an unknown freedom of environment of the whole system. The system is assumed to consist of a measured object, a measuring apparatus as well as an environment. In their theory it is never specified what the environment really is. The Machida-Namiki theory takes into account a concrete, well-specified 'environment' and each particle wave is assumed to experience individual environment. This is an advantageous point of the latter theory.

In order to introduce the Machida-Namiki theory from a pedagogical point of view we prepared a 16 m/m movie[5], composed of two parts. The first part consists of Nakazato's Monte Carlo simulation in the Young interferometry type experimental set-up, giving each particle wave a random phase. The ensemble of superposed waves shows an interference pattern on a screen, which varies depending on the scheme by which those random phases are generated. Just in parallel to the present investigation random numbers are given by the Gaussian distribution in terms of parameter δa. It is shown that the N-dependence in the interference pattern really testifies that the more scatterers exist, the more reduction of wacvepacket is caused.

The latter half of the film is dedicated to the same calculations as in this paper, though rather preliminary. Therein only the dispersion in the interference term is plotted in the case of vanishing phase shift, i.e., $\chi=0$.

In conclusion we showed by simulation that the ensemble averages upto 50 neutrons give a plausible variation of neutron counting rates as a function of given phase shift, where each neutron feels individually a realistic environment. All calculations are based on the time-dependent Schrödinger equation and for each neutron wave environment is taken to be different, i.e., displacement of scatterers to be GAUSSIAN or uniformly RANDOM, the strength of scattering potential to be CONSTANT or RANDOM, and their number to be DEFINITE or INDEFINITE according as the detector is an open system. It was shown by numerical simulation that the criterion given by Machida-Namiki, i.e., $L\overline{\delta k}$ > or < 1 (or more generally, $\overline{\delta(kL)}$ > or < 1) can give a sound basis whether the detecting action is sufficiently comprehensive or not.

ACKNOWLEGMENTS

The authors are very grateful to Professors S. Machida, I. Ohba, H. Nakazato for their discussions and collaborations. Their gratitudes are also due to Professors J.-P. Vigier, F. Selleri, D. Home, and Dr. J. Summhammer for their valuable discussions at the occasion of the Delphi Conference. All simulation was done using a high performance super computer, S-810, of Hitachi, Ltd.

REFERENCES

[1] von Neumann, J., 1932, *Mathematische Grundlagen der Quantenmechanik* (Springer, Berlin); Wigner, E. P., 1963, Am. J. Phys. **31**, 6.

[2] Machida, S and Namiki, M., 1980, Progr. Theor. Phys. **63**, 1457, 1833; for review, Machida, S. and Namiki, M., 1984, *Proc. Int. Symp. on Foundations of Quantum Mechanics*, eds. S. Kamefuchi, et al. (Phys. Soc. Japan, Tokyo) 127, 136; Namiki, M., 1986, *Proc. Int. Conf. on Microphysical Reality and Quantum Formalism*, eds. A. van der Merwe and G. Tarozzi (D. Reidel, Dordrecht) to be published; Namiki, M., Found. Phys., to be published; Namiki, M., *Proc. Int. Workshop on Matter-Wave Interference, 1987, Vienna*, to be published in Physica.

[3] For example, Araki, H., 1987, *Proc. 2nd Int. Symp. on Foundations of Quantum Mechanics*, eds. M. Namiki, et al. (Phys. Soc. Japan, Tokyo) 348.

[4] Tonomura, A., 1986, *Single–electron build–up of a quantum interference pattern*, 16m/m film, 3', produced by Hitachi, Ltd.-Cinesell, Inc.

[5] Murayama, Y., Nakazato, H., Namiki, M., and Ohba, I., 1986, *Numerical simulations of reduction of wavepacket on quantum measurement processes*, 16m/m film, 15', produced by Hitachi, Ltd.-Cinesell, Inc.

[6] For example, Schiff, L. I., 1968, *Quantum Mechanics*, 3rd ed. (McGraw-Hill, New York) Chapter 3.

[7] Rauch, H., 1984, *Proc. Int. Symp. on Foundations of Quantum Mechanics*, eds. S. Kamefuchi, et al. (Phs. Soc. Japan, Tokyo) 277; 1987, *Proc. 2nd Int. Symp. on Foundations of Quantum Mechanics*, eds. M. Namiki, et al. (Phys. Soc. Japan, Tokyo) 3.

[8] Namiki, M., Otake, Y., and Soshi, H., 1987, Progr. Theor. Phys. **77**, 508.

[9] Summhammer, J., private communication.

[10] Murayama, Y. and Namiki, M., to be published.

[11] Wootters, W. K. and Zurek, W. H., 1979, Phys. Rev. **19**, 473; Zurek, W. H., 1981, Phys. Rev. **24**, 1516; 1982, Phys. Rev. **26**, 1862.

DESCRIPTION OF EXPERIMENTS IN PHYSICS: A DYNAMICAL APPROACH

Jan von Plato
University of Helsinki
Department of Philosophy
Unioninkatu 40 B
00170 Helsinki
Finland

ABSTRACT. A way of describing repetitive experiments can be founded on the notions of the theory of dynamical systems. Dynamical invariants of a stationary system decompose its set of trajectories. Asymptotic statistical behaviour is uniform within a component which is not further decomposed by an invariant. The time average probabilities of classical systems become limits of relative frequencies when time and state space are discretized. A complete set of invariants is replaced by a parametric family of probability distributions. The parameters are precisely those factors the control of which specifies statistical laws of the repetitive experiment. Various ways of interpreting stationary probabilities are suggested, these interpretations depending on whether parameters exist, have been identified, or are randomized over.

1. PRELIMINARIES

1.1. Let X be the state space of a classical dynamical system. Usually X is a bounded subset of R^n for some n. This will be assumed to be the case in the sequel. The law of motion for the microstates $x \in X$ has as solutions trajectories in the phase space X: given an initial state x_0 at time t_0, there is for any time t a state x which belongs to the dynamical trajectory passing through x_0. Let us write this basic dynamical relation as $x = T(t_0, t, x_0)$. For a system with a fixed total energy, we have identically $T(t_0, t, x_0) \equiv T(t, x_0)$. The time evolution is only dependent on initial state but not initial time.

1.2. The point-like microstates are not physically observable. Observation tells as best that the microstate of the system is in some set of states $A \subset X$. With $X \subset R^n$, one takes as the algebra A of observable macrostates that of measurable subsets of X. As X was supposed to be bounded, we can have normalized measures m over A. If $m(A) = 0$, A is equated with the empty set. This makes for a non-atomic algebra, and corresponds to the unobservability of the exact microstates.

1.3. Given two systems, we can make a difference between them only so far as they have some permanent properties and so far as they differ with respect to at least one of these properties. For one system, we can only ascribe to it such properties as we would to any other system with identical permanent properties. An example will clarify the matter. Let us assume that we have plain ordinary gas in a container. Everybody believes the molecules collide in all sorts of ways so as to wipe out any other properties except the constant total energy. There is no physical means of saying that the time evolution of the system belongs to some specific set of trajectories, except the set of all trajectories. Specifically, it is physically meaningless to ascribe one particular trajectory to the system's evolution.

On the other hand, there are systems with many permanent properties, such as two-body systems. Otherwise one would not be able to tell the planets apart, so to speak.

Permanent properties are dynamical invariants: let f be a phase function over X. If $f(x) = f(T(t,x))$ holds for any time t, f is by definition an invariant.

In classical mechanics, the behaviour of a system is identified through the identification of values of invariants. For plain ordinary gases, the total energy $H(x) = H$ is the only invariant. Therefore, if one knows how much gas there is and what the volume of the container is, one can estimate the energy from a thermometer attached to the latter. From the reading any other properties the system might have, can be determined.

The classic ideal of deterministic science has been to give a dynamical equation for a system of moving bodies and to solve it completely. But this has been found to work only on very simple cases. A paradigm is given by Kepler motion. We have two gravitating mass points. Six numbers are needed for a description of one moving point, three for position and three for velocity or momentum coordinates. Its state space is a subset of R^6. Each single closed orbit is an ellipse in configuration space. Motion along it is preserved, so numbers characterizing which ellipse the moving body is gravitating on, are values of invariants. Five invariants will suffice for the determination of a particular ellipse. One should think in this order: first take the whole state space, then reduce the set of accessible states one by one into subsets of lower and lower dimension. In the case of Kepler motion, one arrives finally at a single trajectory in state space. To be exact, physical parameters never are determined with exactness. However, Kepler motion is stable so approximate determinations are respected by the system.

If we fix some properties of a system, there will be others left open. These are contingent relative to our specification. If we fix all possible invariants of a system, the remaining (non-permanent) properties are contingent relative to any specification. In Kepler motion, the contingent property is the angular position of the orbiting point. If this is observed at some specific time, the motion is completely determined in the way once was thought possible in mechanics in general.

1.4. Starting from the whole phase space, we determine parts of it via identification of invariants. This is in a way a reverse of the traditional picture where one thinks of the phase space as being formed from its points. With the use of invariants, we shall see that there is no dramatic difference between deterministic and probabilistic descriptions. Specifically, there is nothing contradictory in the idea of determining probabilities for classical systems. This determination proceeds from the physical properties of a system. Often, physicists and philosophers don't see this. They claim instead that probabilities in classical systems only refer to ignorance, basing the argument on the alleged "determinism" of all classical motions.

1.5. Above, we introduced a normalized measure space (X,A,m). The conservativeness or isolatedness of a system is expressed in dynamical terms as dH/dt = 0, where H is the Hamiltonian function of the system. Let us denote by P the Lebesgue measure over the phase space of Hamiltonian coordinates. This is called the natural measure. By what is known as Liouville's theorem, it is shown that the dynamics preserves the measure P over the Hamiltonian coordinate space. Define T(t,A) as the set of all points T(t,x) where x∈A. T is a measure preserving transformation if P(T(t,A)) = P(A). In particular, P(X) = P(T(t,X))=1. The motion is accordingly pictures as an incompressible flow.

Let us call the condition for having a time-independent Hamiltonian the condition of dynamical stationarity. If you require that the measure that you are going to introduce over the state space should be absolutely continuous with respect to the Lebesgue measure over it, there is only one measure which is preserved in the motion. Let us call this preservation a probabilistic stationarity condition. The dynamics gives us probabilities if we assume that events of zero measure have zero probability. In the Hamiltonian formulation we get the right probability measure right away, and what it is is determined by a dynamical condition.

1.6. Next we come to the interpretation of P. Let $I_A(x) = 1$ if x A and 0 in the opposite case. There is a theorem which says that the time average of $I_A(x)$ exists almost everywhere in X:

$$\hat{I}_A(x) = \lim_{t\to\infty} \frac{1}{t} \int_0^t I_A(T(t,x))dt$$

The phase average of the time average $\hat{I}_A(x)$ over X is

$$\bar{\hat{I}}_A = \int_X \hat{I}_A(x)dP = \bar{I}_A = P(A)$$

If we had the possibility of picking states x according to the law P, and followed the time averages $\hat{I}_A(x)$, the average of time averages would reproduce the a priori probability P(A) of A.

1.7. Suppose a system has only one invariant. Define a set of states A invariant if $T(t,A) = A$. It follows that such a system has no invariant sets with measure strictly between zero and one. One the other hand, if we have an invariant phase function f, the set $f^{-1}[x] = \{y | f(x)=f(y)\}$ is invariant. In particular, it contains with each x also the trajectory of x. Invariant sets are in this sense complete sets of trajectories.

If our system accepts only one invariant, it has only one invariant set which is the whole phase space (and the empty set). It is not possible to sort out a set of states (macrostate) that would remain invariant (= the macrostate preserved) under the dynamical evolution.

A theorem shows that in this case, which is the case known as ergodicity, we have constant time averages over X. Specifically

$$\hat{I}_A(x) = \hat{I}_A = P(A).$$

Under ergodicity, the statistical behaviour of our system is uniform, it does not depend on the particular trajectory. This makes it possible to determine uniform statistical laws from dynamics. (More precisely, uniform almost everywhere in the measure-theoretic sense.) On the other hand, one knows that any one time evolution will realize the statistical law of the system. It is asymptotically identifiable from data. With suitable conditions stronger than ergodicity, one does not even have to wait infinitely long to see how the system behaves statistically.

1.8. In a non-ergodic case, it will depend on the values of further invariants what particular statistical law the evolution of the system obeys. There is a theorem which says that any stationary system decomposes in a unique way into ergodic parts. In terms of invariants, this can be seen to be rather obvious for classical systems:

A collection of invariant phase functions $f_1,\ldots,f_k$ is a complete set of invariants if any other invariants can be determined from these. So for example, strictly speaking our plain ordinary gas, referred to above, accepts an infinity of invariants. These are always functionally determined by the single constant invariant the system was supposed to have. They must then be functions of this invariant, the total energy. Time averages are good examples. They are invariants for stationary systems, because it makes no difference what time is taken as zero time in their determination. Therefore probabilities in ergodic systems are determined as functions of total energy. Now, to come back to the case of several invariants, we shall assume that our invariants $f_1,\ldots,f_k$ are independent of each other and complete. If the degree of freedom of our system is n, and if $k = n-1$, each ergodic component is an individual trajectory. Then that trajectory is an invariant set, and we place probability one on it. This is the case of "determinism". If $k < n-1$, there are invariant sets with a continuous number of trajectories that don't decompose. They are the ergodic components of a stationary system. The stationary measure P is parametrized by k real numbers, the values of $f_1,\ldots,f_k$. Each specification of values gives an ergodic measure P_a, where $a = (a_1,\ldots,a_k)$. There is no physical means of making a difference between time evolutions within one component that would be respected by the system. This is the case of genuine statistical behaviour in dynamical systems. We saw that determinism is a special case with k = n-1, so it cannot be seen as somehow being an opposite or contradiction of the genuinely statistical case.

2. DESCRIPTION OF EXPERIMENTS IN CLASSICAL DYNAMICS

2.1. Let us first discretize the above. For the sake of simplicity, divide the phase space X into only two parts X_0 and X_1. These are designated by 0 and 1. Let t be a fixed unit of time, and write $T(t,x) = T(x)$, $T(2t,x) = T^2(x)$, and so on. I_0 and I_1 are the indicator functions of the two components. Under stationarity, we have for the time average $\hat{I}_1(x)$

$$\hat{I}_1(x) = \lim_{n \to \infty} \frac{1}{n} \sum_{i=0}^{i=n-1} I_1(T^i(x))$$

This is the limit of relative frequency of the macroscopic event X_1 in the sample sequence $(x, T(x), T^2(x), \ldots)$. The sampling distributions $p_n(x, \ldots, T^{n-1}(x))$ are determined once the measure P over the underlying continous trajectories is given.

The situation is different from the general statistical case in two respects:

1. The sample sequence is obtained by a physically well defined dynamical process.
2. The probabilistic description of the sampling procedure is based on the physical laws governing the behaviour of the underlying dynamical evolution.

It is a physically well defined question to ask if the sampling process is representative or unbiased. That it is biased can only mean: there is some systematic factor that affects the statistical behaviour. This factor must be, according to what was presented above, an invariant of the system in question.

We saw also that the dynamics singles out the right probability measure P for the system. The central task of the whole approach is the determination of the statistical behaviour of dynamical systems from an underlying basic physical theory. This is the reverse in order of the situation in statistics in general. There, one postulates a hypothetical probabilistic model as a description of the sampling process. The most common assumption is to have probabilistically independent consecutive experiments, with a constant though generally unknown success probability. In this case, we cannot give physical arguments for not being in a situation where some uncontrolled factor (invariant) affects the statistics of our experiments. In conclusion, the only way (so far known) to <u>prove</u> that one's experimental arrangement is statistically representative, is to give a physical description for it, and a dynamical description for the succession of experiments.

2.2. In the above, we have a nice picture of what goes on in the classical case. If two seemingly equivalent series are statistically different, there should exist an invariant hidden somewhere in the experimental arrangement to account for this. In the first series the factor, designated here by λ, had a certain value λ_1, and we were led to a component whose probabilistic law was given by the measure P_{λ_1}, in the second by P_{λ_2}. A possibility in between having an ergodic component and uniform statistical behaviour, and several components with differing behaviour, would be the following. We have an extra invariant λ, but it is uncontrollable in the experimental arrangement. For representing the uncontrollable, random character of the variation of the extra invariant, we assume a probability distribution $\mu(\lambda)$ for it. Then, our experiment randomizes over the parameter λ according to the law μ. The probability law is a mixture of the form

$$\int_\Lambda P_\lambda d\mu$$

where λ varies in Λ. This should be seen as a conditional probability. First a choice of λ is done, then an experiment with the probability law P_λ. As this is repeated at each experiment, the successive results are probabilistically independent. This is what one wants to have with a randomization procedure. Note that the situation would be completely different if we were just given, by Nature's choice so to speak, some one value of λ unknown to us. Then our first experiment would still be described by a probability law of the above mixture type. Note, however, that even though all the individual measures P_λ are ergodic for fixed values of λ, the mixture itself is not ergodic but only stationary. If we take the special case of ergodicity where we in fact have independence between consecutive events, the mixture probabilities don't have this independence. Instead, they are what are known as exchangeable measures. Remaining in one component with an unknown value of the parameter, one modifies the mixture probability by conditionalizing in the usual probabilistic fashion. The law P_{λ_i} is recovered asymptotically from data, but one knows nothing of the statistical laws for other values of λ.

3. REMARKS ON THE QUANTUM MECHANICAL CASE

What was presented above warrants us to say that there are at least <u>some</u> cases where the statistical laws are derived from a physical basis. In practical terms, the picture offered for classical systems is idealized too far. But it gives us an example of how the requirements for a justification of the introduction of probabilistic laws can be met.

It is often said that in quantum physics probability enters on a more fundamental level than in the classical case. The probabilities of observable events are derived from a physical description of the system under study. They are computed from the wave function. The macroscopic statistical behaviour that one observes has no hidden dynamical structure as in the classical systems. But there is another level on which the dynamical character of the latter leads to a justification of probabilistic laws. This concerns the repetition of an experiment, and the probabilistic description of sequences of results from experiments. In quantum mechanics it seems to remain an assumption that the repetitions of the experiment are probabilistically independent. It is a statistical hypothesis about the sampling procedure as one would say in the terminology of statistics. It may not be explicitly stated, but one sees from calculations (e.g., estimation of standard deviation from data) that it is as said.

The claim of the experimenter is that he repeatedly brings his system to an identical quantum state. When a measurement takes place, there appears in common terminology a localization, or reduction of the wave packet. Next time the experiment is performed, there should be no 'memory' of the previous result left in the system. One concludes that the preparation or 'filtering' of the object system into an identical pure quantum state has been achieved.

The assumption of independence leads to a probability of the product type over the sequences of denumerable series of experimental results. But is there a conclusive justification for this step, from the quantum mechanical description of the situation? It seems that the classical side is sometimes better able to justify this independence assumption. We know that there are classical dynamical systems so strongly non-linear that they macroscopically are examples of Bernoulli processes. Of course, hardly anyone believes that, e.g., Aspect's photon pair experiments which give a reading on only about every tenth pair, should be a manifestation of Nature's wicked ways of choosing a subensemble of the original sequence, which latter, if it had been observed in its entirety, would have produced a deviating statistics. The independence postulate appears as an extra assumption. Its role is perhaps partly epistemic, or it is justified methodologically by statistical testing.

We can look at the situation in more general terms. The mathematical results about existence and uniqueness of limits of time averages, and the decomposability of a stationary probability law into ergodic parts, can be established as results of probability theory. Their essential content is independent of the two specific features of classical theories for which they were originally shown to hold. A continuous state space and a continuous time dynamics are not essential, as was indicated above in the comparison of time averages with relative frequencies. For the case of quantum mechanics, one should show that statistical representativity does not appear as a hypothesis supported by theoretical insight, belief, or statistical methodology, but as a physically well founded conclusion.

BIBLIOGRAPHICAL NOTE

Proofs of all the results of ergodic theory referred to, except for the ergodic decomposition theorem, can be found in e.g. Farquhar, <u>Ergodic Theory in Statistical Mechanics</u> (Wiley, New York 1964). For the latter result, see references in Cornfeld, Fomin and Sinai, Ergodic Theory (Springer, Berlin 1982). I have developed further my ideas on probability in dynamical systems in my essays Ergodic theory and the foundations of probability, pp. 257 - 278 in Skyrms and Harper (eds.), Causation, Chance, and Credence, vol. 1 (Reidel, Dordrecht 1988). For Kepler motion see Sternberg, Celestial Mechanics, Part I, chapter II, p. 108 ff. particularly.

THE PHYSICAL QUANTITIES IN THE RANDOM DATA OF NEUTRON INTERFEROMETRY

J. Summhammer
Atominstitut der österreichischen Universitäten
Schüttelstrasse 115
A-1020 Vienna, Austria

ABSTRACT. This work consists of two parts. In the first section a review of typical neutron interferometry experiments is given, in which three different interactions of the neutron are exploited: nuclear, magnetic and gravitational. In all cases the familiar sinusoidal interference pattern is obtained. In the second section we start from the observation, that the primary data in neutron interferometry, as in all of quantum physics, are random "clicks" in detectors. In the case of the interferometer each "click" is just a yes-no answer, since there are only two detectors. The "clicks" are related in a probabilistic manner to the underlying physical quantity, the phase shift. We identify criteria how possible physical quantities can be derived from primary data. Without referring to quantum theory at all we are led step by step to the cosine law between probability and related physical quantity and ultimately to the wavefunctions present at the two detectors.

I. TYPICAL EXPERIMENTS IN NEUTRON INTERFEROMETRY

Since its first successful introduction in 1974 (1), the single crystal neutron interferometer has made possible precision measurements of neutron-nucleus scattering lengths as well as the demonstration of a very non-intuitive feature of quantum mechanics: interference phenomena for massive particles (2,3).
  Neutron interferometers are made from single crystal silicon, and generally cut in such a way, as to leave three parallel slabs standing on a common basis (Fig. 1). Normally the interferometer (IFM) is illuminated with an incident beam of fairly monochromatic neutrons, with a wavelength of about 2 Å (corresponding to neutron velocities of around 2000m/s), and a wavelength spread of less than 1%. At the first crystal slab the incident beam is separated into a transmitted and a reflected beam by the principle of Bragg diffraction in Laue geometry. The wavefunctions of these two beams are coherent. At the second slab,

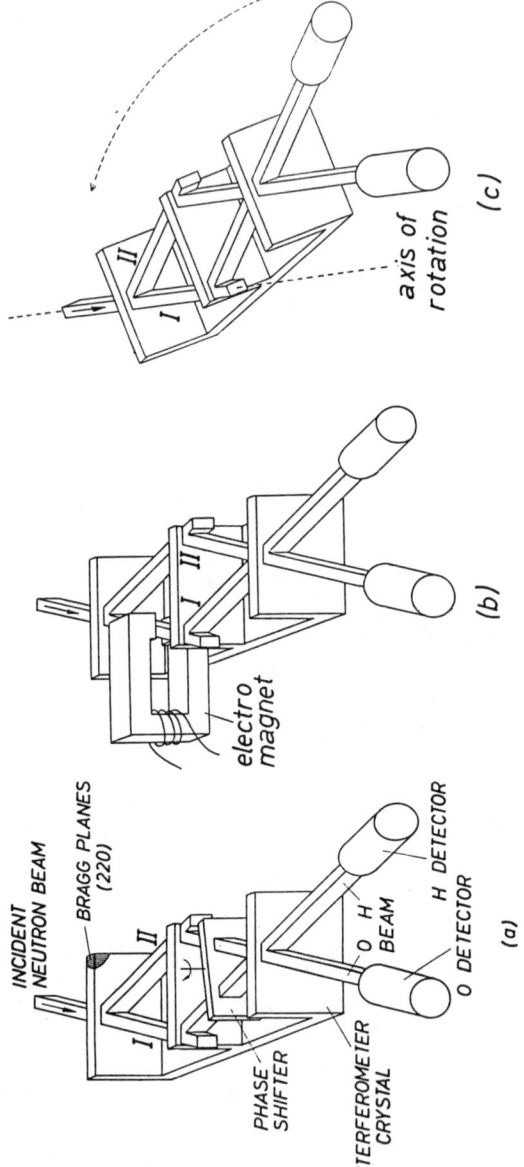

Figure 1: IFM experiments testing the influence of three different interactions on the neutron wavefunction. The IFM crystal is typically 6-10 cm long, 6 cm wide and 6 cm high. The three crystal slabs are usually around 4 cm high and 2-5 mm thick.
a) A slab of nonmagnetic homogeneous material as phase shifter exploits the nuclear potential.
b) A magnetic field as phase shifter uses the potential due to the neutron magnetic moment.
c) Rotation by a certain angle of the whole experimental set up around the incident beam causes paths I and II to experience a different mean gravitational potential. This results in a phase shift.

each of the two beams is again separated. The transmitted components leave the IFM and are lost. But the reflected beams approach each other and meet in the third slab. There again, each of them is split coherently. Thus the two beams emerging from the third slab — they are customarily called 0- and H-beam — each are a superposition of two wavefunctions, where one carries the information of path I, and the other that of path II.

At the second crystal slab, the beams inside the IFM are typically about 3-4 cm apart. Thus one can conveniently set up in one path an interaction potential to which the neutron is sensitive, without affecting the other path. In the simplest experiment, one inserts a plate of a solid homogeneous material as a phase shifter, as shown in Fig. 1a. For practical reasons the phase plate often extends over both paths. By turning the plate the pathlength of the two beams in the plate can be changed. Since the neutrons are subjects to the nuclear potential, the plate represents a region with a refractive index different from that of air, so that the wavefunctions of both paths experience a phase-shift proportional to their pathlength in the plate. But only the relative phase shift is measurable. The wavefunction at the 0-detector can therefore be written as (the wavefunction at the H-detector is formally completely analogous and can thus be neglected here):

$$\Psi_0(\chi) = \Psi_{0I} + \Psi_{0II} \exp(i\chi) \qquad (1)$$

with $\chi = N\lambda b_c \Delta D$. χ is the relative phase shift between the left and the right beam paths. N is the density of nuclei in the phase plate, b_c is the coherent neutron nucleus scattering length, ΔD is the difference of the pathlengths through the phase plate, λ is the de Broglie wavelength of the neutron. Ψ_{0I} and Ψ_{0II} are the wavefunctions in the empty IFM of beams I and II which lead to the 0-detector. They can be taken as plane waves, if the monochromaticity of the neutrons is very good. Furthermore, one has $\Psi_{0I} = \Psi_{0II}$ for an IFM with equal crystal plate thicknesses. The intensity at the 0-detector thus is:

$$I_0(\chi) = |\Psi_0(\chi)|^2 = |\Psi_{0I}|^2 (1 + \cos\chi) \qquad (2)$$

By changing the phase shift in increments, one thus gets the familiar interference pattern (Fig. 2). It should be mentioned, that Ψ_0 is a one particle wavefunction. This is justified, because even at the high flux reactor of the ILL in Grenoble the beam incident on the IFM carries less than 1 neutron per meter. Interferometer experiments are thus one particle experiments.

Aside from the nuclear potential, the neutron is also sensitive to the electromagnetic field, because of its magnetic moment, and of course, to gravity. The phase shift can thus also be accomplished through these interactions. For instance, one can create a small region with a static magnetic field (of the order 100 - 1000 Gauss) over one beam, and observe the intensity as a function of the strength of the field (Fig. 1b). In the field region, the neutrons perform Larmor precession

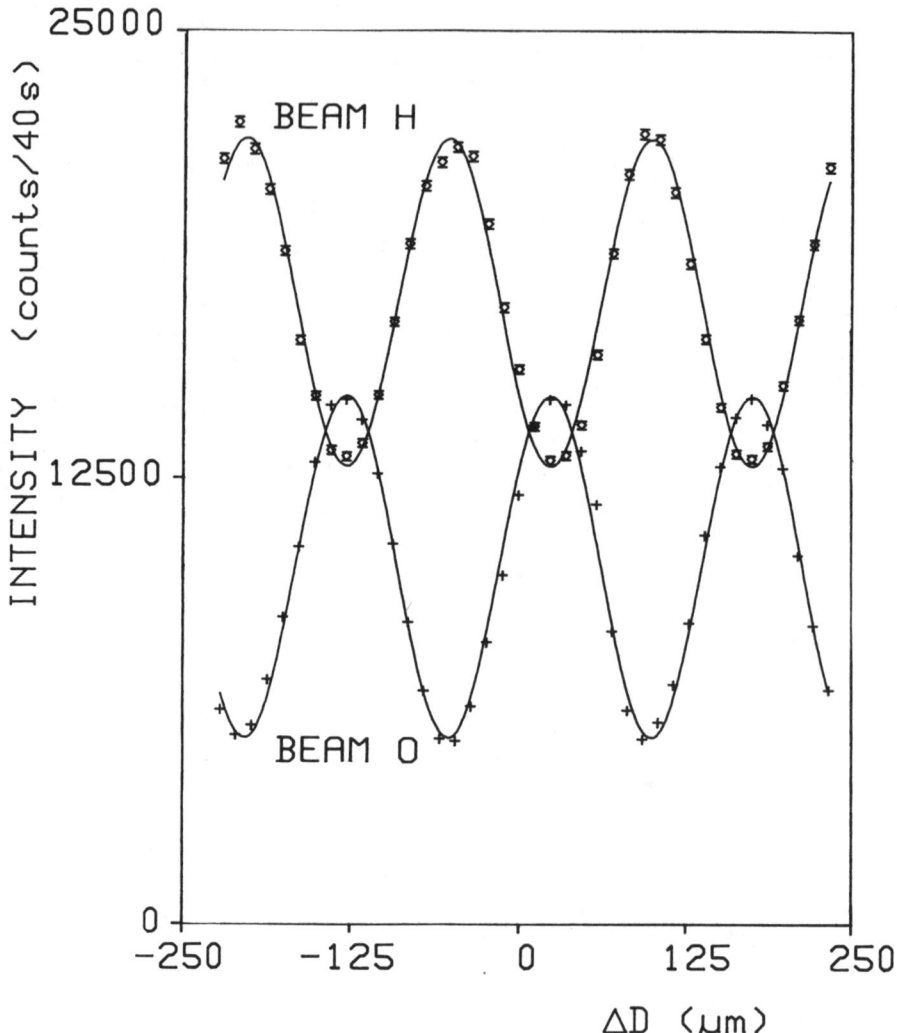

Figure 2: Typical interferogram obtained with unpolarized neutrons by rotating a 5 mm thick AI — phase plate in increments at the IFM instrument of the Institut Laue Langevin, Grenoble. Central wavelength: 1.924(6) Å. Cross section of incident beam: 9 mm vertical, 5.6 mm horizontal.

around an axia parallel to the field direction. It was found, that only when the neutrons were rotated by (716.8∓ 3.8)°, or an integral multiple of this, could the original intensity be restored (4). When increasing the magnetic field from zero to the strength needed to rotate the neutrons by 4π, the intensity only goes through one full oscillation. In this way, the 4π-periodicity of the wavefunction of spin-1/2 particles under rotation, was demonstrated.

Now an example of gravity as the interaction potential shall be given (5). When the IFM is tilted, so that on the mean one path will be farther away from the center of the earth than the other one, this will produce a phase shift, too (Fig. 1c). The reason is, that the average momentum of the neutron, when it is on the higher path, is lower, whereas when it is on the lower path, it is higher, than the momentum in a horizontal interferometer. Of course, at the point of superposition in the third slab, the momentum is equal again. Thus, the intensity at the 0-detector is:

$$I_0 = \left|\Psi_{0I} + \Psi_{0II}\exp[i(\bar{p}_{II} - \bar{p}_I)L/\hbar]\right|^2 = \qquad (3)$$

$$= \left|\Psi_{0I}\right|^2\{1 + \cos[(\bar{p}_{II} - \bar{p}_I)L/\hbar]\}.$$

where $\bar{p}_I$ and $\bar{p}_{II}$ are the average momenta on paths I and II, respectively, and L is the pathlength in the IFM (equal for both paths).

Finally, the quantum mechanical principle of spin superposition shall be considered. It ways that the superposition of two coherent neutron beams with opposite polarization does not result in an unpolarized beam, but in one fully polarized in a direction normal to the original polarizations. In the experiment the IFM is illuminated with neutrons polarized in the +z direction (Fig. 3). Inside the IFM in path I the spin is inverted by means of an oscillating magnetic field with $\omega/2\pi = 55$ kHz. Together with the spin flip the neutrons absorb (or emit) a photon from the field and thus change their energy slightly. One now has a spin up state in one path and a spin down state in the other. With the usual phase plate an additional phase shift can be created. Superposition of the beams results in a state vector for the 0-beam:

$$\Psi_0(\chi,t) = \begin{pmatrix} 0 \\ \Psi_{0I}e^{-i\omega t} \end{pmatrix} + \begin{pmatrix} \Psi_{0II}e^{i\chi} \\ 0 \end{pmatrix} = \Psi_{0I}\begin{pmatrix} e^{i\chi} \\ e^{-i\omega t} \end{pmatrix}. \qquad (4)$$

One notes that although it consists of two beams of opposite polarization the 0-beam is not unpolarized, but is fully polarized in a direction in the x-y plane. This polarization is given by:

$$\vec{P}_0(\chi,t) = \langle\Psi_0|\vec{\sigma}|\Psi_0\rangle / \langle\Psi_0|\Psi_0\rangle = \qquad (5)$$

$$= (\cos(\chi+\omega t), -\sin(\chi+\omega t), 0).$$

The dependence of the polarization direction on time and on χ could be demonstrated by means of polarization analysis of the 0-beam and by stroboscopic detection of the neutrons synchronous to the oscil-

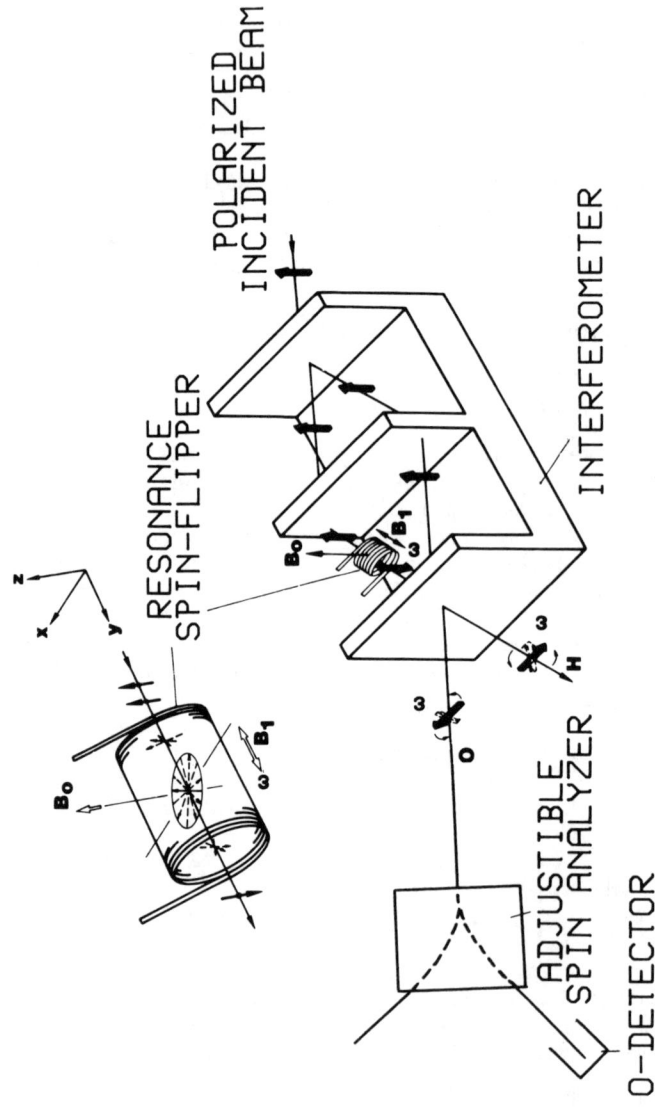

Fig. 3: Schematic of spin superposition experiment with oscillating magnetic field as spin flipper. The whole experimental set up is in a guide field in +z direction with amplitude B_0 = 19 Gauss. The amplitude of the oscillating field B_1 is of the same order, its frequency ω is such that $\hbar\omega = 2\mu B_0$, μ being the neutron magnetic moment. Intensity measurement at the 0-detector is synchronous with the phase of the oscillating field.

lating magnetic field (6). Polarization analysis was such that only particles in the desired spin state of any spatial direction were reflected towards the 0-detector. When one analyzed for a direction in the x-y plane the expected sinusoidal intensity oscillations as a function of t and of χ were found.

II. THE NEUTRON INTERFEROMETRY EXPERIMENT SEEN AS A BINARY ALTERNATIVE

The essence of neutron interferometry experiments is this: one allows a neutron to enter the apparatus, in which it has two possible paths which remain undetermined, and observes at which of the two detectors it ends up. It is a one particle experiment which is repeated sufficiently often to collect statistically significant results. The surprising phenomenon is, that although in the interaction potential the properties of the neutron as a particle enter (neutron-nucleus scattering length, neutron magnetic moment, mass of the neutron), the formal description is by means of complex valued wave functions, and the appearing intensity oscillations seem to corroborate the idea that the neutron has wavelike properties. Can this be understood?

We do not want to investigate this question by operating with the preconceived notions of particle and wave, nor by introducing any other object-like notion. Rather, we want to concentrate on the aspect how we obtain information in an IFM experiment and what conclusions we can draw with ordinary reasoning.

An experimenter will collect data in any quantum physical experiment with the following assumptions in mind:

(A1) A single experiment gives one discrete result out of a denumerable range of possible results. If n results are possible, one experiment is the decision of an n-fold alternative (7).
(A2) The outcome of a single experiment is unpredictable.
(A3) The probability of the outcome of a single experiment is determined by the physical conditions of the experiment.

In neutron interferometry a single experiment would mean to allow one particle to go through the apparatus and to detect it at either the 0- or the H- detector. Since two results are possible, one experiment gives the answer to one binary alternative.

The purpose of experiments is to learn something about nature, in particular about the values of physical quantities. How many independent such quantities can we measure in a neutron interferometry experiment? We have only two detectors. If we perform a time independent experiment, then it is only important at which detector the particle ends up, but not at what time. Thus one can only determine the relative frequency of firings in detectors 0 and H. Therefore, we can only determine one physical quantity.

Now we can ask: what, and how much, do we learn in an IFM experiment when we keep the experimental conditions constant and count a total of K particles in the 0- and H- detector? Let there be L counts in the 0-detector. Then we have found a relative frequency $P = L/K$ in K performances of the experiment. The numbers P and K characterize the information we have obtained. From them it should be possible to infer the probability p that detector 0 will fire in a future performance of the experiment. Because of assumption (A3) p should be a function of the physical quantity to be measured, which we will denote by E. We will now try to extract as much information as possible about E from the experimentally established numbers P and K.

For that purpose it is important to investigate the meaning of P and K. Quantum experiments are probabilistic experiments (assumption (A2) and (A3). Therefore, any quantity we want to derive from them can be specified better, if we have more performances of the experiment. In other words, the amount of experimental information increases strictly monotonously with the number of performances, K. Since we are dealing with identical performances of the experiment we must admit, that each one gives us the same amount of information, simply because it is just one more answer of nature to the same question. Also, the experiments are independent of each other. Therefore, the amount of information acquired is proportional to K (8). K is the answer to the question how much we have learnt. In contrast, P is obviously the answer to the question, what we have learnt in the experiments. It therefore represents the content of information.

This is the complete information we have gained. Here it is worth noting, that the specific series of firings of the 0- and H- detector does not contain information, since it is stochastic (assumption (A2)).

The physical quantity E and its error interval ΔE which we want to derive, will in general be functions of P and K:

$$E = E(P, K) \tag{6}$$
$$\Delta E = \Delta E(P, K). \tag{7}$$

The reason, why we must consider E as well as ΔE becomes apparent, when we think of what it means to specify a number as representing the most likely value for a physical quantity, when all the information from which it is to be derived is stochastic. Before the first experiment we know nothing of E, except that it will be somewhere in the real continuum (we do not consider complex quantities at this point). But after several performances we will be able to single out a specific interval ΔE in the real continuum as containing with a certain probability the true value of E. Obviously the width of this interval should decrease as the amount of information increases (this belief is at the heart of probability theory, otherwise there would be no point in trying to gain information from a stochastic series of yes-no answers). Therefore we can say that for all P

$$\Delta E(P, K+1) < \Delta E(P, K) \tag{8}$$

and
$$\lim_{K \to \infty} \Delta E(P, K) = 0. \tag{9}$$

The latter equation just expresses, that we can know the true value of E arbitrarily well. Now let us investigate the dependence of ΔE on

P. It means, that for a given amount of information the interval can still depend on P. Since the real continuum is everywhere equally dense this implies, that the "number" of points which we isolate with a given amount of information is not fixed. But how can that be? What additional information would allow us to isolate fewer points for one P and more points for another P? P itself is not an amount of information, it just expresses what is contained in that amount. Therefore, as we have pointed out above, there is no additional amount of information and we must conclude, that it is unwarranted that for a given K the interval should sometimes be wider and at other times be narrower, depending on P. Hence we cannot permit ΔE to depend on P. Rather we must have:

$$\Delta E = \Delta E(K) \tag{10}$$

$$\Delta E(K+1) < \Delta E(K) \tag{11}$$

$$\lim_{K \to \infty} \Delta E(K) = 0. \tag{12}$$

Now let us turn to the most likely value of E. It is clear, that it should depend on P, since this is exactly what we learnt in the experiment. But should E also depend on K? E is to approximate the true value of a physical quantity increasingly better as K increases (equ. (11)). Furthermore, we believe a physical quantity to be something objective (assumption (A3)), which should not depend on the number of experiments we perform to determine it. Therefore, the most likely value of E should not depend on K. Thus we must write:

$$E = E(P). \tag{13}$$

Equations (10)-(13) are the requirements we must demand from any physical quantity that is to be derived from interferometry data: the most likely value of this quantity must only depend on the content, whereas its error interval must only depend on the amount of the information acquired in the experiments.

Now we will investigate the functional relation in between E and the probability p which is assumed in (A3). In K performances the relative frequency of firings of detector 0 was found to be P. Then the most likely value of the probability that it will fire in the next performance of the experiment is given by (9):

$$p = P \tag{14}$$

with the error interval of the binomial distribution:

$$\Delta p = \sqrt{p(1-p)/K} \tag{15}$$

Combining equations (13) and (14) we can write

$$E = E(p) \tag{16}$$

For ΔE we now get:

$$\Delta E = |dE/dp|\Delta p = |dE/dp|\sqrt{p(1-p)/K} \qquad (17)$$

In order that equation (10) is fulfilled we must have:

$$|dE/dp|\sqrt{p(1-p)} = C_1 \qquad (18)$$

C_1 is a constant. In principle the l.h.s. could also be a function of K, but this is excluded by requirement (13). Now $\sqrt{p(1-p)}$ is always understood as positive, so that we can drop the absolute value sign in (18) and perform the integration. We get

$$E = -C_1 \arcsin(1-2p) + C_2, \qquad (19)$$

where C_2 is a constant of integration. And for ΔE we find:

$$\Delta E = C_1/\sqrt{K}. \qquad (20)$$

Requirements (10)-(13) are fulfilled, so that E can really be considered a physical quantity. With the particular choice $C_1 = 1$ and $C_2 = \pi/2$ we find the relation:

$$p = \cos^2(E/2). \qquad (21)$$

We see, that experimentally E is only determined up to an additive constant $2\pi N$, N being an integer. This equation is somewhat surprising, because it expresses exactly the peculiar wavelike relation between physical quantities in the IFM (any kind of phase shift) and experimentally determined probabilities as we know it from quantum mechanics. But we have at no point made use of quantum mechanics. All we did was to make clear to ourselves what is our experimental information and what we can infer from it about possible physical quantities by ordinary logical reasoning. It should be pointed out that with similarly simple arguments, but from a different starting point, Wootters has also found this relation (10).

Having obtained E, can we infer other physical quantities from primary data of IFM experiments? Since the content of information is given by the one number P, there can of course not be other independent physical quantities. But are there any related ones? In interferometry E is essentially the phase shift. But this phase shift controls two other quantities, which are the fluxes of particles towards the two detectors. In our time independent consideration we can understand this as the tendencies that detector 0 or H fires, respectively. Let us denote them by s_0 and s_H. Can we find a relation between the physical quantity E and the quantities s_0 and s_H on purely logical grounds? We will only consider s_0, since s_H should be obtainable by analogy. s_0 is to be a physical quantity. Therefore it must conform to requiremnets (10)-(13).

s_0 is a function of P, thus of p, and therefore of E. From (21) it then follows that s_0 must be periodic in E. Also we can reasonably require that it is continuous.

$$s_0(E) = s_0(E+2N\pi) \qquad (22)$$

For the error interval we have:

$$\Delta s_0 = |ds_0/dE|\Delta E = |ds_0/dE|C_1/\sqrt{K} \qquad (23)$$

Because of (10)-(12) Δs_0 should only be a function of K, but because of (13) neither s_0 nor E depend on K. Therefore $|ds_0/dE|$ must be a constant. In order that s_0 be periodic in E as required by (22) $s_0(E)$ could then be taken as a zig-zag function (Fig. 4). But we see, that this function does not fulfill the requirement, that Δs_0 only depend on K. Therefore the zig-zag function is not possible for $s_0(E)$. But this means, that there is no real valued function $s_0(E)$ that conforms to requirements (10)-(13). If we try complex functions we find,

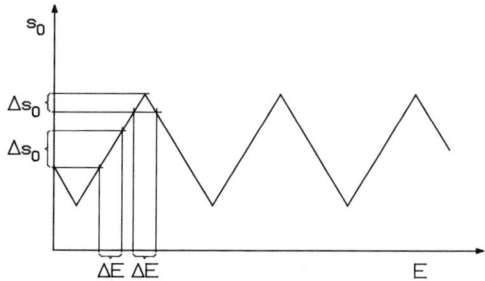

Fig. 4: $s_0(E)$ seen as a zig-zag function. Close to the extrema of s_0 the interval Δs_0 is smaller than for average values of s_0 for any ΔE, which is a constant for a given amount of information K. But according to (10) Δs_0 should be a constant for a given K.

that the only one that is periodic with 2π and conforms to the requirement that $|ds_0/dE|$ is a constant is

$$s_0(E) = A_0 + A_1 e^{iE} \qquad (24)$$

or its complex conugate. Here, A_0 and A_1 are complex constants which we must determine. For this purpose we can define s_0 to be zero when detector 0 never fires, and to be at its maximum, s_0^{max}, when detector 0 always fires. These two cases we have for $p = 0$ and $p = 1$, or equivalently, for $E = (2N+1)\pi$ and $E = 2N\pi$. Then we obtain:

$$s_0(E) = s_0^{max}\cos(E/2)e^{iE/2}. \qquad (25)$$

With similar arguments we find for $s_H(E)$:

$$s_H(E) = -s_H^{max}\sin(E/2)e^{iE/2}, \qquad (26)$$

where s_H^{max} is a constant like s_0^{max}. For the error intervals we obtain:

$$\Delta s_0 = |s_0^{max}|C_1/(2\sqrt{K'}), \qquad (27)$$
$$\Delta s_H = |s_H^{max}|C_1/(2\sqrt{K'}). \qquad (28)$$

Requirements (10)-(12) are fulfilled by Δs_0 and Δs_H Therefore $s_0(E)$ and $s_H(E)$ can justifiably be considered physical quantities. A comparison with quantum mechanics shows, that they are proportional to the wavefunctions at the detectors 0 and H, respectively. But we have nowhere in our deduction appealed to quantum theory or any physical model. Thus it seems, that the very concept of wavefunction can be arrived at by drawing conclusions from the assumptions (A1)-(A3), which are generally accepted in quantum physics. We note that in our arguments we have not referred to the two paths in which the particle can be found inside the IFM. On the contrary, we stressed, that there are two detectors at which it can end up. Therefore (21), (25) and (26) are not an explanation of the interference phenomenon. These equations only show that for any quantum physical experiment that can be understood as giving answers to a time independent binary alternative (e.g. also the Stern Gerlach experiments with spin-1/2 particles), one does not need to stop at the observed relative frequencies and from then on employ quantum mechanics. Rather it is possible to arrive by simple arguments at physical quantities, which are related to the probability by in general complex trigonometric functions. It is only necessary to clarify the meaning of pinpointing a small region in the real or complex continuum as containing the most likely value of a physical quantity by means of a certain amount of information. One can then look for an empirical relation between these physical quantities, in particular the real valued quantity E, and experimentally adjustable parameters like thickness of a phase plate, strength of a magnetic field, angle of a polarizer etc. One finds that these relations are just linear. The wave property then appears as nothing but a logical structure implicitly contained in the way we order stochastic data by means of probability theory. In our special case they are only a structure contained in the binomial distribution, which governs the statistics of time independent interferometry experiments. (But since logical structures in our mind are quite immutable, it is a matter of philosophical preference whether we consider the wave property as just rooted in the way the human mind thinks, or as a property in nature.)

In conclusion, we want to point out, that the present considerations are not limited to the binary alternative. They can easily be extended to the time independent n-fold alternative, whose statistics is given by the multinomial distribution. Then one finds n-1 independent real valued physical quantities. But in analogy to (25) and (26) one can also derive the general n-component complex valued state vector familiar from the Hilbert space formalism. Also, one can extend the considerations to include time dependent observations. The resultant probability distributions are Poissonian. But by applying requiremnets (10)-(13) one again arrives at physical quantities which are related to the detection probability in a certain time interval by the same complex valued wave functions we know from quantum theory in these cases (8). It thus appears that exactly those features of quantum mechanics which are in strongest contrast to our intuition, namely the appearance of wavelike phenomena despite the fact that we always only detect discrete "clicks" can easily be understood by carefully investigating how we analyze and then draw conclusions from stochastic data. By such careful analysis it should ultimately be possible to get an understanding of the idea of space as a sumbol for the properties extendedness and dimensionality, and therefore of the idea of particle. This should then render a true understanding of interference phenomena from simple logical arguments.

REFERENCES

1. H. Rauch, W.Treimer and U. Bonse, Phys. Lett. $\underline{47A}$, 369, (1974).

2. "Neutron Interferometry", ed. U. Bonse and H. Rauch, Clarendon Press, Oxford 1979.

3. The most recent experiment with polarized neutrons is reported in (see also references therein): G. Badurek, H. Rauch and D. Tuppinger, Phys. Rev. $\underline{A34}$, 2600, (1986).

4. H. Rauch, A. Wilfing, W. Bauspiess, U. Bonse, Z. Physik $\underline{B29}$, 281, (1978).

5. J.-L. Staudenmann, S.A. Werner, R. Collela, A.W. Overhauser, Phys. Rev. $\underline{A21}$, 1419, (1980).

6. This experiment is reported in: G. Badurek, H. Rauch and J. Summhammer, Phys. Rev. Lett. $\underline{51}$, 1015, (1983). Its analogue with a static spin flipper is given in: J. Summhammer, G. Badurek, H. Rauch, U. Kischko and A. Zeilinger, Phys. Rev. $\underline{A27}$, 2523, (1983).

7. To view one quantum physical experiment as the decision of an n-fold alternative was proposed by C.F. von Weizsäcker, "Aufbau der Physik", Carl Hanser Verlag, 2. Edition, 1986.

8. It can be shown that this amount of information is equivalent to Shannon's definition of information:
J. Summhammer, Found. Phys. Letters, (issue of July 1988).

9. H. Geiringer: "Probability: Objective Theory". In "Dictionary of the History of Ideas", ed. P.P. Wiener, Vol. III, p.605. Charles Scribner's Sons, New York (1973).

10. W.K. Wootters, Phys. Rev. $\underline{D23}$, 357 (1981).

Part 5

Questions about irreversibility and stochasticity

INTRINSIC IRREVERSIBILITY IN CLASSICAL AND QUANTUM MECHANICS

I. E. Antoniou and I. Prigogine +
Faculté des Sciences , CP 231
Université Libre des Bruxelles
1050 Bruxelles
Belgium

+also in
Center for Studies in Statistical Mechanics
The University of Texas at Austin
Austin TX 78712
U.S.A.

ABSTRACT. Intrinsically Irreversible Dynamical Systems allow for an exact passage to Irreversible Evolution through appropriate non-Unitary change of Representation. The property which characterises such systems is Dynamical Instability expressed by the Kolmogorov Partition and Internal Time or by the non-vanishing of the asymptotic Collision Operator. This leads to an extension of both Classical and Quantum Mechanics. Certain implications of the Kolmogorov Instability and Internal Time for Relativistic Systems as well as of the non-vanishing of the asymptotic Collision Operator for Unstable quantum systems are discussed .

1. INTRODUCTION

It is a pleasure to thank the organizers of this initiative to celebrate E. Schrodinger's anniversary here in Delphi ,the meeting place of the ancient city states.

Schrodinger (1) recognized the unique standing of the second Law of Entropy increase among the Laws of Physics, as well as the probabilistic meaning of Irreversibility as was discussed by Boltzmann . But Schrodinger did not go further than Boltzmann's Stosszahlanzatz in elucidating the problem of Irreversibility : How probability gives rise to the Dissipative Irreversible Evolution of the Conservative Reversible Dynamics ?

The coarse graining averagings over the cells of microstates caused by lack of knowledge or by incomplete observations,as well as the various approximations to the "complete"

Dynamical description being extraneous to Dynamics, cannot be considered as satisfactory answers to the problem. The second Law of Entropy increase and the very notion of Time and Becoming cannot be illusions associated to averagings or to approximations extraneous to the "real" Reversible Evolution. Unstable particles and excited Atomic states as well as the cosmic background radiation indicate the presence of time arrow in elementary processes as well as on the cosmic scale. The self-organization of dissipative structures emerging in far from equilibrium conditions testify to the essential constructive role of Irreversibility at the macroscopic level (2).

In the approach to the problem of Irreversibility developed in Brussels (2), Irreversibility expressed by the Law of Entropy increase, is considered as a fundamental physical principle. Irreversible Evolutions may be understood to arise as appropriate realizations of the Conservative Dynamics. Such realizations are effected by a non-Unitary time Reversal Symmetry breaking transformation Λ acting on the states ρ of the Dynamical System:

Conservative Dynamics
Entropy is constant

$$U_t : \rho \longmapsto U_t \rho$$

$$\Lambda \downarrow \qquad \downarrow \qquad \downarrow$$

Dissipative Dynamics
Entropy is increasing

$$W_t : \Lambda\rho \longmapsto W_t\Lambda\rho = \Lambda U_t\rho$$

Λ intertwines the Unitary Dynamics U_t on the states ρ, being density functions for classical systems or density operators for quantum systems, with the Irreversible Markov semigroup W_t, $t \geqslant 0$, being the prototype of Entropy increasing evolutions. An analogous formulation also holds for the Observables -Heisenberg picture of Dynamics.

This formulation of the problem of Irreversibility allows an investigation of the implications of the second Law to Dynamics. The property which qualifies Intrinsically Irreversible Dynamical Systems and gives rise to a time Reversal Symmetry breaking representation is Instability. Two forms of instability have been found giving rise to two kinds of Λ transformations:
A) The Kolmogorov instability expressed by the presence the time asymmetric K-partitions and Internal Time. The transformation Λ effectively incorporates the limitations to predictability arising from the Dynamics itself.
B) The instability due to the presence of continuous sets of Resonances expressed by the non-vanishing of the asymptotic Collision Operator. Here Λ is a suitable generalization of diagonalization transformations.

Let us also emphasize that the very existence of the Λ transformation is equivalent to the existence of a Liapunov observable M having the meaning of a microscopic Entropy operator. Such observables cannot be ordinary observables ,i.e. phase functions or Hermitian operators, but superobservables i.e. operators acting on the density functions or on the density operators .This was shown by Misra (3,4) who in this way went beyond the limitations of Poincaré Recurrence theorem, namely that "a phase function will infinitely often assume its initial state".A Liapunov observable is given by :

$$M = \Lambda^\dagger \Lambda$$

and $\langle \rho_t, M \rho_t \rangle = h(\rho;t)$ is a monotonically decreasing function

$$\langle \rho_1, \rho_2 \rangle = \begin{cases} \int d\mu \, \rho_1 \rho_2 & \text{for classical systems} \\ \mathrm{tr}\, \rho_1 \rho_2 & \text{for quantum systems} \end{cases}$$

Therefore in Intrinsically Irreversible Systems the Algebra of Observables is extended to include superobservables .

The Entropy observable M ,as well as the Λ transformation have another important property : they do not preserve the purity of states ;the distinction between pure states and mixtures is lost for intrinsically irreversible systems . Pure states ,being phase points or wave functions lose their priviledged position in Dynamics . this is also expressed as non-locality ; Λ -transformed states, cannot be localized on phase points or on wave functions .

2. KOLMOGOROV SYSTEMS AND INTERNAL TIME

Classical Kolmogorov systems are highly unstable dynamical systems characterized by two partitions of the Phase Spa-Space ,which evolve asymmetrically under the dynamical evolution group S_t . Points in the same cell of the (stable) K-Partition ξ have common future in the sense that their trajectories converge and become indistinguishable in the far future ,eventhough they have completely different past ,in the sense that the partition becomes coarser approaching the coarsest partition (one phase cell) in the far past . The behaviour of the unstable K-Partition ξ^u is the reverse : the trajectories of points in the same cell diverge from each other ,eventhough they are as close as we want in the past . The divergence is exponential for systems with finite Kolmogorov -Sinai Entropy . The description in terms of partitions is a generalization of the stable and unstable manifolds of Hyperbolic Systems and arose from the study of non-deterministic stochastic processes .For K-Systems the very concept of trajectory is lost much before Poincaré recurrence time,thus invalidating immediately the traditional argument against Irreversibility .

The behaviour of K-systems is illustrated in the Baker Cascade on the unit square Phase Space $[0,1) \times [0,1)$:

$$B(q,p) = \begin{cases} (2q, \frac{1}{2}p) & , q \text{ in } [0, \frac{1}{2}) \\ (2q-1, \frac{1}{2}p + \frac{1}{2}) & , q \text{ in } [\frac{1}{2}, 1) \end{cases}$$

As the name suggests, Baker's transformation resembles the kneading dough process :

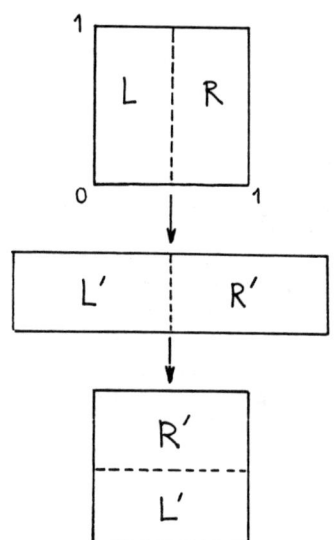

squeeze the 1×1 square to a 2×½ rectangle

cut the 2×½ rectangle in half put the right half piece above the left half to make a new 1×1 square

The cells of the K-partition are the vertical lines of lenth one , and contract in the future , while the unstable K-partition consists of the horizontal lines which expand in the future :

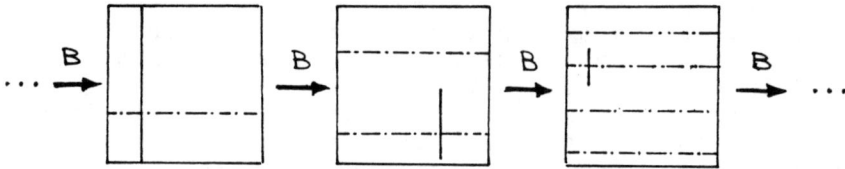

It is clear that it becomes increasingly more difficult to distinguish operationally points in the same K-cell as time goes on .Therefore the notion of a single trajectory ceases to be meaningful and only suitable averagings over the K-cells acquire meaning ,thus incorporating our limitations to predictability .

The averaging projections $\mathbb{P}_\tau$ onto the K-cells $S_\tau \tilde{\mathfrak{S}}$ are the spectral projections of the self-adjoint Internal Time operator

$$T = \int \tau \, d\mathbb{P}_\tau$$

The Time operator is canonically conjugate to the Liouville generator L of the dynamical evolution of the density functions :

$$\tfrac{1}{i}[L,T] = -I$$

$$U_t^+ T U_t = T + tI \quad , \quad U_t = e^{iLt}$$

The Expectation $\langle \rho, T\rho \rangle$ is the Age of the state ρ, keeping step with the external clock time t :

$$\langle U_t\rho, TU_t\rho \rangle = \langle \rho, U_t^+ T U_t \rho \rangle$$
$$= \langle \rho, T\rho \rangle + t$$

The Commutation relation of the Liouville operator L and the Internal Time T shows that there exist two complementary descriptions of Dynamics ; one in terms of frequency eigenstates :

$$\rho = \sum_\omega C_\omega \varphi_\omega \quad , \quad L\varphi_\omega = \omega \varphi_\omega$$

and another in terms of age eigenstates :

$$\rho = \sum_\tau C_\tau \chi_\tau \quad , \quad T\chi_\tau = \tau \chi_\tau$$

The limitations to predictability associated with the refining property of the K-Partition necessitate that the contribution of the higher age components acquires less operational significance. Therefore the operationally attainable evolution corresponds to appropriately weighted density functions

$$\tilde{\rho} = \Lambda \rho \quad , \quad \text{with} \quad \Lambda = \Lambda(T) = \int \Lambda(\tau) d\mathbb{P}_\tau$$

$\Lambda(\tau)$ being a monotonically decreasing, positive, logarithmically concave function with $\Lambda(\tau) \underset{\tau \to +\infty}{\longrightarrow} 0$ and $\Lambda(\tau) \underset{\tau \to -\infty}{\longrightarrow} 1$.

The non-Unitary transformation Λ provides a weight over the precision characterising the progressively refined family of Conditional Expectation Projections $\mathbb{P}_\tau$, and converts the Unitary group U_t to a Markov semigroup W_t approaching equilibrium in the far future. These results are discussed in (6,7,8)

Quantum K-systems are defined in an analogous way. The classical K-partition of the phase cells corresponds to a partition of the pure states (wavefunctions) which becomes refined in the future. This concept has been applied (9) in the Quantum measurement problem.

Many physically interesting systems are known to be K-systems, such as the hard spheres in a box, the convex billiard, the Lorentz gas, geodesic flows in spaces with negative curvature, the infinite ideal gas and hard rods, shifts in certain Ising models, the infinite harmonic lattice, the Mixmaster cosmological model, the classical relativistic fields.

3. THE K-PROPERTY OF CLASSICAL RELATIVISTIC FIELDS

Every classical Relativistic Field Equation, namely the scalar wave, the Electromagnetic Radiation and the linear Gravitational Field on flat spacetime, defines a Gaussian Flow on the space of initial data or equivalently on the space of solutions, which has the Kolmogorov property (10,11,12). The K-property arises from the relativistic transformation properties of the fields. An Internal Time operator can be constructed on the space of solutions or on the space of initial data as a function of the generators of the Relativistic Symmetry Group. The spectral projections of the Time operator allow us to test whether two initial data ψ_1, ψ_2 belong to the same cell of the K-partition.

For the simple case of the Wave Equation:

$$\partial_t^2 \psi = \Delta \psi$$

the Internal Time on the Hilbert space of square integrable solutions of the Wave Equation has the form:

$$T_R = \tfrac{1}{2}(DP_o^{-1}+P_o^{-1}D) = DP_o^{-1}+\tfrac{1}{2}P_o^{-1}$$

with $D=-(x^\nu \partial_\nu +I)$ the generator of Dilatations: $U_\lambda \psi(x) = e^{-\lambda}\psi(e^{-\lambda}x)$ and $P_o = \partial_o$ the generator of the Time Evolution:

$U_t \psi(x^o;\vec{x}) = \psi(x^o+ct;\vec{x})$.

The explicit form of the relativistic Internal Time is:

$$T_R = -x^o - x^a \frac{\partial}{\partial x^a}\frac{\partial}{\partial x^o}\Delta^{-1} - \tfrac{1}{2}\frac{\partial}{\partial x^o}\Delta^{-1}$$

The presence of the inverse Laplace integral operator Δ^{-1} in the expression for T_R is expected since the internal time cannot generate field transformations implementable by point transformations in Minkowski space. The Internal Time T_R is related by similarity to another Internal Time operator T_E

$$T_R = |P_o|^{\tfrac{1}{2}} T_E |P_o|^{-\tfrac{1}{2}}$$

$$T_E = D P_o^{-1} + P_o^{-1}$$

The operator T_E is self adjoint on the space of initial data with finite energy. The spectral projection $P_o^{(E)}$ of T_E

corresponds to the Lax-Phillips (13) incoming waves which vanish inside the past light cone. However the spectral projections of T_R do not admit such a simple physical interpretation.

The K-property of the relativistic fields implies that their evolution is asymptotically unstable, a condition which is not in contradiction with the stability of these fields, expressed as continuous dependence of the solutions upon the initial data, for finite times. The relativistic character of the fields gives the possibility to study the transformation properties of the internal time. The relativistic fields provide concrete representations of the Algebra generated by the Relativistic Poincaré Algebra and the Internal Time.

4. THE RELATIVISTIC INTERNAL TIME ALGEBRA

The Lie Algebra of Relativistic Systems with Internal Time is found to be an infinite algebra $(11,12)_a$. The Internal Time commutes with the generator of rotations J^a but does not commute with the generators P^a and N^a of space translations and Lorentz boosts, thus giving rise to the velocity V^a and to the internal position Q^a:

$$[P_o, T] = -I$$
$$[P^a, T] = -P^a P_o^{-1} = -V^a$$
$$[J^a, T] = 0$$
$$[N^a, T] = TV^a = Q^a$$

The velocity and position generate the infinite monomials $V^a V^b, V^a V^b V^c, \ldots, TV^a V^b, TV^a V^b V^c, \ldots$:

$$[N^a, V^b] = V^a V^b - \delta_{ab} I$$
$$[N^a, Q^b] = 2TV^a V^b - \delta_{ab} T$$
$$\ldots$$

The Relativistic Internal Time Algebra allows us to find the transformation properties of the internal time T and the internal position Q^a under Lorentz Boosts:

$$(T)_B = \frac{T\sqrt{1-\left(\frac{v^1}{c}\right)^2}}{1 - \frac{v^1 V^1}{c^2}}$$

$$(Q^1)_B = \frac{Q^1 - TV^1}{\sqrt{1-\left(\frac{v^1}{c}\right)^2}} \cdot \frac{1}{\left(1 - \frac{v^1 V^1}{c^2}\right)^2}$$

These formulas differ from Einstein formulas and show clearly that the internal spacetime T, Q^a differs from the Minkowski spacetime of localized observations .The presence of a field endows space with new properties such as the age and the field is of course, not necessarily homogeneous. The field is like a landscape and allows a characterization of spacelike points in terms of the field's age, a subject to be discussed in a forthcoming publication .

5. RESONANCE INSTABILITY AND THE ASYMPTOTIC COLLISION OPERATOR

In classical Dynamical Systems the presence of resonances destroys the analytical invariants as shown by Poincaré (14) This is the so called Poincaré Catastrophe (2) .In the case of systems with two degrees of freedom for example, Resonance corresponds to points where the ratio of the frequencies is rational ,while the non-Resonant stable points have an irrational ratio .Obviously Poincaré's theorem is not directly applicable to quantum systems, since they have discrete spectrum .However Poincaré's theorem acquires meaning in the limit of large systems with continuous spectrum ,both classical and quantum ,where two possibilities arise (15) : either we have ,in the limit of large systems ,the appearance of continuous sets of resonances or in this limit we may neglect the role of resonances .The existence of continuous sets of resonances is equivalent to the non vanishing of the asymptotic Collision operator $\Psi(+i0)$,which emerged in the study of the asymptotic behaviour of the Liouville-von Neumann Equation of dynamics using Laplace transform methods (16,17).

The Collision operator of a system with energy $H=H_o+\lambda H_I$ and Liouville operator $L=L_o+\lambda L_I$, λ being the coupling parameter ,is defined with respect to the projection P on the relevant part of dynamics (vacuum of correlations) and its complement $Q=I-P$ as :

$$\Psi(z) = PLQ \frac{1}{QLQ - z} QLP$$

where z is a complex number associated to the Laplace transform .The operators QLP ,QLQ and PLQ create ,propagate and annihilate the correlations correspondingly .

The asymptotic Collision operator $\Psi(+i0) = \lim_{z \to +i0} \Psi(z)$
describes the long time behaviour of the dynamical equations and contains integrations over resonances .All known kinetic theories are characterized by a non-vanishing asymptotic Collision operator. In large quantum systems the asymptotic Collision operator has a well defined meaning in terms of continuous sets of resonances .Quantum systems with non vanishing asymptotic Collision operator are characterized by excited states and finite lifetimes .

As could be expected, there is a close connection between Poincaré's theorem and the existence of canonical transformations leading to a cyclic Hamiltonean for classical systems or of unitary transformations diagonalizing the Hamiltonean operator of quantum systems. For systems with non vanishing asymptotic Collision operator canonical-unitary transformations do not exist, because they diverge (15). For such systems a non Unitary Λ transformation may be constructed as an appropriate diagonalizing transformation, using perturbation techniques.

The construction of Λ involves an appropriate analytic continuation which is effected by a rule in which the direction of time plays the essential role.

The process collision⟶correlation in which plane waves are transformed into spherical waves, is associated with the analytic continuation $z \to -i\varepsilon$

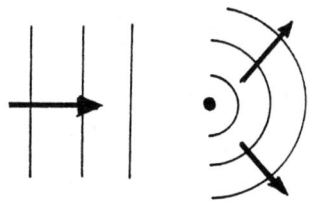

The reverse process correlation⟶collision is associated with the analytic continuation $z \to +i\varepsilon$

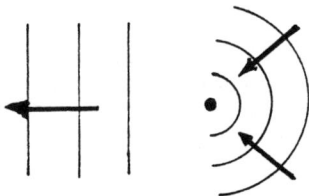

The physical time order is collisions first, correlations then, and leads to equilibrium in the future. The reverse order would correspond to a transformation of spherical waves into plane waves (18) and gives rise to another transformation Λ' leading to equilibrium in the past.

The non local superoperator Λ thus constructed, cures the divergences related to the resonances and leads to an extension of the quantum formalism, as well as to a new type of non-distributive algebra of observables (19,20). The non-distributivity is expressed by the condition:

$$(\widetilde{H})^n \neq \widetilde{H^n}$$

$\widetilde{H}$ being the Λ-transformed Hamiltonean.

6. PREDICTIONS FROM THE FRIEDRICHS MODEL

Let us indicate some of the predictions formulated into the frame of the Friedrichs model of unstable quantum systems (20).
1) The non-distributivity of the algebra of observables leads to the Energy - Lifetime uncertainty relation :
$$E = \frac{\hbar}{2\tau}$$
2) The existence of off-diagonal elements for the energy superoperator and the lack of distributivity lead to a new anomalous Lamb shift which is lifetime dependent .This effect is the quantum analogue of the classical frequency shift due to radiation damping .For long lifetimes the usual expression is recovered .

3) The most interesting new effect is the existence of a new cosmological Lamb shift which comes from the time asymmetry of Λ . Indeed it has been shown (20) that :
$$\Lambda[H] \neq \Lambda'[H]$$
This cosmological Lamb shift involves the time direction of the approach to equilibrium ,which is as far as we know, a global property of our universe .The origin of these effects is due to the radiation damping ,i.e. the cooling of the atom by spontaneous emission ,which is neglected in the semi-classical derivations of the Lamb shift .

7. CONCLUDING REMARKS

For Dynamical Systems with the Kolmogorov property and Internal Time or with non-vanishing asymptotic Collision operator, appropriate non- Unitary Λ-transformations lead to the physical irreversible evolution providing the proper realization of the unstable dynamics .The probabilistic language arises as the appropriate one for unstable dynamical systems and arises intrinsically ,not as a result of assumptions extrinsic to the system .Both K-systems and systems with non vanishing asymptotic Collision operator are intrinsically irreversible and non deterministic .

The significance of classical non deterministic systems is beautifully sketched by Lighthill (21) ,president of the International Union of Applied Mechanics ,who also expressed "the collective apology of the practitioners of mechanics for having misled the general educated public by spreading ideas about determinism of systems satisfying Newton's Laws of motion that ,after 1960 ,were to be proven incorrect " .

The intrinsic character of indeterminism is expressed in

the "metaphysical dream" of K.Popper (22) :
"it is likely that the world would be just as indeterministic as it is, even if there were no 'observing subject' to experiment with it and to interfere with it" .

ACKNOWLEDGEMENTS

It is a pleasure to thank the financial support of the Instituts Internationaux de Physique et de Chimie fondés par E. Solvay .We are grateful to Profs B.Misra and T.Petrosky for many stimulating discussions .

REFERENCES

1. Schrodinger E. (1944) 'The statistical law in nature' Nature 153 , 704

2. Prigogine I. (1980) From being to becoming , Freeman.

3. Misra B. (1978) 'Non equilibrium entropy, Liapunov variables and ergodic properties of classical systems' P.N.A.S. U.S.A. 75 , 1627

4. Misra B. ,Prigogine I. ,Courbage M. (1979) 'Liapounov variable, entropy and measurement in quantum mechanics' P.N.A.S. U.S.A. 76 , 4768

5. Misra B. ,Prigogine I. (1983) 'Irreversibility and non-locality' Lett. Math. Phys. 7 421

6. Misra B. ,Prigogine I. ,Courbage M. (1979) 'From deterministic dynamics to probabilistic descriptions' Physica 98A ,1

7. Misra B. ,Prigogine I. (1982) 'Time, Probability and Dynamics' in Long Time Predictions in Dynamic Systems ed. by Horton E.W. et als , Wiley N.Y.

8. Goodrich R.K. ,Gustafson K. ,Misra B. (1986) 'On K -Flows and Irreversibility' J. Stat. Phys. 43 317

9. Lockhart C. ,Misra B. (1986) 'Irreversibility and Measurement in Quantum Mechanics' Physica 136A 47

10. Misra B. (1987) 'Fields as Kolmogorov flows' J. Stat. Phys. 48 1295

11. Antoniou I.E. (1988) Internal Time and Irreversibility of Relativistic Dynamical Systems , Thesis Free University of Brussels .

12. Antoniou I.E. ,Misra B. to appear
13. Lax P. Phillips R. (1967) Scattering Theory Academic Press
14. Poincaré H. (1892) Les Methods Nouvelles de la Mecanique Celeste Dover Reprint 1957
15. Petrosky T. Prigogine I. (1988) 'Poincaré's theorem and Unitary transformations for classical and quantum theory' Physica 147A 459
16. Prigogine I. (1962) Non Equilibrium Statistical Mechanics Wiley
17. Prigogine I. ,George C. ,Henin F. ,Rosenfield L. (1973) 'A unified formulation of Dynamics and Thermodynamics' Chem. Scr. 4 5
18. George C. ,Mayné F. ,Prigogine I. (1985) 'Scattering theory in superspace' ,in Adv.Chem.Phys. 61 ,Wiley
19. Prigogine I. ,Petrosky T. (1987) 'Intrinsic Irreversibility in quantum theory' Physica 147A 33
20. Prigogine I. ,Petrosky T. (1988) 'An alternative to quantum theory' Physica 147A 461
21. Lighthill J. (1986) 'The recently recognized failure of predictability in Newtonian Dynamics' Proc. R. Soc. London A407 35
22. Popper K. (1982) Quantum theory and the schism in Physics from the Postscript to the Logic of the scientific discovery Rowman and Littlefield ,Otowa

A NEW CHALLENGE FOR STATISTICAL MECHANICS

W.T. Grandy, Jr.
Department of Physics and Astronomy
University of Wyoming
Laramie, Wyoming 82071 USA

ABSTRACT. Traditionally we understand the need for probability in statistical mechanics to arise primarily from insufficient knowledge regarding initial conditions. But nonlinearities in the equations of motion can introduce irregular behavior that may lead to new phenomena, so that it is necessary to investigate how the role of probability in statistical mechanics may be affected by this observation. We conclude that the important effects are related to macroscopic equations *derived* from statistical mechanics.

For over a century the role of statistical mechanics has been thought to lie entirely with the explication of the behavior of systems consisting of very large numbers of microscopic particles. In a sense, the job has been to amplify these complicated microscopic motions into a macroscopic form suitable for observation, and the enormous number of particles involved has necessitated a statistical description. The methods developed for this purpose, which were summarized so beautifully by Schrödinger (1960), have been extraordinarily successful, to the point that they have been adopted in the analyses of problems which are basically not at all statistical in nature. Examples are the traveling salesman problem (e.g., Baskaran, *et al*, 1986), and the analysis of Feynman diagrams contributing to very-high-order perturbation terms in quantum field theory (Bender and Wu, 1976). The character of these problems is such that an 'average' behavior turns out to make sense, thereby alleviating the difficulty of dealing with large numbers of alternatives.

In the recent past it has become clear that nonlinearities in equations of motion —which seem to be the rule both microscopically and macroscopically—introduce the possibility of new phenomena in the form of chaotic or irregular behavior. As a consequence, it is important to re-examine the role of probability itself in the many-body problem, and in what follows we discuss the beginnings of a program to carry out such a re-examination.

1. Conventional Statistical Mechanics

Aside from the fact that one must deal with an enormous number of particles, the fundamental problem of statistical mechanics is one of missing information: there is no way to know the microscopic initial conditions needed to solve for the particle motions. Generally, we possess only a small amount of macroscopic information about the system and are content to ask only macroscopic questions in return, a strategy dictated by our ignorance of those microscopic initial conditions. This ignorance forces the use of a probability distribution over the possible microstates of the system consistent with that macroscopic information, and the only requirement the particle dynamics places upon us is that of being able to specify those possible alternatives, or states.

But how is one to construct such a distribution? When the known information can be stated in the form of an expectation value, the most effective means of construction was first stated by Gibbs: maximize the entropy of the probability distribution subject to the constraints provided by the given information. That is, for a number of mutually exclusive and exhaustive discrete alternatives, and a given datum $\langle f \rangle$, the entropy

$$S = -\kappa \sum_i P_i \ln P_i, \quad \kappa > 0, \tag{1}$$

is maximized over $\{P_i\}$ subject to $\sum_i P_i = 1$ and

$$\langle f \rangle \equiv \sum_i P_i f_i. \tag{2}$$

The immediate solution to this variational problem is

$$P_i = \frac{1}{Z} e^{-\lambda f_i}, \tag{3a}$$

where the *partition function* is defined as

$$Z \equiv \sum_i e^{-\lambda f_i}, \tag{3b}$$

and the Lagrange multiplier λ is determined by substitution into the constraint equation (2):

$$\langle f \rangle = -\frac{\partial}{\partial \lambda} \ln Z(\lambda). \tag{4}$$

That is, λ is determined such that the known *number* $\langle f \rangle$ is reproduced, and that is all that is logically required of $\{P_i\}$ if that is all we know about the problem. The expectation value of any other function $g(x)$ is then evaluated as in Eq.(2), and the extension to numerous pieces of data is straightforward.

This maximum-entropy procedure has been outlined in great detail elsewhere (e.g., Grandy, 1987), which for this kind of problem is simply a means for selecting an optimum distribution from a large set that might also reproduce the given data. Viewed in this way we are faced with an inverse problem, and the distribution (3) just corresponds to the solution that can be obtained in the greatest number of ways, a maximization process often expressed in terms of a multinomial coefficient. Indeed, thinking of the problem in this way leads us to Boltzmann's method of most probable values, an approach championed so articulately by Schrödinger (1960). In the physical context of a gas in thermal equilibrium, for example, specification of the spectrum of system energy levels $\{E_i\}$ leads to

$$P_i = \frac{1}{Z} e^{-\beta E_i}, \quad Z = \sum_i e^{-\beta E_i}, \tag{5}$$

and β is identified as the reciprocal of the absolute temperature T in units of Boltzmann's constant. Other thermodynamic functions are then evaluated as expectation values.

The essential point to be emphasized here is that one has only macroscopic information about the system, necessitating a probabilistic description, and thus we only make predictions of macroscopic quantities associated with the system. There are essentially no conceptual changes when considering a system in a nonequilibrium state, for such a state is still identified by macroscopic data. By definition, however, a number of parameters are changing in time, and maybe in space, so that this information is generally specified over some space-time region. The probability distribution then constructed merely provides us with an optimal estimate of the initial nonequilibrium state of the system, and the subsequent behavior is to be deduced fundamentally from the known microscopic equations of motion. The analogue to Eq.(5) can be expressed most simply in terms of the statistical operator, or density matrix,

$$\hat{\rho} = \frac{1}{Z[\lambda]} \exp\left\{-\int_R d^3x\, dt\, \lambda(\mathbf{x},t) \hat{f}(\mathbf{x},t)\right\}, \tag{6}$$

where $\hat{f}$ is a Heisenberg operator whose expectation value over the space-time region $R(\mathbf{x},t)$ provided the initial data, and we now must consider the *partition functional*:

$$Z[\lambda(\mathbf{x},t)] = \text{Tr} \exp\left\{-\int_R d^3x\, dt\, \lambda(\mathbf{x},t) \hat{f}(\mathbf{x},t)\right\}. \tag{7}$$

As expected, the Lagrange-multiplier function $\lambda(\mathbf{x},t)$ is obtained from the functional-differential equation

$$\langle \hat{f} \rangle = -\frac{\delta}{\delta \lambda(\mathbf{x},t)} \ln Z[\lambda(\mathbf{x},t)], \quad (\mathbf{x},t) \in R, \tag{8}$$

and again the predicted value of another Heisenberg operator $\hat{J}$ at *any* space-time point is just

$$\langle \hat{J}(\mathbf{x},t) \rangle = \text{Tr}[\hat{\rho}\, \hat{J}(\mathbf{x},t)]. \tag{9}$$

Details of this nonequilibrium theory have been discussed at length elsewhere (Grandy, 1988), where applications to hydrodynamics are also carried out. The underlying microscopic description of a hydrodynamic state is given in terms of five conserved density operators $\hat{e}_i$, which are related to a corresponding set of current operators $\hat{\jmath}_i$ through equations of continuity: $\partial_t \hat{e}_i + \nabla \cdot \hat{\jmath}_i = 0$. These, in turn, generate a similar set of *macroscopic* conservation laws in terms of nonequilibrium expectation values:

$$\partial_t \langle \hat{e}_i(\mathbf{x},t) \rangle + \nabla \cdot \langle \hat{\jmath}_i(\mathbf{x},t) \rangle = 0. \tag{10}$$

Initial information defining an initial nonequilibrium state is given in terms of expectation values such as these, and the associated Lagrange-multiplier functions are identified in terms of the standard functions of fluid flow. The usual equations of hydrodynamics are thus seen to be equations of motion for the Lagrange multipliers. For example, the multiplier associated with the system stress tensor is the fluid velocity $\mathbf{v}(\mathbf{x},t)$, for which we are able to derive an equation of motion (Grandy, 1988). When the fluid is incompressible, $\nabla \cdot \mathbf{v} = 0$, the result is just the well-known Navier-Stokes equations:

$$mn_0\big(\partial_t v_i + [(\mathbf{v}\cdot\nabla)\mathbf{v}]_i\big) = -\partial_j P \delta_{ij} + \partial_k p_{ik} + mn_0 F_i, \tag{11}$$

nonlinear in the fluid velocity. The notation is as follows: n_0 is the equilibrium number density of the system of particles of mass m, P is the pressure, F_i is a possible external force, and the shear tensor p_{ik} depends on the viscosity as well as the velocity. Statistical mechanics also provides explicit expressions for the transport coefficients, such as viscosity, in terms of the behavior of the microscopic constituents of the system.

At this point the standard many-body theory is conventionally thought to have completed its task, resulting in macroscopic deterministic equations such as those of Eq.(11). One would think that the role of probability in describing these systems was now over, but recent investigations into the dynamics of classical systems give one reason to pause.

2. Irregular Dynamical Motion

It has been known since the work of Poincaré that the number of integrable dynamical systems is severely limited, and that nonlinear equations of motion possess solutions exhibiting highly irregular behavior for given parameter values. Not until relatively recently, however, with the enormous advances in computational ability, has it been possible to study this behavior in any detail. Examples of deterministic dynamical systems in which irregular or 'chaotic' motions can occur are now commonplace. An early, and important, physical model was presented by Hénon and Heiles (1964) in connection with the distribution of stellar velocities within the galaxy. One is compelled to ask what bearing, if any, these developments might have on the statistical-mechanical description of many-body systems.

Let us first consider the implications if the microscopic equations of motion are nonlinear and can exhibit irregular solutions. As we have emphasized strongly above, by 'observable' we mean 'macroscopic' in this context, for it is ignorance of microscopic initial conditions and subsequent motions which dictates employment of a probability distribution over the possible microscopic states. The only role of the microscopic equations of motion, therefore, is to specify those alternative states. But even for relatively simple Hamiltonians this last task is not generally trivial, and requires thoughtful approximation. Nevertheless, this is precisely where the detailed physics enters the problem, and converts it from an exercise in applied statistics to one in physical analysis.

Note carefully that the formal theory only requires that it be *possible in principle* to enumerate the possible solutions to the microscopic equations—actual calculations, of course, require more than this. Some introspection suggests that nothing really changes here if the particle equations are highly nonlinear, for the procedures remain the same. Were we unable to enumerate the possible states of the system *in principle*, then we would no longer be discussing deterministic physics. It may not be possible to follow the microscopic motion, but that is not desired in any event. As long as the spectrum can be presented in principle, however difficult in practice, then it matters little how irregular that microscopic motion may be—that, after all, is just the point of statistical mechanics!

Entirely different conclusions can be reached, however, with respect to *macroscopic* motions of a system if the governing equations are nonlinear, for we usually wish to, and often can follow the trajectories in this case. Let us continue to use the example of conventional hydrodynamics for an incompressible fluid, so that the equations of motion are just the Navier-Stokes equations (11). As derived from the microscopic theory, these are actually a linear approximation in the Lagrange-multiplier function $\mathbf{v}(\mathbf{x},t)$, and higher-order equations can be obtained in a straightforward way from the perturbation expansion of the nonequilibrium expectation values. The non-linearity in Eq.(11) arises solely from the convection term in the derivative on the left-hand side. It is useful to rewrite these equations in terms of dimensionless variables by introducing a characteristic speed u of the fluid, and a characteristic length ℓ. Then, in vector notation, and without the external-force term, Eq.(11) is written

$$\partial_t \mathbf{v} + R\left(\mathbf{v}\cdot\nabla\right)\mathbf{v} = -\nabla p + \nabla^2 \mathbf{v}, \tag{12}$$

where $R \equiv u\ell/\nu$ is the *Reynolds number*, and ν is called the kinematic viscosity (the ratio of viscosity coefficient to the density). Clearly, the effect of nonlinearity is controlled completely by R, and when $R = 0$ these are known as the Stokes equations.

As R increases from zero the experimentalist observes in some systems a series of spectacular instabilities, and eventually complete turbulence in the fluid flow emerges for sufficiently large Reynolds numbers. Many of these stages in the progression to fully-developed turbulence for various systems have been captured photographically, and can be observed in the beautiful collection of Van Dyke (1982). One believes that this progression is described theoretically by Eq.(12) as R varies, but these equations are very difficult either to solve or to analyze in general. Nevertheless, in some applications it is possible to approximate them without destroying the essential nonlinearity.

The now-classic example of this latter procedure begins with the attempt by Saltzman (1962) to model the Rayleigh-Bénard instability in two dimensions by Fourier expansion in Eq.(12) and truncation into a set of ordinary differential equations. Shortly thereafter these equations were adopted by Lorenz (1963) as a model for the unpredictable behavior of the weather, and studied extensively by him—with remarkable results. These reduced equations for convection of the fluid are

$$\frac{dx}{dt} = \sigma(y - x),$$
$$\frac{dy}{dt} = rx - y - xz,$$
$$\frac{dz}{dt} = xy - bz, \tag{13}$$

where $x(t)$ is proportional to the amplitude of convective motion, $y(t)$ and $z(t)$ are proportional to two temperature modes, σ is called the *Prandtl number* (the ratio of kinematic viscosity to the thermal diffusivity), r is the *Rayleigh number* in units of its critical value (the convective analog of the Reynolds number), and b is a constant related to the wavenumber of the fundamental mode.

The set of equations (13) is completely deterministic, so that we can study (via computer) the trajectories generated from various initial conditions by fixing σ and b at the values adopted by Lorenz, say, and varying r. For $r < 1$ all trajectories are attracted to a stable solution at the origin of the variables in Eq.(13): $x = y = z = 0$. If r exceeds unity by much the model is no longer physically realistic, but nevertheless still worth studying. One finds two stable solutions for $1 < r < 13.9$, to which all stable trajectories are attracted, and in the region $13.9 < r < 24.1$ a complicated transition begins to take place. For $r > 24.1$ all trajectories are attracted toward a subspace in which they wander 'chaotically' forever. That is, the motion is highly irregular and essentially unpredictable. This subspace is called a *strange attractor*.

We hesitate to use the word 'chaos' here, because it tends to convey an impression of motion which is not deterministic. In reality, the motion here is no more chaotic than that of particles in an equilibrium gas—they all obey well-defined equations of motion. But 'chaos' now assumes a more technical meaning—namely, the result of an extraordinary sensitivity to initial conditions. In the chaotic regime it is virtually impossible to specify initial conditions precisely enough to be sure of the ensuing trajectory, and it is in this sense we employ the above phrase 'essentially unpredictable'.

The importance of these results lies with the possibility of being able to describe turbulence in some detail as a solution of the Navier-Stokes equations, say. As noted above, it has been known for many years that smooth laminar flow will become unstable and cascade into turbulence when R exceeds some critical value. This is a *macroscopic* phenomenon, and so would seem to be outside the purview of statistical mechanics. That is, the role of the latter should cease with the complete derivation of the macroscopic equations of motion and provision for calculation of the relevant parameters.

But completely developed turbulence is more than just 'chaotic' motion, and the phenomena uncovered by study of the Lorenz equations only provide us with a beginning. The onset of chaos may well signal the approach to a turbulent state, which is intrinsically nonequilibrium and collective in nature. As the parameters of the macroscopic equations continue to change, and full turbulence develops, one realizes that the number of *macroscopic* degrees of freedom has increased enormously. There are now a great many possible trajectories available to the system, but it is very difficult to know which is taken owing to the extreme sensitivity to initial conditions. Although the system state may well be described by only a few macroscopic variables—or 'supermacroscopic variables—just what that state may be is difficult to determine exactly. It is as if one did not really know the precise initial conditions.

At this point everything begins to sound familiar, as if statistical mechanics were emerging anew, but on a higher level than previously. In the problems of hydrodynamics it appears that the elementary volumes associated with the velocity field $\mathbf{v}(\mathbf{x},t)$ play the role of the basic units, or 'particles', with laminar flow being analogous to the equilibrium state. Some systems can then pass through a number of 'second-order phase transitions', corresponding to the hydrodynamic instabilities, and for a given range of parameter values the various states are both stable and reproducible.

In order to verify such notions, however, there are a number of questions which must be addressed and resolved to a degree that has not yet been achieved. These questions are suggested by the earlier discussion of the microscopic theory.

(a) What, exactly, are the experimentally reproducible phenomena of the macroscopic theory? Can a definite catalog be constructed? To put it another way, what are the 'macroscopic' quantities that we can either measure or observe, both on and off the strange attractor? One known class of such quantities, for example, consists of power spectra of the velocity field—but it is not clear that this class is sufficient to characterize the phenomena.

(b) More generally, what kind of information can be obtained and utilized in constructing a probability distribution over possible trajectories? For example, in attempting to predict the climate one is really forced to develop a 'statistics of the solutions' to the basic fluid equations of motion (Lorenz, 1964). Years ago Hopf (1952) attempted to construct a statistical theory of turbulence employing much the same point of view advocated here, and statistical theories of fully-developed turbulence were subsequently stimulated by this work. Unfortunately, these studies lacked the perspectives which were only to emerge from the more recent computer-assisted understanding of chaos.

(c) Finally, precisely how does the notion of 'insensitivity to initial conditions' arise in a specific real problem, and what role does it play? The suggestion is strong that, as R increases and the nonlinearities become increasingly more important, the observed instabilities signal a breakdown in the severely rigid uniformity of laminar flow. The onset of turbulence is characterized by 'insensitivities' which are analogous to the ignorance of microscopic initial conditions leading to the statistical description of many-body systems discussed earlier.

These comments serve to outline briefly a program in need of a great deal of development. As these points raised above are clarified, one hopes that there will emerge a 'canonical' form of probability distribution every bit as effective as that of Gibbs in describing ordinary thermodynamics.

REFERENCES

Baskaran, G., Y. Fu, and P.W. Anderson: 1986, 'On the Statistical Mechanics of the Traveling Salesman Problem', *J. Stat. Phys.* **45**, 1.

Bender, C.M., and T.T. Wu: 1976, 'Statistical Analysis of Feynman Diagrams', *Phys. Rev. Letters* **37**, 117.

Grandy, W.T., Jr.: 1987, *Foundations of Statistical Mechanics, Volume I: Equilibrium Theory*, Reidel, Dordrecht.

Grandy, W.T., Jr.: 1988, *Foundations of Statistical Mechanics, Volume II: Nonequilibrium Phenomena*, Reidel, Dordrecht.

Hénon, M., and C. Heiles: 1964, 'The Applicability of the Third Integral of the Motion: Some Numerical Experiments', *Astron. J.* **69**, 73.

Hopf, E.: 1952, 'Statistical Hydromechanics and Functional Calculus', *J. Rat. Mech. Anal.* **1**, 87.

Lorenz, E.N.: 1963, 'Deterministic Nonperiodic Flows', *J. Atmos. Sci.* **20**, 130.

Lorenz, E.N.: 1964, 'The Problem of Deducing the Climate from the Governing Equations', *Tellus* **16**, 1.
Saltzman, B.: 1962, 'Finite Amplitude Free Convection as an Initial Value Problem–I', *J. Atmos. Sci.* **19**, 329.
Schrödinger, E.: 1960, *Statistical Thermodynamics*, Cambridge Univ. Press, Cambridge.
Van Dyke, M.: 1982, *An Album of Fluid Motion*, The Parabolic Press, Stanford, CA.

QUANTUM STOCHASTIC CALCULUS AS A UNIFYING FORCE IN PHYSICS AND PROBABILITY

R. L. Hudson
University of Nottingham
Mathematics Department
University Park
Nottingham NG7 2RD
England

ABSTRACT. A survey is made of the theory of quantum stochastic calculus in Fock space developed by the author and K. R. Parthasarathy, together with some of its applications. Comparison with the classical Ito stochastic calculus of Brownian motion is emphasised, as is the power of the quantum theory to unify aspects of the classical theory of stochastic processes, Boson and Fermion second quantisation and dilations of quantum dynamical semigroups.

1. INTRODUCTION

As an outside who is neither a philosopher nor a physicist, I am sympathetic towards the aspirations for a realist interpretation of quantum theory expressed by many contributors to these proceedings. But in so far as I have understood the proposals to achieve these aspirations, I must confess to finding them Ptolemaic in character, reminiscent of the attempts to retain the geocentric theory of the universe by the introduction of epicycles. Could it be that quantum theory means what it appears to say, and that it does require a modification of our understanding of what is real? I invite my audience to suspend their misgivings, and to lie back and enjoy some of the insights that the mathematical formalism has to offer.

I shall describe the theory of quantum stochastic calculus which I have developed in collaboration with K. R. Parthasarathy [6,7,8]. It may be thought of as a theory of quantum noise, indeed as a mathematical rigorisation and extension of the quantum noise of Lax and Senitzky [9,12]. I emphasise that it is a *quantum* or noncommutative, operator theoretic noise and will thus be found of no use in the construction of random ethers, stochastic mechanics or other edifices of the Ptolemaic gallimaufry. But it offers powerful unifying perspectives, both in the classical mathematical theory of stochastic processes and in the physics of dissipative systems. These I will describe.

2. CLASSICAL AND QUANTUM PROBABILITY

There is no antithesis between classical and quantum probability; the former is merely an incomplete version of the latter. To see this at its simplest we consider a finite classical probability space, namely a sample space

$$\Omega = \{\omega_1, \omega_2, \ldots, \omega_n\}$$

and a probability measure $\mathbb{P}$ on Ω characterised by the n numbers $p_j = \mathbb{P}(\{\omega_j\})$, $j = 1,\ldots,n$, satisfying

$$p_j \geq 0, \quad \sum_{j=1}^{n} p_j = 1. \tag{2.1}$$

A random variable ξ is represented classically by a real valued function on Ω, taking values $\xi_j = \xi(\omega_j), j = 1, \ldots, n$. The expected value of ξ is

$$\mathbb{E}[\xi] = \sum_{j=1}^{n} p_j \xi_j. \tag{2.2}$$

Now let us introduce the Hilbert space $\mathbb{C}^n$ of complex n-tuples with the usual inner product $\langle , \rangle$. Then $\psi = (\sqrt{p_1}, \ldots, \sqrt{p_n})$ is a unit vector in view of (2.1). The expected value (2.2) can be expressed in the quantum mechanical form

$$\mathbb{E}[\xi] = \langle \psi, T\psi \rangle,$$

where T is the self-adjoint operator which has the diagonal matrix

$$\begin{bmatrix} \xi_1 & 0 & \ldots & 0 \\ 0 & \xi_2 & \ldots & 0 \\ & & & \\ 0 & 0 & \ldots & \xi_n \end{bmatrix}$$

with respect to the canonical basis of $\mathbb{C}^n$. Quantum probability removes the restriction to diagonal matrices and allows arbitrary self-adjoint operators to represent random variables.

3. BROWNIAN MOTION FROM EINSTEIN AND SMOLUCHOWSKY TO THE WIENER–SEGAL ISOMORPHISM

In the Einstein [3]–Smoluchowsky [13] theory of Brownian motion, each Cartesian coordinate $X(t)$ of the Brownian particle at time t is a Gaussian random variable of mean zero and (with suitable normalisations of physical constants) variance t. The stochastic process $X = (X(t): t \geq 0)$ has the further (Markov) property that it begins anew at every time independently of the past. Thus, if $0 \leq s \leq t$, $X(t) - X(s)$ is normally distributed of mean 0 and variance $t-s$.

Wiener [14] demonstrated the mathematical existence of such a stochastic process. Specifically, he proved the existence of a probability measure $\mathbb{P}$, *Wiener measure*, on the space Ω comprising all real valued, continuous functions ω on $\mathbb{R}_+ = [0, \infty)$, such that the evaluations

$$X(t)(\omega) = \omega(t)$$

realise X. Among the more notorious properties of Brownian motion is its non-differentiability; specifically the probability, according to Wiener measure $\mathbb{P}$, that the Brownian path ω have a derivative, or more generally be of bounded variation, on any finite interval, is zero.

Wiener also discovered a property which, after its refinement by the latter, is known as the *Wiener–Segal isomorphism*. The Hilbert space $L^2(\Omega, \mathbb{P})$ of square-integrable complex-valued random variables is isomorphic in a natural way to the *Boson Fock space* $\Gamma(L^2(\mathbb{R}_+))$ over the *one-particle space* $L^2(\mathbb{R}_+)$ comprising (Lebesgue) square-integrable functions on the half-line $\mathbb{R}_+$. The Fock space may be defined as the infinite direct sum

$$\Gamma(L^2(\mathbb{R}_+)) = \mathbb{C} \oplus L^2(\mathbb{R}_+) \oplus (L^2(\mathbb{R}_+) \otimes L^2(\mathbb{R}_+))_{\text{sym}} \oplus (L^2(\mathbb{R}_+) \otimes L^2(\mathbb{R}_+) \otimes L^2(\mathbb{R}_+))_{\text{sym}} \oplus \ldots.$$

Here sym indicates that only the symmetric part of the full Hilbert space tensor product is to be taken. In the Wiener–Segal isomorphism

$$L^2(\Omega, \mathbb{P}) \sim \Gamma(L^2(\mathbb{R}_+)),$$

the identity function $1 \in L^2(\Omega, \mathbb{P})$ corresponds to the Fock space *vacuum vector*

$\psi_0 = (1, 0, 0, \ldots) \in \Gamma(L^2(\mathbb{R}_+))$. $X(t) \in L^2(\Omega, \mathbb{P})$ corresponds to the vector $(A^\dagger(t) + A(t))\psi_0$. Here $A^\dagger(t)$ and $A(t)$ are the *creation* and *annihilation operators* corresponding to the function $\chi_{[0,t]} \in L^2(\mathbb{R}_+)$, which is 1 on $[0, t]$ and 0 on (t, ∞). Thus $A^\dagger(t)$ acts on $\Phi = (\phi_0, \phi_1, \ldots) \in \Gamma(L^2(\mathbb{R}_+))$ as

$$(A^\dagger(t)\Phi)_n = \sqrt{n} P_{\text{sym}}^{(n)}(\chi_{[0,t]} \otimes \phi_{n-1}),$$

where $P_{\text{sym}}^{(n)}$ is the symmetrising projection, and $A(t)$ is adjoint to $A^\dagger(t)$. More generally, for arbitrary $n = 0, 1, 2, \ldots$ and $t_1, t_2, \ldots, t_n \geq 0$, $X(t_n) \ldots X(t_1) \in L^2(\Omega, \mathbb{P})$ corresponds to $(A^\dagger(t_n) + A(t_n)) \ldots (A^\dagger(t_1) + A(t_1))\psi_0 \in \Gamma(L^2(\mathbb{R}_+))$. Since the former vectors are total in $L^2(\Omega, \mathbb{P})$ and the latter are total in $\Gamma(L^2(\mathbb{R}_+))$, this correspondence defines the Wiener–Segal isomorphism. From it we see also that the operator in $L^2(\Omega, \mathbb{P})$ of multiplication by $X(t)$ corresponds to the operator $A^\dagger(t) + A(t)$.

4. THREE BASIC PROCESSES OF QUANTUM STOCHASTIC CALCULUS

In addition to the creation and annihilation operators $A^\dagger(t)$ and $A(t)$, we also consider the operator $\Lambda(t)$, which may be defined as the sum $\sum_j A^\dagger(f_j) A(f_j)$, where $\{f_j\}$ is an orthonormal basis of the subspace $L^2[0, t]$ of $L^2(\mathbb{R}_+)$, and $A^\dagger(f_j)$ and $A(f_j)$ are the creation and annihilation operators corresponding to f_j. Alternatively, $\Lambda(t)$ is the differential second quantisation of the projector onto $L^2[0, t]$.

Formally, the operators $A^\dagger(t)$, $\Lambda(t)$ and $A(t)$ may be expressed in terms of the improper point creation and annihilation operators $a^\dagger(t)$ and $a(t)$, which satisfy the formal commutation relation

$$[a(t), a^\dagger(s)] = \delta(t-s),$$

as

$$A^\dagger(t) = \int_0^t a^\dagger(s)\, ds \tag{4.1}$$

$$\Lambda(t) = \int_0^t a^\dagger(s) a(s)\, ds \tag{4.2}$$

$$A(t) = \int_0^t a(s)\, ds \tag{4.3}$$

respectively. The families of operators $A^\dagger = (A^\dagger(t), t \in \mathbb{R}_+)$, $\Lambda = (\Lambda(t), t \in \mathbb{R}_+)$ and $A = (A(t), t \in \mathbb{R}_+)$ are the basic processes of quantum stochastic calculus. They are called, respectively, the *creation*, *gauge* (or *conservation*, or *number*) and *annihilation processes*. K.R. Parthasarathy likens them to the three Hindu deities Brahma, the creator, Vishnu, the preserver and Shiva the destroyer.

As we saw in §3, the family

$$Q(t) = A(t) + A^\dagger(t), \quad t \geq 0$$

is essentially a Brownian motion, transforming into multiplication by Brownian motion under the Wiener–Segal isomorphism. In particular the $Q(t)$, $t \geq 0$ are mutually commuting essentially self-adjoint operators, and for $t \geq s$ we have, expressing in Fock space language the defining probabilistic properties of a Brownian motion

$$\langle \psi_0, T\exp(iu(Q(t)-Q(s)))\psi_0 \rangle = \langle \psi_0, T\psi_0 \rangle \exp(-\tfrac{1}{2}(t-s)u^2), \tag{4.4}$$

where $u \in \mathbb{R}$ and T is any operator belonging to the algebra of operators generated by the $Q(v)$ with $0 \leq v \leq s$. It is not difficult to show that *exactly the same is true* of the family of operators

$$P(t) = i(A^\dagger(t) - A(t)), \quad t \geq 0.$$

These do not commute with the $Q(t)$ and thus cannot be diagonalised simultaneously with them. Thus, in addition to the Wiener–Segal isomorphism $\sim_Q$, there is a second Hilbert space isomorphism $\sim_P$ from $L^2(\Omega, \mathbb{P})$ to $\Gamma(L^2(\mathbb{R}_+))$, under which each $P(t)$ corresponds to multiplication by $X(t)$. ($\sim_Q$ and $\sim_P$ are related by

$$\sim_P = \sim_Q \cdot \mathscr{F},$$

where $\mathscr{F}$ is the Fourier–Wiener transformation from $L^2(\Omega, \mathbb{P})$ to itself.) More generally, exactly the same is true of the family of operators

$$e^{i\theta} A^\dagger(t) + e^{-i\theta} A(t), \quad t \geq 0$$

for each fixed $\theta \in [0, 2\pi)$. Thus Fock space contains not one but many mutually incompatible Brownian motions no two of which can be measured simultaneously according to orthodox quantum theory.

More remarkably, the same Fock space $\Gamma(L^2(\mathbb{R}_+))$ plays host not only to Brownian motions but also to realisations of the other fundamental class of classical stochastic processes, namely Poisson processes. For each fixed $l \in \mathbb{C}$, the operators

$$\pi_l(t) = \Lambda(t) + l A^\dagger(t) + l^\dagger A(t) + |l|^2 t, \quad t \geq 0, \tag{4.5}$$

are a commuting, essentially self-adjoint family whose joint probability distributions in the vacuum state are those of a Poisson process of intensity $|l|^2$. That is, for arbitrary $t_1, \ldots, t_n \in \mathbb{R}_+$ and $u_1, \ldots, u_n \in \mathbb{R}$, we have

$$\left\langle \psi_0, \prod_{j=1}^n \exp(i u_j \pi_l(t_j)) \psi_0 \right\rangle = \prod_{j=1}^n \{\exp(e^{i u_j} - 1)|l|^2 (t_j - t_{j-1})\}.$$

Moreover repeated action of the $\pi_l(t), t \in \mathbb{R}_+$ on the vacuum vector yields a total family of vectors in $\Gamma(L^2(\mathbb{R}_+))$. Thus, in analogy with the Wiener–Segal isomorphism, we can construct a Hilbert space isomorphism from $\Gamma(L^2(\mathbb{R}_+))$ to the Hilbert space of square integrable complex-valued random variables for a canonical realisation of the Poisson process. Fock space is thus the natural home of the fundamental classical stochastic processes.

Rewriting (4.5) in the case when $l \in \mathbb{R}_+$ as

$$\frac{\pi_l(t) - l^2 t}{l} = l^{-1} \Lambda(t) + Q(t) \tag{4.6}$$

we obtain, in an amusing way, the well known limiting approximation of the compensated Poisson process by Brownian motion, in the limit as $l \to \infty$. The joke is that, in (4.6), no two of the Poisson process, the Brownian motion and the error process $l^{-1} \Lambda(t)$ commute with each other. Perhaps this is the pons asinorum of quantum probability theory.

5. ITO STOCHASTIC CALCULUS

The Einstein–Smoluchowsky–Wiener process X is actually rather a poor model of the physical phenomenon of Brownian motion, despite the fact that it has appropriated the name "Brownian motion", at least among mathematicians. Construction of better models, such as the Ornstein–Uhlenbeck process, as well as the constructive theory of Markov diffusion processes requires the formulation of integrals in which the process X occurs as the integrator. An instructive example is

$$\int_0^t X \, dX. \tag{5.1}$$

In *defining* such an integral it is not permissible to replace dX by $\frac{dX}{dt} dt$ since, as we have seen, the derivative $\frac{dX}{dt}$ fails to exist with probability 1. Nor, because X is also of unbounded variation, can the integral be defined as a Riemann–Stieltjes one.

In 1941 Ito devised a stochastic integral of sufficient generality to include the example (5.1). The method is to start by defining the integral of a step function, say

$$F(t) = \begin{cases} 0 & 0 \leq t < t_0 \\ F(t_j) & t_j \leq t < t_{j+1}, \quad j = 0, 1, \ldots, n-1 \\ 0 & t_n \leq t \end{cases}$$

for which a moments thought will reveal that there can be no argument - the integral *must* be

$$I_t[F] = \int_0^t F \, dX = \sum_{j=1}^n F(t_{j-1})(X(t \wedge t_j) - X(t \wedge t_{j-1})),$$

where $t \wedge s$ denotes the lesser of t and s. Notice that

$$\mathbb{E}[|I_t(F)|^2] = \sum_{j=1}^n |F(t_{j-1})|^2 (t \wedge t_j - t \wedge t_{j-1})$$

$$= \int_0^t |F(s)|^2 \, ds.$$

Using the fact that X begins anew independently of the past at each time t_j, it may be seen that this identity remains true, in the form

$$\mathbb{E}[|I_t(F)|^2] = \int_0^t \mathbb{E}[|F(s)|^2] \, ds \tag{5.2}$$

for a *random* step function F, provided that F is *adapted*, that is each $F(t)$ depends only on the past at t; equivalently $F(t)$ is an element of the function algebra generated by the $X(s)$ with $s \leq t$. The Ito integral of more general adapted random functions F (including the case $F = X$) is defined by taking the isometric extension to the Hilbert space completion of the map I_t, whose action on adapted step functions is isometric according to (5.2).

When this procedure is applied to the example (5.1) one finds that

$$\int_0^t X \, dX = \tfrac{1}{2} X(t)^2 - \tfrac{1}{2} t. \tag{5.3}$$

The unexpected term $-\tfrac{1}{2} t$ is called an *Ito correction*. It arises as an instance of the *Ito product formula*. Let us agree to denote by the stochastic differential notation

$$dM = F \, dX + G \, dt,$$

the circumstance that each $M(t)$ differs from $\int_0^t (F \, dX + G \, ds)$ by a constant. Then, given two such stochastic differentials

$$dM_j = F_j \, dX + G_j \, dt, \quad j = 1, 2,$$

the differential of the product is found to be

$$d(M_1 M_2) = (M_1 F_2 + F_1 M_2) \, dX + (M_1 G_2 + G_1 M_2 + F_1 F_2) \, dt.$$

We may write this as

$$d(M_1 M_2) = M_1 \, dM_2 + dM_1 \, M_2 + dM_1 \, dM_2,$$

where the third term, the Ito correction, is found by bilinear extension of the multiplication table for the basic processes dX and dt

	dX	dt
dX	dt	0
dt	0	0

(5.4)

Thus (5.3) arises by taking $M_1 = M_2 = X$.

6. QUANTUM STOCHASTIC CALCULUS

We shall generalise the Ito stochastic calculus so as to define operator-valued stochastic integrals in Fock space

$$M(t) = \int_0^t (E\, d\Lambda + F\, dA^\dagger + G\, dA + H\, ds). \tag{6.1}$$

Here the integrands E, F, G, H are adapted operator valued processes, and *adaptedness* means that, for example, $E(t)$ is an element of the operator algebra generated by the $\Lambda(s)$, $A^\dagger(s)$ and $A(s)$ with $s \leq t$ (or, equivalently, that generated by the two Brownian motions $Q(s)$ and $P(s), s \leq t$).

Our task is made easier by the fact that, in a formal sense, unlike Brownian motion the processes Λ, $A^\dagger$ and A are differentiable. We introduce the so-called *coherent states*

$$\psi(f) = (1, f, (2!)^{-\frac{1}{2}} f \otimes f, (3!)^{-\frac{1}{2}} f \otimes f \otimes f, \ldots), f \in L^2(\mathbb{R}_+).$$

As is well known, these are eigenvectors of annihilation operators, in particular

$$A(t)\psi(f) = \int_0^t f(s)\, ds\ \psi(f).$$

It is tempting to differentiate this relation, using the fundamental theorem of calculus, to get

$$\frac{dA}{dt}\psi(f) = f(t)\psi(f). \tag{6.2}$$

The operator $\dfrac{dA}{dt}$ (in fact it is the point annihilation operator $a(t)$) is only well-defined on those $\psi(f)$ for which f is continuous ($f(t)$ has no meaning for a general element f of the Hilbert space $L^2(\mathbb{R}_+)$), and is not a closeable operator. Nevertheless, writing (6.2) in the form

$$dA\psi(f) = f(t)\, dt\ \psi(f)$$

and using similar relations for the adjoint action of $A^\dagger(t)$ and of $\Lambda(t)$, we expect that, for the stochastic integral (6.1)

$$\langle \psi(f), M(t)\psi(g) \rangle = \int_0^t \langle \psi(f), (\bar{f}(s)g(s)E(s) + \bar{f}(s)F(s) + g(s)G(s) + H(s))\psi(g) \rangle\, ds. \tag{6.3}$$

Thus quantum stochastic integrals should reduce to Riemann–Lebesque integrals. (6.3) is easily verified directly when the adapted processes E, F, G and H are operator-valued step functions, for which $M(t)$ can be given the obvious meaning. In this case a slightly more complicated argument, using the canonical commutation relations, shows that

$$\|M(t)\psi(f)\|^2 = \int_0^t \{2\operatorname{Re}\langle M(s)\psi(f), (|f(s)|^2 E(s) + \bar{f}(s)F(s) + f(s)G(s) + H(s))\psi(s)\rangle$$
$$+ \|f(s)E(s) + F(s)\|^2\} \, ds, \tag{6.4}$$

and hence, after some manipulations (see [6] for details), leads to the estimate

$$\|M(t)\psi(f)\|^2 \leq \int_0^t \alpha(s,t,f)\{3\|f(s)E(s)\psi(f)\|^2 + 3\|F(s)\psi(f)\|^2$$
$$+ \|G(s)\psi(f)\|^2 + \|H(s)\psi(f)\|^2\} \, ds \tag{6.5}$$

where $\alpha(s,t,f) = \exp\{t - s + \int_s^t |f|^2\}$. This estimate permits a unique extension by continuity of stochastic integration from the case of step function integrands to integrands which are measurable adapted processes satisfying the *local square integrability conditions*

$$\int_0^t \{\|f(s)E(s)\psi(f)\|^2 + \|F(s)\psi(f)\|^2 + \|G(s)\psi(f)\|^2 + \|H(s)\psi(f)\|^2\} \, ds < \infty$$

for all $t > 0$ and $f \in L^2(\mathbb{R}_+)$. The extended integral continues to satisfy (6.3), (6.4) and (6.5), which are the basic identities and estimate of quantum stochastic calculus.

By polarising the relation (6.4) and comparing with (6.3) we can read off a *quantum Ito product formula*: for sufficiently well behaved stochastic differentials

$$dM_j = E_j \, d\Lambda + F_j \, dA^\dagger + G_j \, dA + H_j \, dt, \quad j = 1, 2,$$
$$d(M_1 M_2) = (M_1 E_2 + E_1 M_2 + E_1 E_2) \, d\Lambda + (M_1 E_2 + F_1 M_2 + E_1 F_2) \, dA^\dagger$$
$$+ (M_1 G_2 + G_1 M_2 + G_1 E_2) \, dA + (M_1 H_2 + H_1 M_2 + G_1 F_2) \, dt;$$

equivalently

$$d(M_1 M_2) = M_1 \, dM_2 + dM_1 \, M_2 + dM_1 \, dM_2,$$

where the Ito correction $dM_1 \, dM_2$ is evaluated by the convention that the basic differentials $d\Lambda$, $dA^\dagger$, dA and dt commute with adapted processes, together with the multiplication table

	$d\Lambda$	$dA^\dagger$	dA	dt
$d\Lambda$	$d\Lambda$	$dA^\dagger$	0	0
$dA^\dagger$	0	0	0	0
dA	$d\Lambda$	dA	dt	0
dt	0	0	0	0

(6.6)

(6.6) contains the Ito table (5.3) for the Brownian motions Q and P, as well as the corresponding table for the Poisson process

	$d\pi_l$	dt
$d\pi_l$	$d\pi_l$	0
dt	0	0

Students of the Christian theological doctrine of the Trinity may note that the avatars $dA^\dagger$, $d\Lambda$ and dA of Hindu deities may be united in the single formula

$$dJ \, dJ = dJ,$$

where dJ is the 2×2 matrix of differentials

$$dJ = \begin{pmatrix} dt & dA \\ dA^\dagger & d\Lambda \end{pmatrix}$$

7. BOSON–FERMION UNIFICATION [8]

We saw that the (Boson) Fock space $\Gamma(L^2(\mathbb{R}_+))$ is the natural carrier space for realisations of Brownian motions and of Poisson processes, and thus unifies the fundamental classical stochastic processes. We turn now to physics and show that $\Gamma(L^2(\mathbb{R}_+))$ can be regarded also as a *Fermion Fock space*, in such a way that the Fermion and Boson n-particle space structures coincide, as also do the time filtrations determined by Fermion and Boson field operators.

We introduce the *reflection process* which is the adapted process R such that $R(t)$ acts on exponential vectors by

$$R(t)\psi(f) = \psi(-\chi_{[0,t]}f + \chi_{[t,\infty)}f).$$

Then (see [8] for details) the solution of the stochastic differential equation

$$db(f) = \overline{f(s)}R(s)\,dA(s), \quad b(f)(0) = 0,$$

where $f \in L^2(\mathbb{R}_+)$, is a bounded-operator-valued adapted process, satisfying the Fermion field anticommutation relations

$$\{b(f), b(g)\} = 0, \quad \{b(f), b^*(g)\} = \langle f, g \rangle 1.$$

Furthermore each $b(f)$ annihilates the vacuum ψ_0, and ψ_0 is cyclic, in so far as the set of vectors

$$b^*(f_n)\ldots b^*(f_1)\psi_0, \quad n = 0, 1, \ldots, \quad f_1, \ldots, f_n \in L^2(\mathbb{R}_+)$$

is total in $\Gamma(L^2(\mathbb{R}_+))$. Thus the triple $(\Gamma(L^2(\mathbb{R}_+)), b, \psi_0)$ realises the Fock representation of the Fermion anticommutation relations, which is characterised to within unitary equivalence [1] by the existence of a cyclic vacuum vector. The coincidence of the Fermion and Boson n-particle structures in $\Gamma(L^2(\mathbb{R}_+))$ follows from the identities

$$b^*(f_n)\ldots b^*(f_1)\psi_0 = \int_{0 \leq t_1 \leq \ldots \leq t_n} \det(f_i(t_j))\,dA^\dagger(t_1)\ldots dA^\dagger(t_n)$$

$$a^\dagger(f_n)\ldots a^\dagger(f_1)\psi_0 = \int_{0 \leq t_1 < \ldots \leq t_n} \mathrm{per}(f_i(t_j))\,dB^*(t_1)\ldots dB^*(t_n),$$

where $\mathrm{per}(f_i(t_j))$ is the permanent $\sum_{\pi \in S(n)} f_1(t_{\pi(1)})\ldots f_n(t_{\pi(n)})$, B^*, B are the *Fermion creation and annihilation processes* given by

$$B^*(t) = b^*(\chi_{[0,t]}), \quad B(t) = b(\chi_{[0,t]})$$

and $a^\dagger(f)$ is the Boson creation operator corresponding to $f \in L^2(\mathbb{R}_+)$. B and B^* satisfy

$$dB = R\,dA, \quad dB^* = R\,dA^\dagger.$$

Because $R^2 = 1$, this relation between Boson and Fermion is symmetrical; we also have

$$dA = R\,dB, \quad dA^\dagger = R\,dB^*.$$

From these relations it is clear that the operator algebras generated by the $A(s)$ with $s \leq t$, and by the $B(s)$ with $s \leq t$, coincide for each $t \in \mathbb{R}_+$. Thus the Bosonic and Fermionic notions of adaptedness coincide. The theory of quantum stochastic calculus can be developed from the beginning using Fermions instead of Bosons.

The Fermionic approach to quantum stochastic calculus can be used to develop a very beautiful quantum generalisation of the convergence of random walks to diffusion processes [11].

8. UNITARY PROCESSES AND DILATIONS

We turn to the main application of quantum stochastic calculus in physics, which is the construction of dilations of irreversible evolutions.

We suppose that there is given an *initial Hilbert space* $\mathcal{H}_0$, which is thought of as carrying the physics of a dissipative system. By identifying the processes Λ, $A^\dagger$ and A with their "ampliations" $1 \otimes \Lambda$, $1 \otimes A^\dagger$ and $1 \otimes A$, we regard quantum stochastic calculus as living in the Hilbert space tensor product $\mathcal{H}_0 \otimes \Gamma(L^2(\mathbb{R}_+))$. Given bounded operators L_0, L_1, L_2 and L_3 on H_0, which we similarly identify with their ampliations $L_j \otimes 1$ to $\mathcal{H}_0 \otimes \Gamma(L^2(\mathbb{R}_+))$, we consider the stochastic differential equations for an adapted process U

$$dU = U(L_0\, d\Lambda + L_1\, dA^\dagger + L_2\, dA + L_3\, dt), \quad U(0) = 1. \tag{8.1}$$

A variation of the Picard iterative method based on the estimate (6.5) shows that this equation has a unique solution [6]. Necessary conditions for the solution to be unitary are found from the quantum Ito formula (6.6). Indeed if U is unitary then $U^*U \equiv 1$ and so

$$0 = d(U^*U) = U^*\, dU + dU^*U + dU^*\, dU$$
$$= U^*U(L_0\, d\Lambda + L_1\, dA^\dagger + L_2\, dA + L_3\, dt)$$
$$+ (L_0^*\, d\Lambda + L_2^*\, dA^\dagger + L_1^*\, dA + L_3^*\, dt)U^*U$$
$$+ L_0^*U^*UL_0\, d\Lambda + L_0^*U^*UL_1\, dA^\dagger + L_1^*U^*UL_0\, dA$$
$$+ L_1^*U^*UL_1\, dt.$$

Using again the fact the $U^*U = 1$, and comparing coefficients of the basic differentials $d\Lambda$, $dA^\dagger$, dA and dt, we find that

$$L_0 + L_0^* + L_0^*L_0 = 0$$
$$L_1 + L_2^* + L_0^*L_1 = 0$$
$$L_3 + L_3^* + L_1^*L_1 = 0.$$

These, together with similar identities derived from the condition that $UU^* = 1$, imply that L_0, L_1, L_2 and L_3 must take the form $W-1$, L, $-L^*W$, $iH - \frac{1}{2}L^*L$ respectively, where W, L, H are bounded operators on H_0 with W unitary and H self-adjoint. Thus (8.1) becomes

$$dU = U((W-1)\, d\Lambda + L\, dA^\dagger - L^*W\, dA + (iH - \tfrac{1}{2}L^*L)\, dt), \quad U(0) = 1 \tag{8.2}$$

It can be shown [6] that these conditions are in fact sufficient; the solution of (8.2) is indeed unitary.

Two interesting semigroups are associated with the solution of (8.2) [6]. The first is formed by taking the *vacuum conditional expectation* $T(t) = \mathbb{E}_0[U(t)]$ of $U(t)$, that is, the unique bounded operator $T(t)$ on $\mathcal{H}_0$ such that

$$\langle u, T(t)v \rangle = \langle u \otimes \psi_0, U(t)v \otimes \psi_0 \rangle.$$

$(T(t): t \in \mathbb{R}_+)$ is a semigroup of contractions on $\mathcal{H}_0$ of which the infinitesimal generator K is given by

$$K = iH - \tfrac{1}{2}L^*L,$$

as follows from the polarised form of (6.4). The second comprises the maps $\mathcal{T}(t)$ from $B(\mathcal{H}_0)$ to itself defined by

$$\mathcal{T}(t)(S) = \mathbb{E}_0[U(t)S \otimes 1 U(t)^{-1}].$$

$(\mathcal{T}(t) = t \geq 0)$ is a semigroup of completely positive maps, whose infinitesimal generator is found from (6.4) to be

$$\mathcal{L}(S) = i[H, S] - \tfrac{1}{2}(L^*LS - 2L^*SL + SL^*L).$$

That $(T(t)$ and $(\mathcal{T}(t))$ are indeed semigroups, that is that

$$T(t)T(s) = T(t+s), \quad \mathcal{T}(t)\mathcal{T}(s) = \mathcal{T}(t+s) \quad (t, s \geq 0),$$

follows from the fact that each of the fundamental processes Λ, $A^\dagger$ and A, like classical Brownian motion, begins anew independently of the past at each time t.

Gorini, Kossakowski and Sudarshan [2] in the case when $\mathcal{H}_0$ is finite dimensional, and Lindblad [10] in the general case, have shown that the general form of infinitesimal generator of a uniformly continuous semigroup of completely positive maps on $B(\mathcal{H}_0)$ is

$$\mathcal{L}(S) = i[H, S] - \tfrac{1}{2}\sum_j (L_j^*L_j S - 2L_j^* SL_j + SL_j^*L_j),$$

where $H \in B(\mathcal{H}_0)$ is self-adjoint, and the $L_j \in B(\mathcal{H}_0)$ are such that $\sum_j L_j^*L_j$ converges strongly. By solving a generalisation of (8.2), for example the equation

$$dU = U(\sum_j (L_j \, dA_j^\dagger - L_j^* \, dA_j) + (iH - \tfrac{1}{2}\sum_j L_j^*L_j)\, dt) \quad U(0) = 1,$$

where the $A_j^\dagger$ and A_j are independent creation and annihilation processes, the corresponding semigroup can be expressed in the form

$$e^{t\mathcal{L}}(S) = \mathbb{E}_0[U(t)S \otimes 1 U(t)^{-1}].$$

Thus the general uniformly continuous quantum dynamical semigroup admits a *stochastic dilation* [7].

The unitary processes U satisfy the cocycle identity

$$U(t) = 1 \otimes \Gamma^\dagger(s) U^\dagger(s) U(s+t) I \otimes \Gamma(s),$$

where $\Gamma(s)$ is the isometry which maps each $\psi(f)$ to $\psi(f_s)$ when $f_s(t) = \chi_{(s,\infty)} f(t-s)$. A *group dilation*, in which

$$U(s)U(t) = U(s+t) \quad (s, t \in \mathbb{R})$$

is obtained by enlarging the Fock space to $\Gamma(L^2(\mathbb{R}))$ and replacing each $U(t)$ by $U(t) 1 \otimes \Gamma(t)$.

Quantum dynamical semigroups model dissipative dynamics. For a physical system whose algebra of observables is a proper C^*-subalgebra $\mathcal{A}$ of $B(\mathcal{H}_0)$, the Lindblad formula for the infinitesimal generator of a quantum dynamical semigroup may fail to hold if $\mathcal{A}$ has non trivial cohomology spaces. In this case the theory of *quantum diffusions* [4,5] may be used to construct dilations. A one dimensional quantum diffusion over $\mathcal{A}$ is a family $j = (j_t = t \geq 0)$ of *-algebra homomorphisms from $\mathcal{A}$ into $\mathcal{A} \otimes B(\Gamma(L^2(\mathbb{R}_+)))$, such that the $j_t(S)$, $S \in \mathcal{A}$, are adapted processes satisfying

$$dj_t(S) = j_t(\lambda(S)) \, d\Lambda + j_t(\alpha(S)) \, dA^\dagger + j_t(\alpha^\dagger(S)) \, dA$$
$$+ j_t(\tau(S)) \, dt, \quad j_0(S) = S \otimes 1.$$

Here λ, α, $\alpha^\dagger$ and τ are maps from $\mathcal{A}$ to itself which satisfy certain cohomological identities following from the quantum Ito formula. In cases when the cohomology is trivial every such diffusion is of form

$$j_t(S) = U(t)(S \otimes 1)U(t)^{-1}$$

where U is a unitary process of the form (8.2), and the Lindblad formula holds for the infinitesimal generator of the corresponding quantum dynamical semigroup

$$\mathcal{T}_t(S) = \mathbb{E}_0 j_t(S) \quad (S \in \mathcal{A}).$$

REFERENCES

[1] O. Bratteli and D.W. Robinson, *Operator algebras and quantum statistical mechanics* II, Springer, Berlin (1979).
[2] V. Gorini, A. Kossakowski and E. C. G Sudarshan, *J. Math. Phys.* **17**, 1298 (1976).
[3] A. Einstein *Annalen der Physik* **17**, 549 (1905).
[4] R.L. Hudson, 'Quantum diffusions and cohomology of algebras', in *Proceedings of Ist World Congress of Bernoulli Society, Tashkent 1986*.
[5] R.L. Hudson and M. Evans, 'Multidimensional quantum diffusions', to appear in *Quantum Probability (Oberwolfach) Proceedings* 1987, ed. L. Accardi, Springer LNM.
[6] R.L. Hudson and K.R. Parthasarathy, *Commun. Math. Phys.* **93**, 301 (1984).
[7] R.L. Hudson and K.R. Parthasarathy, *Acta Applicandre Math.* **2**, 353 (1984).
[8] R.L. Hudson and K.R. Parthasarathy, *Commun. Math. Phys.* **104**, 457 (1986).
[9] M. Lax, *Phys. Rev.* **145**, 111 (1965).
[10] G. Lindblad, *Commun. Math. Phys.* **48**, 119 (1976).
[11] J.M. Lindsay and K.R. Parthasarathy, 'The passage from random walk to diffusion in quantum probability II', preprint.
[12] I.R. Senitzky, *Phys. Rev.* **119**, 670 (1960).
[13] M. Smoluchowsky, *Annalen der Physik* **25**, 205 (1908).
[14] B Wiener, *J. Mathematics and Physics.* **2**, 131 (1923).

ON THE SEARCH OF THE TIME OPERATOR SINCE SCHRÖDINGER

Z. Marić
Institute of Physics
P.O.B. 57
11000 Beograd
Yugoslavia

ABSTRACT. The conceptual and mathematical problems concerning the time operator and its operational meaning in quantum mechanics and the "inner-time" operator in classical statistical physics are reviewed. The common point is the use of non-unitary similarity transformation. In the quantum case the time operator is related to the non-orthogonal resolution of identity. It is argued that the similar results for the notion of entropy obtained for a certain class of classical dynamical systems in Hilbert-space and phase-space representation require further elucidations.

1. INTRODUCTION

Historically, in the explicit form, the problem of time in quantum mechanics was connected with the time energy uncertainty relation, first discussed by Bohr[1] and further by Mandelstam and Tamm[2], Fock and Krylov[3] and Aharonov and Bohm[4] The discussion concerns mainly the operational meaning of a time-measurement related to the traditional interpretation of the uncertainty relations as the limit imposed on the simultaneous measurement of the time and energy. In the formal sector the notion of a time operator originates from the view that all classical quantities have to be "translated" into Hermitian operators, the measured values of which are then interpreted as their spectral values. This view immediately conflicts with the standard interpretation of the time as a parameter in the Schrödinger equation. Schrödinger[5] himself was the first one who tried to treat, in the relativistic case, all four coordinates (x,y,z,t), as Hermitian multiplicative operators and arrived at the conclusion that they are without physical meanings. Later, Pauli[6], in treating the non-relativistic case, remarked that if an inner-time operator T obeys the equation of motion

$$\frac{dT}{dt} = \frac{1}{i\hbar}[T,H] = 1 \tag{1}$$

i.e. if we keep on one side the parametric nature of t and on the other side by introduce the internal time properties of the system considered through the operator T, then no Hermitian operator T exists since the spectrum of the Hamiltonian H has a lower bound. The explicit expressions for the time operator were obtained by Paul[7] and Engelmann and Fick[8], but simple inspection shows that they are not Hermitian with respect to the set of functions usually used in physical problems.

It seems that there are three measurement problems which ask for a well defined time operator concept. They are: the time of arrival, the time of passage and the time of sojourn, the last one being studied in connection with a decay experiment. A systematical study of the concept of time of arrival has been carried out by Allcock[9]. He concludes that, both in the Hilbert space of a free particle ($0<E<\infty$) as well as within a wider family of wave functions given via the non-relativistic free-space Schrödinger equation with arbitrary sources ($-\infty<E<\infty$), there exists no self-adjoint operator whose expectation value predicts an outcome of the time of arrival measurements; here three primary requirements of a measurement connected with resolution, time translation and overall probability conservation have to be fulfilled. In analyzing decay experiments Ekstein and Siegert[10] have found an expression for a sojourn-time operator proposing a correspondence between a classical observable and its quantum-mechanical counterpart which is not deducible from any form of the Weyl rule.

In Sec.2. a review of the properties of a time operator is presented as it has been analyzed by Holevo[11] in the framework of the statistical theory of quantum measurement. Here the limitations regarding the estimation of inner-time properties are derived following an algorithm common to all quantum measurements in which Galilean invariance is required for the shifted measurements in question. The central feature of the time measurement lies in the fact that it is described via a non-orthogonal resolution of identity. Closely connected with this is the problem of Weyl-like canonical commutation relations. It is easily observed that the energy shift operators constitute a semi-group of isometric operators on a positive-valued energy axis, i.e. the transformations are one-sided unitary. By enlarging the analysis to the whole energy axis, it is possible formally to arrive at an expression for a time-operator by which one elucidates the relationship between this operator and parametric time as it appears in the Schrödinger equation. This has been done by Bauer[12] using a method proposed by Newton[13] for the pair of angle-action variables.

An analogous situation characterizes the attitude of the Brussels school towards the genuine problem of non-equilibrium statistical mechanics, i.e. the problem of irreversibility. Prigogine et al.[14] have developed the view that one can reconcile dynamical and thermodynamical evolution by using a non-unitary similarity transformation acting on the distribution function ρ so that the original deterministic Liouville equation is transformed into a dissipative evolution equation. Here the problem of a lower bound for the Hamiltonian operator is circumvented by using the Liouville equation. In the same spirit, Misra[15] has revisited the old no-go theorem on entropy for the Hamiltonian classical system formulated by Poincaré[16], and found for mixing dynamical system and K-flow systems an entropy-like operator (the Lyapunov variable) which via the Weyl commutation relations admits the construction of a time-operator. The analysis is done in the special Hilbert space representation of classical mechanics originally advocated by Koopman. However, Burić[17] in a not yet published work has shown that the similar result can be obtained in the standard phase-space representation of the classical statistical mechanics. This topic is discussed in Sec. 3.

2. STATISTICAL THEORY OF QUANTUM MEASUREMENTS AND THE TIME OPERATOR IN THE HARDY CLASS SPACE. THE FORMAL PROPERTIES OF THE TIME OPERATOR IN AN ENLARGED SPACE

A) In the framework of the statistical theory of quantum measurements[18] the central topic of investigation is a map, more precisely an affine map, from a quantum state S (belonging to the convex set of density operators $\mathcal{J}(\mathcal{K})$) into the set of all probability distributions defined in a measurable space, say U, which is the space of <u>outcomes</u> of a measurement. Its domain in R^n is naturally <u>supplied</u> with a Borel σ-field. In short, a quantum measurement is a map

$$S \rightarrow \mu_S(dU) \tag{2}$$

where the probability distribution of the results of measurement performed on a state S, $\mu_S(dU)$, is interpreted in terms of a resolution of the identity. By this one understands a collection $M = \{M(B); B \in \mathcal{A}(U)\}$ of Hermitian operators in a Hilbert space $\mathcal{K}$, B belonging to a subset $\mathcal{A}(U)$ of the σ-Borel field of measurable space U, with the properties:

(i) $M(\emptyset) = 0$; $M(U) = 1$

(ii) $M(B) \geq 0$; $B \in \mathcal{A}(U)$

(iii) for any at most countable decomposition $\{B_j\}$ of $B \in \mathcal{A}(U)$ the relation $M(B) = \Sigma M(B_j)$ holds. By adding the requirement:

(iv) $M(B_1)M(B_2) = 0$ if $B_1 \cap B_2 = \emptyset$

we have an particular but very well understood class of the <u>orthogonal</u> resolution of the identity.

For this case one easily establishes the expression for $_S(B)$ which reads

$$\mu_S(B) = TrSM(B), \qquad B \in \mathcal{A}(U) \qquad (3)$$

for a pure state $S_\Psi = |\Psi\rangle\langle\Psi|$ it is transformed into the familiar formula

$$\mu_{S_\Psi}(B) = \langle\Psi|M(B)|\Psi\rangle \qquad (4)$$

Hermitian operators are quantum analogues of classical random variables. Their eigenvalues are real, and the generalization of a spectral representation of an Hermition operator X for the finite-dimensional case $X = \Sigma \lambda_k E_k$ (E_k being the projection onto the invariant subspace corresponding to the eigenvalue λ_k; $\Sigma E_k = I$; $E_j E_k = \delta_{jk} E_j$), to its continuous analogue is traightforward if one deals with a bounded subset λ on R, so that $E(\lambda) = I$. It reads

$$X = \int \lambda E(d\lambda); \qquad E(d\lambda) = [\Sigma_k \delta(\lambda - \lambda_k) E_k] d\lambda \qquad (5)$$

Then, for any Ψ the probability distribution $\mu_\Psi(d\lambda)$ reands

$$\mu_\Psi(d\lambda) = TrS_\Psi E(d\lambda) = \langle\Psi|E(d\lambda)\Psi\rangle \qquad (6)$$

The integrals

$$\int \lambda \langle\Psi|E(d\lambda)|\Psi\rangle = \int \lambda \mu_\Psi(d\lambda) \qquad (7)$$

converge; their convergence is the consequence of the boundedness. The integral (7) corresponds to the Hermitian operator X

$$\langle\Psi|X|\Psi\rangle = \int \lambda \langle\Psi|E(d\lambda)|\Psi\rangle = \int \lambda \mu_\Psi(d\lambda) \qquad (8)$$

This relation establishes one-to-one correspondence between the Hermitian operator X and the orthogonal resolution of the identity $E(d\lambda)$ on bounded subsets of R. It can be directly applied to a function $f(X)$ of a Hermitian operator X if that f is a bounded measurable function. Its generalization for an unbounded self-adjoint operator needs care in determining the domain D; for those Ψ in the domain D for which one has

$$D = \{\Psi: \int_{-}^{+} \lambda^2 <\Psi|E(d\lambda)\Psi> <\infty\}, \qquad (9)$$

the correspondence between a self-adjoint operator X and the orthogonal resolution of the identity is analogous to Eq. (8). The property (9) establishes also that

$$||X\Psi||^2 = \int \lambda^2 <\Psi|E(d\lambda)\Psi>; \qquad \Psi \in D \qquad (10)$$

$||\cdot||$ being the norm.

In contrast to these cases, for a symmetric operator X, non-extendible to the self-adjoint one, defined in a dense domain D, there exists an in general non-unique (i.e. non-orthogonal) resolution of the identity $M(d\lambda)$ such that the following holds:

$$D(X) = \{\Psi: \int \lambda^2 <\Psi|M(d\lambda)\Psi> <\infty\}$$

$$<\Psi|X|\Psi> = \int \lambda <\Psi|M(d\lambda)\Psi> \qquad \Psi \in D(X) \qquad (11)$$

$$||X\Psi||^2 = \int \lambda^2 <\Psi|M(d\lambda)|\Psi> \qquad \Psi \in D(X)$$

The inclusion of the symmetric operators into the scheme enlarges the standard quantum-mechanical measurement estimations to those situations where the non-orthogonal resolution of the identity replaces the orthogonal ones.

In order to characterize statistically the probability distribution $\mu_S(dx)$ one introduces the mean value $\mathcal{E}_S(M)$ and the variance $\mathcal{D}_S(M)$ of the map $M:S$ $\mu_S(dx)$, which read

$$\mathcal{E}_S(M) = \int X \mu_S(dx)$$
$$\mathcal{D}_S(M) = \int (X - \mathcal{E}_S(M))^2 \mu_S(dx) \qquad (12)$$

One supposes that these quantities are well defined, i.e. μ_S has a finite second moment. These definitions are valid for the non-orthogonal resolution of the identity also. For a pure case $S = |\Psi\rangle\langle\Psi|$ one has

$$\mathcal{E}_{S_\Psi}(X) = \langle\Psi|X\Psi\rangle \tag{13}$$

$$\mathcal{D}_{S_\Psi}(X) = ||X - {}_S(X)\Psi||^2 = ||X\Psi||^2 - \mathcal{E}_{S_\Psi}(X)^2$$

For a pair of observables X_1, X_2 defined on a common domain one has

$$\mathcal{D}_{S_\Psi}(X_1)\mathcal{D}_{S_\Psi}(X_2) > |\mathrm{Im}\langle X_1\Psi|X_2\Psi\rangle|^2 \tag{14}$$

For bounded X_1 and X_2, the uncertainty relations (14) become

$$\mathcal{D}_{S_\Psi}(X_1)\mathcal{D}_{S_\Psi}(X_2) > \tfrac{1}{4}|\mathcal{E}_{S_\Psi}([X_1,X_2])|^2; \quad [X_1,X_2] = X_1X_2 - X_2X_1 \tag{15}$$

B) The constraints on the statistical model of quantum measurements are introduced by the requirement that the statistics of any measurement be the same in any inertial reference frame. In non-relativistic mechanics this reduces to invariance with respect to the Galilean group, i.e. for the position and temporal shifts, Galilean boosts and rotations. For any element of this group g which transforms the measurement M into gM and the state S into gS we require

$$\mu^{gM}_{gS} = \mu^M_S \tag{16}$$

In order to translate this constraint into a more transparent expressions, let us consider the projective unitary representation of a shift $\theta \rightarrow V_\theta$. By Stone's theorem

$$V_\theta = \exp(i\theta A) \tag{17}$$

where A is a self-adjoint operator in $\mathcal{H}$. The action of this one-parameter group on a quantum state S is

$$S \to S_\theta = e^{i\theta A} S e^{-i\theta A} \qquad (18)$$

For a pure state Ψ ($\Psi_\theta = \exp(i\theta A)\Psi$) one finds from (15) the Mandelstam-Tamm inequality for any bounded operator X:

$$\mathcal{D}_\theta(X)\mathcal{D}_\theta(A) > \tfrac{1}{4} \left|\tfrac{d}{d\theta}\mathcal{E}_\theta(X)\right|^2 \qquad (19)$$

This result, which can be generalized to arbitrary states and measurements, gives in principle a lower bound for the accuracy of quantum measurements. This is a part of the quantum estimation problem. To see this, suppose that the state S is shifted to the state S_θ, the value of θ is unknown, and one makes the measurement on the object to determine the value θ. The measurements leads to real numbers and the estimation is done via an observable X. In principle any observable X can serve as a statistical estimator for θ. The quality of the estimate X is measured by the mean square deviation $\mathcal{E}_\theta[(X-\theta)^2] = \mathcal{D}_\theta(X) + (\mathcal{E}_\theta(X)-\theta)^2$. The best estimates are the s.c. unbiased estimates for which $\mathcal{E}_\theta(X) = \theta$. For such measurements the inequality (19) translates into

$$\mathcal{D}_\theta(X) > [4\mathcal{D}_\theta(A)]^{-1} \qquad (20)$$

By observing that A commutes with the unitary group $\{\exp(i\theta A)\}$ one has $\mathcal{D}_\theta(A) = \mathcal{D}_S(A)$. One concludes that for an unbiased estimate of the shift θ, $\mathcal{D}_\theta(X)$ is bounded below by a quantity inversely proportional to the uncertainty of the observable A in the basic state S. A simple calculation shows that the variable which is canonically conjugate to A, say B, provides, up to a constant, an unbiased estimate of the shift θ. Denoting the spectral resolution of B by $B(\Lambda)$ for all Borel sets $\Lambda \subset R$, one gets from the Weyl commutation relation

$$e^{-i\theta A} B(\Lambda) e^{i\theta A} = B(\Lambda_{-\theta}); \quad \Lambda_{-\theta} = \{\lambda-\theta, \lambda \in \Lambda\} \qquad (21)$$

A resolution of the identity satisfying this condition is called <u>covariant</u> with respect to the unitary group of shift operators θ. It follows that the probability distribution $\mu_\theta^B(d\lambda)$ of the measurement $B = \{B(d\lambda)\}$ with respect to the pure state $S = |\Psi\rangle\langle\Psi|$ satisfies the condition

$$\mu_\theta^B(\Lambda) = \mu_o^B(\Lambda_{-\theta}) \qquad (22)$$

From this one finds

$$\mathcal{E}_\theta(B) = \int_{-\infty}^{+\infty} \lambda \mu^B(d\lambda) = \int_{-\infty}^{+\infty} (\lambda+\theta)\mu_0^B(d\lambda) = \mathcal{E}_0(B)+\theta \qquad (23)$$

It means that B is, up to a constant, an unbiased estimate of θ. In passing we note that here the Heisenberg uncertainty relations $\mathcal{D}_S(A)\mathcal{D}_S(B) \geqslant 1/4$ follow.

C) Let us consider the family of states

$$S_t = V_t S V_t^*$$

where t is a parameter describing the time shift for the state preparation. Then for un unbiased measurement M(dt) of the parameter t one has the lower bound (20) which reads

$$\mathcal{D}_t\{M\} > \frac{\hbar^2}{4\mathcal{D}_t(E)} \quad ; \quad E = \hbar H \qquad (24)$$

The covariance conditions (21) and (22) for the case of the time variable read

$$V_t M(B) V_t^* = M(B_{-t}) \qquad t \in R \qquad (25)$$

$$\mu_t^M(B) = \mu_0^M(B_{-t}) \qquad t \in R; \quad B \in \mathcal{A}(R)$$

For the system whose Hamiltonian is bounded from below one easily arrives at Pauli's objection in a new language: these is no orthogonal resolution of identity which satisfies the above covariance conditions. Therefore, the time measurement differs from the measurements of other "standard" variables such as position, energy etc. However, there exists a non-orthogonal resolution of identity which satisfies the covariance conditions (21) and (22). In order to show how this comes about, let us consider the free-particle case in the s.c. energy representation, where $E=p^2/2m$. In this representation, V_t is multiplication by $\exp(-itE/\hbar)$, the energy operator is just multiplication with E and the wave functions are given by two-dimensional vectors $\hat{\Psi}_E \in \mathbb{C}^2$. The space is $\mathcal{L}_{\mathbb{C}^2}^2(0,\infty)$, called the Hardy class. The transformation from the momentum to the energy representation is given by

$$\hat{\Psi}_E = \sqrt{\frac{m}{2E}} \begin{vmatrix} \Psi(\sqrt{2mE}) \\ \Psi(-\sqrt{2mE}) \end{vmatrix} \qquad (26)$$

with the norm $\int |\hat{\Psi}_E|^2 dE$. The operator T and its domain D(T) are

$$T = i\hbar \frac{d}{dE} \quad ; \quad D(T) = \{\hat{\Psi}_E : \hat{\Psi}_o = 0 ; \int_o |\frac{d}{dE}\hat{\Psi}_E|^2 dE < \infty \} \qquad (27)$$

The spectral measure of T is the non-orthogonal resolution of identity M(dt),

$$<\Psi|M(dt)|\Psi> = [\int_o^\infty \int_o^\infty \Psi_E^* \Psi_{E'} e^{i(E'-E)t/\hbar} dE dE'] \frac{dt}{2\pi\hbar} , \qquad (28)$$

for which the covariance properties (25) hold. In momentum representation, the time operator T and its domain are given by

$$T = m\hbar i \, sgn(p) \frac{1}{\sqrt{|p|}} \frac{d}{dp} \frac{1}{\sqrt{|p|}} ; \quad D(T) = \{\Psi(p) : \int_{-\infty}^{+\infty} \frac{d}{dp} \frac{\Psi(p)}{\sqrt{|p|}} \frac{dp}{|p|} < \infty \} \qquad (29)$$

The shift operators P_e in $C^2(o,)$ are defined as:

$$P_e \Psi_E = \begin{vmatrix} \Psi_{E-e} & ; & E \geqslant e \\ 0 & ; & E < e \end{vmatrix} \qquad (30)$$

This family constitutes a semi-group of isometric operators in which $P_e^* P_e = I$ but $P_e P_e^* < I$ for $e > o$. The fact that the P_e are not unitary but one-sided unitary has as a consequence the nonself-adjointness of T. The one-sided unitary transformations can also be studied by passing, via canonical transformation, from continuous to discrete spectral measures.[19]

The problem of estimation of a time shift in the family (25) is constrained only by the imequality (24). The further elucidation of an operational meaning of the operator T requires the analysis of a specific non-orthogonal resolution of the identity connected with well defined concrete physical situation.

D) The Naimark theorem[20], which states that any resolution of the identity in a Hilbert space $\mathcal{H}$ can be delated to an orhogonal resolution of the identity in a larger Hilbert space has been implicitelly used in ref. 12) to study the relationship between the parametric time and a mean value of a time operator defined in an enlarged space including the negative energy states. There are two ways to enlarge the space defined by eigenfunctions $|\Psi(E)_o>$, $E>0$, of an Hermitian Hamiltonian H. One of them is called the extended space, the orther the pseudospin extension. The main feature of the result is seen fully in the first one. In the extended space all operators and functions are written with a curet $\hat{}$. By means of the operator $\hat{Q} = \begin{pmatrix} 1 & 0 \\ 0 & -1 \end{pmatrix}$, the Hamiltonian $\hat{H}$, $\hat{H} = H\hat{Q}$, verifies the eigenvalue equation:

$$\hat{H}|\hat{\Psi}(E)> = E|\hat{\Psi}(E)> \qquad -\infty < E < +\infty \qquad (31)$$

where $|\hat{\Psi}(E)> = \begin{vmatrix} |\Psi(E)_o> \\ 0 \end{vmatrix}$ for $E \geq 0$ and $|\hat{\Psi}(E)> = \begin{vmatrix} 0 \\ \Psi(-E)_o \end{vmatrix}$ for $E<0$. With the shift operators P_e and P_e^* defined in the original space as:

$$P_e|\Psi(E)_o> = h(E-e)|\Psi(E-e)_o> \ ; \quad P_e^*|\Psi(E)_o> = |\Psi(E+e)_o> \qquad (32)$$

where $h(x)$ is the step function, and with the projection $\pi(e)$ which is defined as

$$\pi(e) = \int_{-}^{+} dE |\Psi(e-E)_o> h(e-E)h(E)<\Psi(E)_o| \qquad (33)$$

one constructs the shift-operators in the extended space, $\hat{P}_e$, which read

$$\hat{P}_e|\hat{\Psi}(E)> = \begin{vmatrix} P_e & 0 \\ \pi(e) & P_e^* \end{vmatrix} |\hat{\Psi}(E)> \qquad (34)$$

with properties

$$\hat{P}_e|\hat{\Psi}(E)> = |\hat{\Psi}(E-e)> \ ; \quad \hat{P}_e^*|\hat{\Psi}(E)> = |\hat{\Psi}(E+e)> \qquad (35)$$

$\hat{P}_e$ and $\hat{P}_e^*$ verify the unitarity and group properties. Therefore, by the Stone theorem, one has:

$$\hat{P}_e = \exp[-ie\hat{T}] \qquad (36)$$

The commutation relation between $\hat{T}$ and $\hat{H}$ is $[\hat{T},\hat{H}] = e\hat{P}_e$ and for the small e one has:

$$[\hat{T},\hat{H}] = i \qquad (37)$$

The continuous spectrum of $\hat{H}$ implies the continuous spectra of $\hat{T}$. From the general dynamical law

$$\frac{d}{dt}<\hat{\Psi}|\hat{A}|\hat{\Psi}> = \frac{1}{i}<\hat{\Psi}|[\hat{A},\hat{H}]|\hat{\Psi}> + <\hat{\Psi}|\frac{\partial \hat{A}}{\partial t}|\hat{\Psi}> \qquad (38)$$

there follows

$$d<\hat{\Psi}|\hat{T}|\hat{\Psi}> = dt, \qquad (39)$$

whence by integrating one gets

$$<\hat{\Psi}|\hat{T}|\hat{\Psi}>_t = t + <\hat{\Psi}|\hat{T}|\hat{\Psi}>_o \qquad (40)$$

The index o refers to the origin of the parameter t. The relationship between t and the mean value of the operator $\hat{T}$ in $\hat{T}$-representation is seen by looking at the time evolution of the wave packet. One finds that in the Schrödinger equation these two quantities exchange their roles. The fundamental properties (40) remain after the projection to "physical states" has been transformed.

The formal derivation of the time operator is not oriented towards any operational use in time-measuring experiments.

3. TIME-OPERATOR IN CLASSICAL STATISTICAL MECHANICS. THE ENTROPY FOR MIXING SYSTEMS DERIVED IN THE PHASE SPACE

As stated above, for the classical statistical mechanics of the Gibbs form, Misra[15] has obtained a results on the entropy of an isolated system which is analogous to the Poincaré theorem for Hamiltonian system, forbiding the existence of a mechanical quantity like entropy. More precisely, it states that for classical dynamical system $(\Omega, d\Omega, g_t)$, (where Ω is a differentiable manifold, g_t an one-parameter group of diffeomorphisms on Ω, and $d\Omega$ an invariant measure on Ω under the action of g_t), there exists no functional of the form

$$F[\rho] = \int_\Omega f\rho d\Omega$$

where $f \in C^1(\Omega)$

and ρ belongs to the space $V_1(\Omega)$ of all Borel probability measures on Ω. In other words, there are no distribution functions, such that:

$$a) \quad \frac{\partial \rho_t}{\partial t} = 0 \rightarrow \frac{dF(\rho_t)}{dt} > 0$$

$$b) \quad \frac{\partial \rho_t}{\partial t} = 0 \rightarrow \frac{dF(\rho_t)}{dt} = 0$$

(41)

where $\rho_t \stackrel{def}{=} \rho \circ g_t^{-1}$ and the arrow is used to denoted the implication. The proof is based on the assumptions that the integral curves of the system are almost non wandering in the sence that they are subject to the recurrence theorem, and further, only functions defined uniquely on the whole space Ω are allowed as kernels for the functional F.

The construction of a microscopic representative for the non-equilibrium entropy is done in a special Hilbert space in which the quadratic form $<\Psi|\hat{M}|\Psi> = \int \Psi \hat{M} \Psi d\Omega$, [for $\Psi \in \mathcal{L}^2(\Omega, d\Omega)$ and M belonging to the set $B_a(\mathcal{L}^2)$ of bounded self-adjoint operators on $\mathcal{L}^2(\Omega, d\Omega)$} is a single-valued functional on the set of distribution functions ρ , $\{\Psi : |\Psi_\rho|^2 = \rho; \Psi_\rho \in \mathcal{L}^2\}$. The operator M commutes with operators of multiplication by a function. The Lyapunov variable is defined as follows. let $\hat{L}$ be the self-adjoint generator of a one-parameter group $U(t)$ of unitary operators on $\mathcal{L}^2(\Omega, d\Omega)$ given by $U(t)\Psi = \Psi \circ g_t^{-1}$, where g_t characterizes the dynamical system $(\Omega, d\Omega, g_t)$. For Ψ in the domain of $\hat{L}, D(\hat{L})$, one has:

$$\hat{L}\Psi = \text{weak-lim}_{t \to 0} \frac{\hat{U}(t) - 1}{t} \Psi \qquad (42)$$

The operator M $B_a(\mathcal{L}^2)$ is called a Lyapunov variable if the following is valid:

i) $<\Psi|\hat{M}|\Psi> \geq 0 \quad \forall \Psi \in \mathcal{L}^2(\Omega, d\Omega)$

ii) $\Psi \in D(\hat{L}) \rightarrow \hat{M}\Psi \in D(\hat{L})$

iii) $[\hat{L}, \hat{M}] \subseteq \hat{D}, \quad <\Psi|\hat{D}|\Psi> \geq 0, \Psi \in \mathcal{L}^2$

iv) $<\Psi|\hat{D}|\Psi> = 0$ is almost every where (a.e);
   i.e. $|\Psi> = |1>$.

The existence of the operator M for classical dynamical system is stated as follows:

α) If $\hat{L}$ given by (42) has absolutely continuous spectrum of uniform multiplicity on the orthogonal complement to the space spanned by $|1\rangle$, i.e. $\mathcal{L}^2 \ominus |1\rangle = H_o^\perp$, extended over the whole real line, then there exists Lyapunov variable $\hat{M}$.

β) If the system admits Lyapunov variable $\hat{M}$, than $\hat{L}$ restricted on $H_o^\perp$ has absolutely continuous spectra in which case the dynamical system is mixing.

Then by using the Weyl canonical commutation rule for $\hat{L}$ and $\hat{M}$ one obtains the time operator.

The closer analysis shows that both statements α) and β) are connected with the wandering orbits of the group U. An orbit $\{\hat{U}(t)|\Psi_o\rangle, t \in R\}$ is wandering if there exists a ball $K_\epsilon \subset \mathcal{L}^2$ and $T \in R^+$ such that for each $t' > t+T$: $U(t')K_\epsilon(|\Psi_o\rangle) \cap U(t)K_\epsilon(|\Psi_o\rangle) = \emptyset$ holds. Then the statement α) can be demonstrated using the bases of wandering orbits in $H_o^\perp$ and β) is a consequence of the boundedness of $\hat{M}$ and the wandering character of orbits.

The analogous result has been obtained in ref. 17) by observing:

1) A simple consequence of the Poincaré theorem on the local exactness of any closed one-form is that for every atlas $\mathcal{A} = \{(U_\alpha, \varphi_\alpha)\}$ on Ω, on which the dynamical system $(\Omega, d\Omega, g_t)$ is defined, there exist functions $f_\alpha \in C_R^\infty(U_\alpha)$ such that:

$$\frac{d}{dt} f_\alpha(\varphi_\alpha(x(t))) = 1, \quad \forall x \in U_\alpha; \qquad (43)$$

2) If the function $f(\vec{x}) \in C_R(G)$ in some domain $G \subset R^n$ satisfies the equation

$$\sum_{k=1} a_k(\vec{x}) \frac{\partial f}{\partial x_k} = b(\vec{x}, f)$$

$b(\vec{x}, f)$ being a continuous function of their arguments, then the value of f on any point $\vec{x}(t)$ of the integral curve of the system $d/dt\, \vec{x}(t) = \vec{a}(\vec{x})$, for which $|f|$ is bounded, is uniquely determined by its value at any point $\vec{x}(t_o)$ on the same curve.

Therefore, one can cover the space Ω with open subsets U_α homeomorphic by $\varphi_\alpha : U_\alpha \to R^{2n-1}$ to subsets $\varphi_\alpha(U_\alpha) \subset R^{2n-1}$. On each $\varphi_\alpha(U_\alpha)$ one takes the integral of the solution $f_\alpha(\varphi_\alpha(x))$ of Eq. (43) with measure $\rho(\varphi_\alpha(x))d\Omega$. This gives the quantity

$$F(\rho) \overset{def}{=} \sum_\alpha \int_{\varphi_\alpha(U)} f_\alpha(\varphi_\alpha(x))\rho(\varphi_\alpha(x))d\Omega; \quad \rho\, V_1(\Omega) \qquad (44)$$

Here $V_1(\Omega)$ is the space of all Borel probability measures on Ω, defined above. The time evolution of this quantity is given trough the evolution group which defines the dynamical system:

$$F_t(\rho) \stackrel{def}{=} \sum_\alpha \int_{\varphi_\alpha(U)} f_\alpha(\varphi_\alpha(x))(g_t\rho)(\varphi_\alpha(x))d\Omega \qquad (45)$$

Locally, one easily finds that for each t R and any open interval $\delta t \subset R^+$

$$(F_{t+\delta t} - F_t)(\rho) > 0, \qquad \rho \neq \rho_{eq}. \qquad (46)$$

Globally, the following theorem is valid.

a) Let $(\Omega, d\Omega, g_t)$ be a mixing dynamical system. Then the map $F_t(\rho)$ of eq. (45) is uniquely defined and continuous in R and $V_1(\Omega)$, with the properties

i) $t'' > t' \in R$, $\rho \neq \rho_{eq}$: $(F_{t''} - F_{t'})(\rho) > 0$ $\qquad (47)$

ii) $\max\{F_t, t \in R\} = F(\rho_{eq})$

b) If the map F_t is continuous in R and $V_1(\Omega)$ and if it satisfies i) and ii) above, then the dynamical system is mixing.

The proof is based on the assertion[21] that a dynamical system is mixing iff for all Borel probability measures $d\mu_\rho$ absolutely continuous with respect to $d\Omega$, i.e. iff for all $d\mu_\rho = \rho d\Omega$, $\rho \in V_1(\Omega)$, we have $\lim_{t \to \infty} g_t \circ \rho \stackrel{a.e.}{=} 1$. The simple consequence is that for a mixing dynamical system all orbits $\{g_t\rho\}$, $\rho \in V_1(\Omega)$, except for $\rho = \rho_{eq}$, are wandering. The similarity with the bases of the Misra proof is obvious. In both cases the expressions for the entropy are, by assumption, bounded. Also, in both cases the maps from $V_1(\Omega)$ into R^+ are multivalued but the difference for the two values t_1 and t_2 can be made unique by fixing one of them.

Naturally, in the phase-space analysis the time t is a parameter. As a consequence the deeper physical meaning of the time operator in the Misra-representation is needed.

4. CONCLUSION

Our understanding of the formal properties of the time operator since the first Schrödinger attempt has considerably improved. In the interpretational sector we are confronted with the non-standard situations related to the use of one-sided unitary transformations and/or non-orthogonal resolutions of the identity. With respect to the ontological questions, i.e. those concerning the nature of time, no profound changes have taken place in the fields considered above.

References

1. N. Bohr, Phys. Rev. 48 (1935). See also N. Bohr, "Atomic Physics and Human Knowledge", Wiley, New-York 1958, pp. 32-66
2. L. Mandelstam and I. Tamm, J. Phys. U.S.S.R. 9 (1945)249
3. V. Fock and N. Krylov, J. Phys. U.S.S.R. 11 (1947) 112
4. Y. Aharonov and D. Bohm, Phys. Rev. 122 (1961) 1649;
5. E. Schrödinger, Sitzungsber. der Preuss. Akad. der Wissenschaften, 1931 (12) p. 238
6. W. Pauli, "Encyclopaedia of Physics" (ed. S.Flügge) vol. 5/I; p.60; (1958)
7. H. Paul, Annal. der Physik 9 (1962) 252
8. F. Engelmann and E.Fick, Phys. 178 (1964) 551
9. G.R. Allcock, Ann. Phys. (N.Y.) 53 (1969) 253
10. H. Ekstein and A.J.F. Siegert, Ann. Phys. (N.Y.) 68 (1971) 509
11. A.S. Holevo, Rep. Math. Phys. 13 (1978) 287
12. M. Bauer, Ann. Phys. (N.Y.) 150 (1983) 1
13. R.G. Newton, Ann. Phys. (N.Y.) 124 (1980) 327
14. I. Prigogine, C. George, F. Henin and L. Rosenfeld, Chem. Scr. 4 (1973) 5
15. B. Misra, Proc. natl. Acad. Sci. USA, 75 (1978) 1627
16. H. Poincare, Com. Ren. Hebd. Seances Acad. Sci. 108 1889) 550
17. N. Burić, "New Aspects of Poincare's Theorems on Entropy", (in Serbian), M. Sc. Thesis, University of Belgrade 1986
18. A.S. Holevo, "Probabilistic and Statistical Aspects of Quantum Theory", North Holland 1987
19. B. Leaf, J. Math. Phys. 10 (1969) 1971
20. M.A. Naimark, Ižv. Acad. Nauk SSSR Ser. Math. 4 (3) (1940) 277
21. P. Walters, An Introduction to Ergodic Theory, Springer-Verlag, 1982, p. 146.

STOCHASTIC OPTICS: A WAVE THEORY OF LIGHT BASED ON CLASSICAL PROBABILITIES

Trevor W. Marshall[1] and Emilio Santos[2]
[1] Dept. of Mathematics, Univ. of Manchester, Manchester, United Kingdom;
[2] Dept. de Física Moderna, Universidad de Cantabria, Santander, Spain

ABSTRACT. Quantum optics is based centrally, on the concept of the photon, a particle which is considered to belong to the family of elementary particles. We believe that this has led to serious errors, resulting from a failure to understand the concept of statistical independence, in the interpretation of photoelectron counts. We show that a purely wavelike description of the electromagnetic field is capable of explaining all of the phenomena which are said to exhibit the "wave-particle duality" nature of the photon. The resulting theory, which we call stochastic optics, is a natural by-product of Planck's radiation theory; it may also be considered as a variety of semiclassical radiation theory in which there is always a random zeropoint electromagnetic field present in addition to any additional sources of radiation.

1. WAVES AND/OR PARTICLES

The wave and particle behaviours of light and matter give rise to the main difficulties for the understanding of the microscopic world. At first sight, both aspects seem well integrated in quantum theory, but a deeper study shows that this is at the price of abandoning either the classical laws of probability or the space-time description of the phenomena. Consider, for instance, the celebrated two-slit experiment with electrons (see Fig. 1). If the electron is a particle travelling from the source S to the detecting screen D through either slit A or B, according to a naive view of probability theory, the arrival at a given point C, when the two slits are open, should be the sum of the probabilities when only one slit is open. This is not the prediction of quantum theory, but rather a probability distribution with several maxima, corresponding to a sum of "probability amplitudes", each one associated with a possible path. This might suggest a wave nature for the electron, were it not for the strong evidence that the electron is a point (or very small) particle. In the first place, each electron is detected at a point of the screen, but there are also more convincing arguments. For instance, detailed calculations of the electronic structure of atoms, in particular helium, have shown the need of considering point-like electrons in order to interpret correctly the observed spectra. In summary, according to quantum theory the electron (or the photon, the neutron, etc.) seems to be a particle, but its motion contradicts the classical laws of probability.

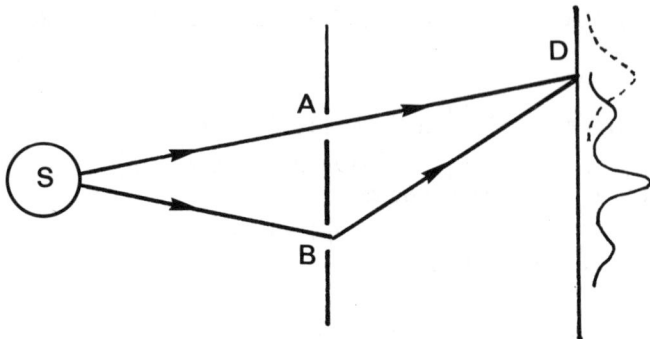

Figure 1. Two-slit experiment. Electrons from the source S arrive at the screen D after going through slit A and/or B. The dashed (continuous) line on the right represents the arrival probability with only slit A (both slits) open.

However, this is not the interpretation of a mature probability theory, which recognizes that an electron passing through one slit has a behaviour which may depend on the presence or absence of the other slit. Such a dependence would be characteristic of a wave-particle duality picture of the de Broglie type, and it is also characteristic of a theory like stochastic electrodynamics which acknowledge the existence of real zero-point fields.

Similar problems appear when one tries to reconcile the wave and particle behaviour of light. Consider for instance, the following experiment (see Fig. 2). A light beam coming from a source S is sent to a beam splitter BS1, where it is divided into two beams, each one carrying half the incoming intensity. Each beam is reflected in an ordinary mirror (M1 or M2) and both are recombined in a second beam splitter BS2. A photomultiplier PM1 put after, will be activated more or less frequently depending on the difference in length between the paths BS1-M1-BS2 and BS1-M2-BS2, the maximum (minimum) activation frequency corresponding to a length difference which is an integer (an integer plus one half) number of wavelengths. This behaviour is attributed to the interference between the two beams and proves the wave nature of light. The experimental setup can be modified by removing the beam splitter BS2 and adding a new photomultipliers PM2, with PM1 and PM2 connected to a coincidence electronic device C . No coincidences (or very few) are observed, which is interpreted as a proof that light is composed of particles ("photons"), each one going to onephotomultiplier. The above interference and anticorrelation experiments can be performed in a more dramatic fashion if the source is a quasar many million light years away, and the role of the beam

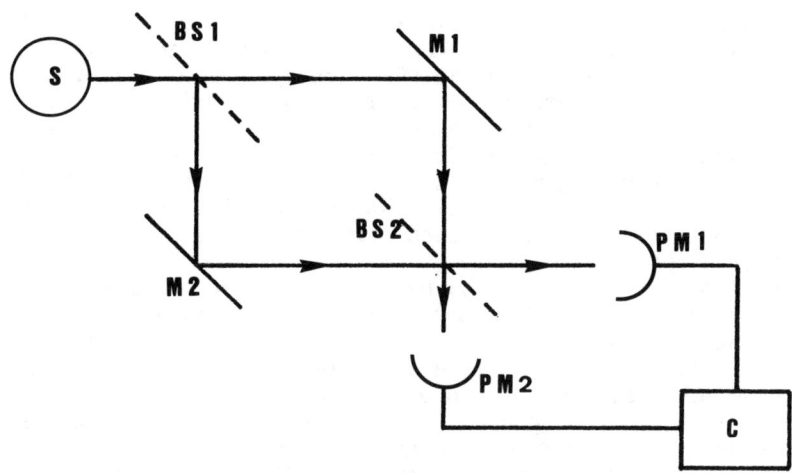

Figure 2. Recombination and anticorrelation experiments. A light signal ("photon") from the source S is divided at the beam splitter BS1 and can be recombined at BS2, showing interference (wave behaviour). Alternatively, the beam splitter BS2 can be removed and the signal is detected either in PM1 or PM2 but not in both (particle behaviour).

splitter is played by a galaxy acting as a gravitational lens. In this case the "photon" must have decided to come through the upper or the lower path (in the anticorrelation experiment) or through both paths at the same time (in the interference experiment) well before the experimenter decides which experiment to perform. This "delayed choice experiment", as Wheeler[1] calls it, illustrates again the kind of difficulty encountered when one attempts to understand the quantum phenomena.

For any physicist not willing to accept the Copenhagen interpretation only one solution seems to exist namely the dualistic picture proposed long ago by Louis de Broglie. In this view it is assumed that both waves and particles are real, interconnected entities. Waves guide the motion of particles, but only particles carry energy and momentum and, therefore, can be detected. We think that this is probably true for material particles, like electrons or neutrons, but not for light. In the following we shall show that anticorrelation, as well as interference experiments, can be explained with a purely wave theory of light. There are people who may reject this solution "a priori" because it treats asymmetrically light and material particles. They consider it more appealing to assume - as quantum theorist do - that every entity has a wave and a particle aspect. We do not agree with this opinion. We think that it is simpler to return to the prequantal, Maxwellian world view. According to it, light is a purely (electromagnetic) wave phenomenon, whilst particles (e.g. electrons) are real particles, but they are always accompanied by a wave partner, namely, the fields created by the particle. However, in contrast with the prequantal view, we think that the fields (waves) have a much richer structure than previously assumed due to an inherently stochastic element. We are convinced that this world view, when properly developed, will result in a realist local alternative to quantum theory. Obviously, we have in mind not only the electromagnetic field but also those associated with strong, weak and gravitational forces and, possibly, others. With this picture, we hope, for instance, to be able in the future to interpret neutron interference experiments. For the moment, we shall deal exclusively with the behaviour of light. That is, we are trying to develop just a local realist alternative to quantum optics, which we call stochastic optics.

After this general discussion, let us return to the experiments considered above. If light is waves, the interference experiment has no mystery, so we shall discuss only the anticorrelation experiment. There are frequent mentions in the literature of experiments of this kind, which, it is claimed, have shown a very small number of coincidences, in agreement with the particle ("photon") view of light. The words "very small", however, are nonsense if a quantitative discussion is not made. To do that, it is appropriate to introduce the following "correlation parameter"

$$\alpha = N N_c / N_r N_t , \qquad (1)$$

where $N_t (N_r)$ is the number of counts in the transmitted (reflected) beam per unit time - that is the counting rate of the photomultiplier PM1 (PM2), see Fig. 2 - N_c the number of coincidence counts and N the number of "photons" in the incident beam. A naive particle ("photon") theory predicts $\alpha = 0$ because $N_c = 0$, when the incoming beam is so weak that only one (or zero) "photon" will arrive at the beam splitter within a coincidence window. Alternatively, a naive wave theory, where the activation rates of the two photomultipliers by the reflected and transmitted beams are proportional to the incoming intensities, and these are uncorrelated, predicts $\alpha = 1$ because the detection probabilities fulfil

$$p_c = p_t p_r, \quad N_c/N = (N_t/N)(N_r/N), \quad \alpha = 1 . \qquad (2)$$

In consequence, an experiment will provide an argument for a particle behaviour of light only if it gives $\alpha \ll 1$. The problem is that the total number, N, of incoming "photons", and hence α, cannot be measured. The escape from this difficulty is to compare simultaneous with delayed coincidences. That is, whenever a count occurs in PM1, say at time t, it is recorded whether another count takes place in PM2 within $t+\tau$ and $t+\tau+w$, w being a suitable (small) coincidence window. The measurable coincidence probability with τ delay, $p_c(\tau)$ will certainly fulfil eq. (2) for high enough τ. Therefore, the parameter α of (1) can be calculated through

$$\alpha = N_c(0) / N_c(\infty) , \qquad (3)$$

that is, as the ratio between the number of coincidences at zero delay ($\tau = 0$) and the number of coincidences with a very large delay ($\tau \to \infty$).

The measurement of the correlation parameter α (eq. (3)) with the experiment illustrated in Fig. 2 was made for the first time by Hanbury Brown and Twiss in 1956[2] using thermal light. They obtained $\alpha = 2$, a phenomenon called "photon bunching", instead of the naive "photon" prediction $\alpha \ll 1$. Therefore, no particle behaviour was shown. This proves that Wheeler's delayed choice experiment which is essentially a repetition of the Brown-Twiss experiment poses no problem for a purely wave theory of light. The result $\alpha = 2$ has been shown to be a characteristic of all thermal light <u>no matter how weak</u>. That is, the value of α depends on the coherence properties of the incoming light and not on its intensity. (We suppose this independence is difficult to understand for people thinking in terms of only particles or in terms of waves plus particles). With laser light, the value obtained is $\alpha = 1$, so that again no particle behaviour is shown. Only with light coming from single atom emission is it possible to obtain values of α smaller than one. The first experiment reporting such a small value of the correlation parameter (1) has been made very recently (1985) by Grangier et al.[3]. They were able to estimate N using an atomic source emitting "photons" in pairs (from atomic cascades) and opening the window for the photomultipliers PM1 and the PM2 when one count is recorded in an auxiliary photomultiplier (not shown in Fig. 2). A less conclusive experiment with a similar idea had been made by Clauser[4] in 1974. There are other experiments made in the last two decades, showing "photon antibunching", which have been claimed to support a particle description of light. We show elsewhere[5,6] that they can be easily explained by a stochastic wave theory of light. (See also next section). The discussion of the anticorrelation experiment will be made in section 4.

2. "PHOTON" COUNTING STATISTICS

We want here to discuss briefly "photon" counting statistics in order to show that most of the experiments in this area are easily compatible with a purely wave theory of light that we are developing. It belongs to a class of theories usually termed semiclassical, instead of classical, because they do not contain a detailed explanation of the detection process. In "semiclassical" theories, the probability of activation of a photomultiplier within a small time interval - say between t and t+w - is assumed to be proportional to the light intensity I(t) arriving at time t (Here intensity really means energy per unit time).

We shall consider first a simple experiment consisting of a light source and a photomultiplier, both working in a stationary regime. The observable quantity of interest is the number, $N(\tau)$, of coincidences with time delay τ, produced in a long run. We shall focus on the "correlation function"

$$\alpha(\tau) = N(\tau) / N(\infty), \qquad (4)$$

This function cannot be measured for small values of τ due to memory effects in the photomultiplier, e.g., dead time. For this reason, most experiments have been performed with the setup of Fig. 2, i.e. measuring coincidences between detectors in two channels. In the case of a single photomultiplier, the assumption that the detection probability is proportional to the incoming intensity leads to

$$\alpha(\tau) = <I(t)\,I(t+\tau)> / <I(t)>^2 , \qquad (5)$$

where <> means time average. The denominator corresponds to the autocorrelation for $\tau \to \infty$. In this case the two intensities involved are uncorrelated, so that the average of the product becomes the product of averages. In the case of two photomultipliers (arrengement of Fig. 2), the correlation function is

$$\alpha_{12}(\tau) = <I_1(t)\,I_2(t+\tau)> / <I_1(t)> <I_2(t)> , \qquad (6)$$

which is not the same as (5). However, the implicit assumption is always made that the fraction of the light transmitted by the beam splitter is a time independent constant F and similarly for the reflected beam. With this assumption it is easy to see that $\alpha_{12}(\tau)$ should be identical with (5).

Three possibilities exist, namely $\alpha(\tau)$ being a constant, $\alpha(\tau)$ decreasing with τ ("photon" bunching) and $\alpha(\tau)$ increasing with ("photon" antibunching). (Obviously, $\alpha(\infty) = 1$ always, see (4)). The first two cases can be easily explained within a semiclassical (see above) wave theory of light. The third one presents problems which we discuss elsewhere[5,6]. The case of constant $\alpha(\tau)$ occurs when the light source is a laser. This is very easy to explain, because a laser has a high coherence, that is, constant intensity,

$$I(t) = I_0 . \qquad (7)$$

Combining (7) with (5) we get $\alpha(\tau) = 1$. A decreasing $\alpha(\tau)$ appears if the light source is thermal, where the function decreases from 2, for $\tau = 0$, to 1 for large τ. Again the explanation is simple. Thermal light corresponds to Gaussian fluctuations of the (electric and magnetic) fields of the light beam, which implies an exponential distribution for the intensity. That is, the probability that the intensity I(t) lies between I and I+dI is

$$dP = I_0^{-1} \exp(-I/I_0)\,dI \qquad (8)$$

Hence we get

$$<I(t)> = \int I\,dP = I_0, \quad <I(t)^2> = \int I^2 dP = 2\,I_0^2, \qquad (9)$$

whence the result $\alpha(0) = 2$, follows.

It is interesting to compare the simple explanation just given, with the quantum interpretation of the phenomena. As usual, the quantum formalism gives the correct prediction, but the interpretation is counterintuitive. The laser correlation function $\alpha = 1$ is currently interpreted as due to the fact that "photons" in a laser beam are uncorrelated and so the arrival probability at the photomultiplier PM2 (see Fig. 2) is independent of the time elapsed from the arrival of a given "photon" at PM1. The "photon" bunching of thermal light, in contrast, is attributed to the Bose-Einstein statistics of "photons", that have a tendency to arrive in "bunches". This interpretation (not really an explanation) is very strange because it implies that "photons" in thermal, incoherent light are more organized that in a highly coherent laser.

3. REMARKS ON STATISTICAL INDEPENDENCE

Since this conference has the concept of probability as its central theme, we include here some general observations on statistical independence. This is a concept which is important in science, both in developing theoretical models and in subjecting such models to experimental test.

We believe that, in both of these areas, theoretical and experimental, the modelling of an atomic light signal by a "photon", especially the concept of the latter as an elementary particle, has led to ways of thinking, and of analysing experiments, which would be quite unaceptable in any other branch of science. We remarked in the previous section on the implications of "photon" bunching; it amounts to saying that thermal light is more organized than laser light. This is one example from the theoretical area.

We have remarked elsewhere[7] on a related experimental practice. In analysing the counts of a photomultiplier PM1 activated by a light signal which has passed through a linear polarizer P1 (See Fig. 3) it has become customary, on the basis of the photon hypothesis, to assume that these two devices act independently. Hence, in analysing pairs of light signals originating at a common source, one naturally begins by making the same assumption about PM2 and P2 (Again see Fig. 3). As a working hypothesis this is unexceptionable, but, on the basis of the coincidence rate observed at PM1 and PM2, it is now claimed that a statistical dependence has been demonstrated between the actions of the two polarizers P1 and P2. Such a dependence, if truly observed, would be amazing, because it would violate Einstein's Principle of Local Action[8], which is the basis of both Special and General Relativity, and is the modern form of the Principle of Causality. Nevertheless, there is a quite astonishing resistance in the physics community to accept the alternative proposition; that the observed coincidece rates demonstrate a statistical dependence between the actions of P1 and PM1. The explanation lies, we believe, in an inability to dispense with the photon concept. We shall be disscussing experiments of this type further in section 5.

Now we address the question: what are the independent constituents of the electromagnetic field?. Semiclassical theories, in the optical frequency region at least, reply that the independent constituents are the Fourier components: the statement that, for arbitrary normalization volume, the Fourier components are independent, is equivalent to specifying that the corresponding stochastic process is Gaussian. We discuss the analysis of photoelectron counts on the basis of such a description in the next two sections.

But "photon" theories give no clear answer to the question we have

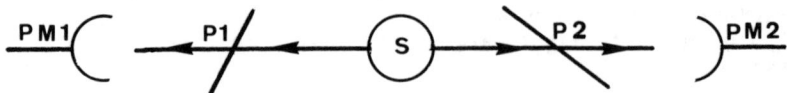

Figure 3. Experiment for the test of Bell's inequalities measuring the polarization correlation of "photon" pairs. The source emits pairs of light signals, one going to the detector PM1 through the polarizer P1, and the other to PM2 through P2.

just posed. They start by asserting that the "photons" are the independent entities, but then they tell us (a) that these elementary particles of the electromagnetic field are indistinguishable and (b) that their total number is not conserved. We arrive in this way at Bose-Einstein Statistics with zero chemical potential, that is the Planck distribution. As we have seen in the previous section, this forces us to describe the "photons" emitted from a thermal source as being organized in "bunches". Even more curiously, experiments involving a pair of carefully tuned laser sources indicate that the "photons" from two independent sources are emitted in some kind of collusive manner[9,10].

Some advocates of the "photon" concept have judged that the particle nature of the photon is not consistent with the indistinguishability which is used in deriving the Planck distribution. They have therefore proposed[11] to use Boltzmann counting but with a different weighting of the cells in phase space. In this way it is possible to derive a Bose-Einstein distribution. There are two criticisms of this procedure. The first is that it leads not to the Planck distribution but to a Bose-Einstein distribution with a definite number of particles - something more appropriate to describe a gas of helium atoms. A "photon" gas is not merely composed of indistinguishable particles; the "particles" are not conserved either.

The second criticism is even more serious, and is precisely the objection Eherenfest made to Einstein[12] immediately after the birth of the Bose-Einstein statistics. In order to obtain the Bose-Einstein statistics for distinguishable particles, Tersoff and Bayer[11] proposed that, for a phase space divided into M cells, instead of giving each cell the a priori weight

$$W_i = M^{-1} \quad (i = 1, 2m \ldots M) \tag{10}$$

we should instead take $0 < W_i < 1$ and

$$\langle W_i \rangle = 1 , \tag{11}$$

and then randomise the weights subject to this constraint. But this latter constraint is precisely the "mysterious interaction" to which Ehrenfest objected, and which Einstein acknowledged. It is now possible to see, more clearly than either Ehrenfest or Einstein were able to see at the time, that the "mysterousness" of the interaction lies precisely in its nonlocalilty[13]; it violates Einstein's own Principle of Local Action[8]. In a sense, therefore, Ehrenfest in 1925, and indeed, as we shall see in the next section, Planck in 1911, anticipated Einstein; the concept of the "photon" seems to lead inevitably to both confusion about statistical independence and to a nonlocal description of phenomena.

We now proceed to discuss the alternative, purely wave description of the electromagnetic field.

4. THE PRINCIPLES OF STOCHASTIC OPTICS

A purely wave theory of light can, obviously, explain diffraction and interference experiments. With the addition of a reasonable hypothesis about the activation of photomultipliers, it is also able to explain most of the "photon" statistics, including "photon bunching". There remain essentially only two apparent difficulties to be explained:
  1) "Photon" anticorrelation or "photon" antibunching in a beam splitter, which is usually interpreted as a corpuscle behaviour.
  2) The violation of the Bell inequalities in the "two photon" polarization correlation experiments.
In the last section, we have shown that anticorrelation experiments can be easily explained if we assume a stochastic action of beam splitters, and a similar explanation is possible for "photon" antibunching, as shown elsewhere[5,6]. On the other hand, in order to explain the violation of Bell's inequalities, it is unavoidable to assume the property of "enhancement", that is, the possibility that the activation probability of a light signal may increase when a polarizer is inserted between the source and the detector. (This is dealt with in the next section). Then the two difficulties above stated are reduced to explaining:
  1) Stochastic action of beam splitters.
  2) The enhancement feature of polarizers.
In the following we show that both facts are straightforward consequences of the principles of stochastic optics.

Stochastic optics[14,15] is just a part of stochastic electrodynamics, which in turn is a quite natural modification of classical electrodynamics. This is not a recent theory since its origin can be traced back to the beginning of the century, when Planck discovered his "second radiation law", which modified the first one by adding to the blackbody spectrum the term

$$(\nu) \, d\nu = (4\pi/c^3) \, h\nu^3 \, d\nu, \tag{12}$$

usually called zeropoint radiation. This term has the undesirable feature of giving rise to an infinite energy density (ultraviolet catastrophe) and it is ignored in modern times for this reason. In quantum field theory the removal of the zeropoint is accomplished by the socalled "renormalization technique", which is an efficient calculational trick, but a clearly unsatisfactory procedure from both the mathematical and the physical point of view. Then, why not consider the zeropoint field as real, attributing the (non-physical) divergence simply to our ignorance of the laws of nature at short distances (or high frequencies)?. Some unknown mechanism might well provide a modification (cut-off) of the high part of the spectrum (12) so removing the divergence. If we consider seriously the zeropoint field, some of the most grossly wrong predictions of classical electrodynamics change drastically, becoming much closer to the experiments, without any need of quantization.

The systematic development of stochastic electrodynamics - that is, classical electrodynamics with the assumption of a real random background radiation having the spectrum (12) - began in the fifties, and a number of interesting results have been obtained since. It is out of the scope of this paper to review the theory, but two conclusions can be obtained from these 30 years of research effort:

1) Contrary to early expectations, the theory is unable to explain all quantum electrodynamical phenomena. We think that the theory is incomplete rather than incorrect. Probably we should include in the physical "vacuum" many other things, for example, an electron-positron sea besides the sea of zeropoint radiation assumed in stochastic electrodynamics.

2) The theory has had a number of sucesses that cannot be attributed to chance, in particular the behaviour of linear systems and the explanation of phenomena like the Casimir effect, involving the interaction of the zeropoint radiation with macroscopic bodies.

The failure of stochastic electrodynamics to give a correct description of the interaction between radiation and charged particles (e.g., emission and absorption of light by atoms) and the success for phenomena involving the interaction of radiation with macroscopic bodies, has encouraged us to develop stochastic optics, i.e. the classical theory of propagation of light through macroscopic bodies (mirrors, lenses, polarizers, etc.), but taking into account the zeropoint radiation. We consider that the theory for emission and absorption of radiation by atoms is outside the scope of stochastic optics. However, in order to be able to interpret real experiments we must have some model for the emission and absorption. This is why we include two "ad hoc" hypotheses as follows. We stress again that these principles constitute just a first approximation, used because we still lack a more complete form which might be specific for each kind of instrument. We assume that theory would change the details of the calculation, but not the general implications.

P_1. Atoms emit radiation in the form of needles, i.e., classical wavepackets with shape similar to a cylinder having a diameter of several wavelengths and a length of the order of meters. The wavevectors and the polarization of the needles are random, with a distribution to be defined later. The total energy of a needle is of the order $h\nu$ (where h is Planck's constant) <u>additional to</u> the energy contained in a similar volume and the same radiation mode in the zeropoint (vacuum) field, which is again of order $h\nu$ ($1/2\ h\nu$ for each polarization). The needles correspond to the "one-photon" states of quantum optics.

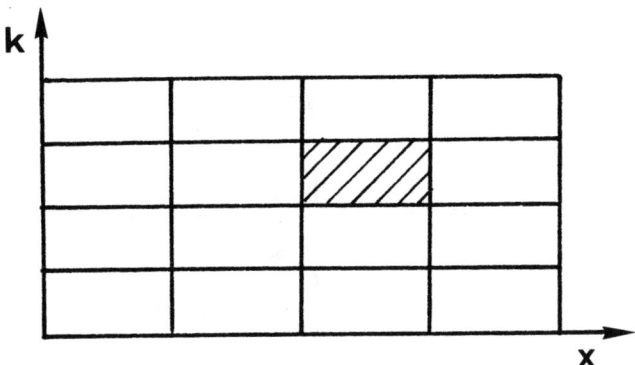

Fig. 4. Schematic representation of the sixdimensional space associated to the electromagnetic radiation. Each cell (represented by a rectangle) contains noise with total energy $h\nu$, except one cell containing a signal with energy $2h\nu$.

The picture that emerges from this is the following. The electromagnetic vacuum contains a set of normal modes or cells in six-dimensional space (three spacial dimensions and three dimensions in wavevector space). For our convenience, we choose the spacial shape of the cells as needles. We may imagine any sample of radiation as a superposition of plane waves, corresponding to the usual Fourier analysis of the electromagnetic field. Then we may cut each plane wave in pieces, each having the form of a needle. Note that, in a volume V in sixdimensional space we must have $V/8\pi^3$ cells. In classical electrodynamics the vacuum corresponds to all cells having zero energy but in stochastic electrodynamics each cell contains an average energy $h\nu$. A "one-photon" state differs from the vacuum state in that one of the cells has energy $2h\nu$ instead of $1h\nu$ (see Fig. 4).

P_2. Photodetectors are insensitive to the zeropoint radiation. They can be activated only when one of the cells in sixdimensional space contains energy above $3/2\ h\nu$ (we choose 3/2 as intermediate between 1 for the zeropoint alone and 2 for the one-photon state). The probability of activation per cell of radiation entering the detector is

$$P = \xi(I/h\nu - 3/2)_+ , \qquad (13)$$

where I is the total energy in the cell, ξ is an efficiency parameter ($\xi \ll 1$) and ( )$_+$ means putting zero if the quantity inside de bracket is negative. Note that the bracket takes the value - 1/2 for each vacuum cell and + 1/2 for a cell carrying one "photon".

This postulate is clearly somewhat ambiguous because it does not specify how to count the cells in sixdimensional space. We hope to be able to make things better in the future when we understand how a photodetector can discriminate between a needle of radiation amongst the infinity of "vacuum" cells given that there is no difference in nature between the radiation of the "vacuum" and that of a signal, and

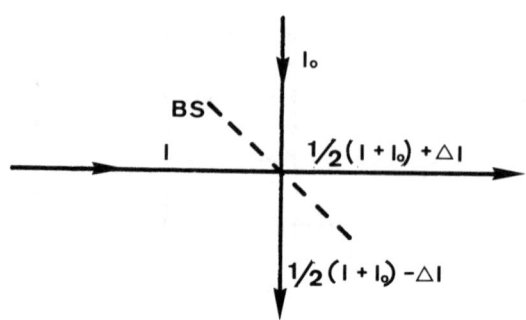

Fig. 5. Stochastic action of a beam splitter BS. The incoming signal, with energy I, interferes with the relevant part of the noise, with energy I_0, giving two beams, one above and the other below the detection threshold.

one signal has just twice the energy of a vacuum cell. We call this the problem of the needle in the haystack!. We shall show immediately how to use this postulate by the simple procedure of ignoring all cells except one or two in each case.

P_3 The transmission, refraction and reflection of radiation (including the zeropoint) in macroscopic bodies follows the laws of classical optics.

This completes the postulates of stochastic optics. Now we illustrate how they work in the interpretation of the interference and anticorrelation experiments of Fig. 2. We begin with the behaviour of a single beam splitter that transmits half the incoming intensity and reflects the other half. A light signal (with energy I, see Fig. 5) coming in the form of a needle of radiation is divided in two half-needles (remember that our needles are not particles but classical wave packets which, therefore, can be divided). On the other hand, some amount of zeropoint radiation (with intensity I_0, Fig. 5) coming in the perpendicular direction is also divided, going to the same cells of the half-needles. However, the intensities transmitted and reflected are not $(I_0+I)/2$, because we should add amplitudes, not intensities. By conservation of energy, if one channel has an excess energy, ΔI, the other channel must have a defect, $-\Delta I$. But the value of ΔI is random due to the random character of the zeropoint field (the size and sign of the inteference term depend esentially on the relative phases of the signal and the noise). In a long run, the relative phases of signal and noise are changing, this been the origin of the stochastic action of a beam splitter, needed to explain "photon" antibunching.

In order to get a more quantitative view, we must find the value of the intensity I_0, because I is $2h\nu$ according to postulate P_1. Is I_0 infinity in view of the infinite energy of the zeropoint radiation?. Clearly it is not, because most of the zeropoint noise emerging from the beam splitter is unable to interfere with the two outgoing half-needle signals. In order to interfere, signal and noise must have the same wave-vector and occupy the same region of space, that is, they should correspond to the same cell in sixdimensional space. In consequence we must only take into account four cells for our problem: the cell of the incoming needle, the two cells of the outgoing half-needles and the cell of incoming noise that goes to the same cells as the two half-needles. This is what we call "the relevant part of the noise". Its energy is, clearly, $1h\nu$. In consequence the transmitted half-needle plus the relevant zeropoint has energy $3/2\ h\nu + \Delta I$ and the reflected one $3/2\ h\nu - \Delta I$. According to Eq. (13), therefore, only the transmitted (reflected) signal plus zeropoint can activate a photodetector if ΔI is positive (negative). This is the anticorrelation property of one-photon signals in a beam splitter, clearly shown for the first time in the recent anticorrelation experiment of Grangier et al.[3] and currently interpreted as a particle behaviour of light (see Fig. 2).

Stochastic optics, on the other hand, is able to explain interference experiments as the one also represented in Fig. 2. In order to analyze the action of the second beam splitter BS2, it can be realized that we must take into account only the radiation incoming in the two cells carrying a "half-needle plus zeropoint". It is rather obvious that if the lenghts of the paths BS1-M1-BS2 and BS1-M2-BS2 are equal, or differ in an integer number of wavelenghts, the action of the beam splitter BS2 is exactly the reverse of the action of BS1, and the full initial needle is reconstructed in the ray going to the photomultiplier PM1. On the other hand, the needle is reconstructed in the lower channel if the path difference is a half-integer number of wavelengths. It can be shown by explicit calculation that the prediction for other situations agrees quite well with the predictions of quantum optics and experiments. In this way, stochastic optics is able to explain quite naturally both the interference and the anticorrelation experiments. In comparison with the "empty wave" and the "wave plus corpuscle" of the dualistic school, we have just two classical waves similar in nature but having different energy, one below and the other above the threshold for the activation of photodetectors. The subthreshold wave carries the latent order that manifests in the recombination experiment.

The careful reader will have realized that our previous model contains one difficulty, namely, that the interference of zeropoint alone can produce signals above threshold. In fact, if we put I_0 instead of I in Fig. 4, the outgoing energy in one channel is $h\nu + \Delta I$, but ΔI may be as high as $h\nu$, so surpassing the threshold $3/2\ h\bar{\nu}$. Of course real detectors produce a dark rate, which fits very well with the general ideas of stochastic optics. However, the dark rate predicted by our eq. (4) is too high. This is a real problem that we have not yet been able to solve. We shall not discuss the difficulty in detail (it is related to the "needle in the haystack" problem), but we simply recognize that our model must be somewhat modified. For the moment, stochastic optics can be considered to be in a preliminary stage and it is able to give only qualitative and semiquantitative interpretations of experiments.

5. OPTICAL TESTS OF THE BELL INEQUALITIES

We shall not discuss here the general aspects of the Bell theorem, which proves the incompatibility of local realism and quantum mechanics for some ideal experiments[16,17]. We shall simply give an idea of the performed optical tests of local realism versus quantum theory and the reason why they are not conclusive. The experimental setup is represented in Fig. 3. An atomic source S emits "photon" pairs going towards the polarizer P1 and P2 through appropriate lens systems ("photons" traveling in other directions are not recorded). The coincidence detection rate $R(\theta_1,\theta_2)$ is measured as a function of the polarizers axes, given by the angles θ_1, θ_2. Quantum optics predicts

$$R(\theta_1,\theta_2)/R_0 = 1/4 \left[1 + \cos 2(\theta_1 - \theta_2)\right] , \qquad (14)$$

where R_0 is the coincidence rate with the polarizers removed.

In a naive classical theory we may assume that the source will emit pairs of light signals both linearly polarized in the same plane, say at an angle λ with the vertical. When the signal goint to the left crosses the polarizer P1, its intensity is reduced by the Malus factor $\cos^2(\theta_1 - \lambda)$. Similarly the reduction of the signal going to the right will be $\cos^2(\theta_2 - \lambda)$. If we assume that λ is uniformly distributed, the predicted reduction is the following average over λ,

$$R(\theta_1,\theta_2)/R_0 = \pi^{-1} \int d\lambda \cos^2(\theta_1 - \lambda) \cos^2(\theta_2 - \lambda) = \qquad (15)$$
$$= 1/4 + 1/8 \cos 2(\theta_1 - \theta_2) .$$

We see that the correlation predicted is smaller than the quantum one (14). It is possible to assume an elliptical, instead of linear, polarization for the emitted light signals. A reasonable distribution of elliptical polarizations gives

$$R(\theta_1, \theta_2)/R_0 = 1/4 + 1/12 \cos 2(\theta_1 - \theta_2) , \qquad (16)$$

that is, an even smaller correlation. We might relax the concept of "classical" rejecting Malus' law, i.e., putting in (15) an arbitrary probability $P(\theta_1, \lambda)$ instead of $\cos^2(\theta_1 - \lambda)$. Again, the model so constructed cannot agree with the quantum prediction (14). This illustrates Bell's incompatiblity theorem.

The formal proof of the theorem is made by defining the most general local realist (or local hidden variables) model by

$$R(\theta_1,\theta_2)/R_0 = \int Q_1(\theta_1,\lambda) Q_2(\theta_2,\lambda) \rho(\lambda) d\lambda . \qquad (17)$$

This expression generalizes (17) in several respects: λ may be a set of parameters instead of just one, $\rho(\lambda)$ is an arbitrary probability distribution and, finally, Q_1 and Q_2 may be different functions of their arguments. (Actually, an even more general expression than (16) is used, but we have simplified it for the sake of clarity). Note that

$Q_1(\ )$ is not allowed to depend on θ_2 nor Q_2 on θ_1, nor $\rho(\lambda)$ on θ_1 or θ_2, which we, following Bell, consider to be the locality condition. It is assumed that Q_1, Q_2 fulfil

$$0 < Q_1, Q_2 < 1 . \tag{18}$$

The first inequality cannot be questioned. The second condition is the "no enhancement" hypothesis mentioned in section 4 above. It states that the detection probability of a light signal with the polarizer in place is not greater than the probability with the polarizer removed. (The ratio of these probabilities is Q_j). The no enhancement hypothesis is quite natural in a theory where light signals are <u>identical</u> "photons" because then Q is the probability that the photon crosses the polarizer. It is also obvious in classical optics where a polarizer always reduces the intensity of a signal crossing it. However, in stochastic optics it is not true because we are dealing with an open system containg zero point radiation besides "photon" signals. Therefore, the energy of a signal may increase, as well as decrease, in crossing a polarizer.

From (17) and (18) (and also from slightly more general assumptions) it can be derived the Bell type inequality

$$\delta \equiv |R(\theta_1-\theta_2=\pi/8) - R(\theta_1-\theta_2=3\pi/8)| / R_0 < 1/4 \tag{19}$$

This is fulfilled by our previous local realist models. For example (15) predicts $\delta = \sqrt{2}/8 \simeq 0.17$ and (16) $\delta = \sqrt{2}/12 \simeq 0.11$. However the quantum expression (14) violates (19), predicting $\delta = \sqrt{2}/4 \simeq 0.35$. The inequality has been also violated in performed experiments, this being the basis for the widespread state of opinion that local realism (or local hidden variables models) has been refuted empirically. It should be pointed out that the inequality originally derived by Bell did <u>not</u> make use of the questionable second inequality (18), and it is not violated by stochastic optics (or any other local realist theory). However, that inequality contradicts quantum mechanics for ideal experiments only, and cannot be tested in real ones.

Stochastic optics predicts quite straightforwardly a violation of the second inequality (18) and also of (19). For instance, let us consider that a light signal of intensity I, linearly polarized at angle φ, arrives at a polarizer at angle θ. According to classical optics the transmitted intensity should be $I\cos^2(\varphi - \theta)$. However, in stochastic optics, an incoming zeropoint radiation must also be considered, whose relevant part has intensity I (see Fig. 5). If this zeropoint radiation is linearly polarized at angle φ_0, the intensity in the outgoing channel will be $I_0 \sin^2(\varphi - \theta)$. (In the second channel of the polarizer the outgoing intensities of signal and noise will be $I \sin^2(\varphi - \theta)$ and $I_0 \cos^2(\varphi_0 - \theta)$, respectively). The detection probability of that signal plus noise will be (see (13) and the estimate of I and I_0 in the previous section)

$$P = \xi \left(2 \cos^2(\theta - \varphi) + \sin^2(\theta - \varphi_0) - 3/2\right)_+ \tag{20}$$

This probability, divided by the detection probability with the polarizer removed (i.e. $\xi/2$) and averaged over φ_0 gives just the function Q introduced in (17). For the sake of simplicity we approximate $\sin^2(\theta - \varphi_0)$ by its average 1/2, which gives

$$Q(\theta, \varphi) = 4\cos^2(\theta - \varphi) - 2 \text{ if } |\theta - \varphi| < \pi/4 \text{ or } |\pi - \theta + \varphi| < \pi/4,$$

$$0 \qquad \text{if } \pi/4 < |\theta - \varphi| < \pi/4. \tag{21}$$

This function violates the second inequality (18) for $|\theta - \varphi| < \pi/3$, so showing the enhancement feature of stochastic optics. Putting this function $Q(= Q_1 = Q_2)$ in eq. (17), with $\rho = \pi^{-1}$, it is easy to obtain the value $\delta = 0.71$, clearly violating the (Bell) inequality (19).

The restriction to linear polarization is not reasonable. If we allow elliptical polarization both for signal and zeropoint, we must use, instead of the parameter λ, two polarization parameters $\{\varphi, \psi\}$, for which we assume the distribution

$$\rho(\lambda) d\lambda \rightarrow (2\pi)^{-1} \sin 2\varphi d\varphi d\psi, \tag{22}$$

with $0 < \varphi < \pi/2, 0 < \psi < 2\pi$. The function Q is now (see Ref. 18 for a detailed derivation)

$$Q(\varphi, \psi, \theta) = 2 \int_0^{\pi/2} \sin 2\Psi_0 \, d\Psi_0 (2 \cos^2\Psi + \sin^2\Psi_0 - 3/2)_+ \tag{23}$$

where

$$2\cos^2\Psi \equiv 1 + \cos 2\varphi \cos 2\theta + \sin 2\varphi \sin 2\theta \cos \psi,$$

$$2\sin^2\Psi_0 \equiv 1 - \cos^2\varphi_0 \cos 2\theta - \sin 2\varphi_0 \sin 2\theta \cos \psi_0$$

and it can be shown that the average over φ_0 and ψ_0 is equivalent to the average over Ψ_0 written in (23). Putting (21) in (17) we obtain $\delta = 0.45$, much closer to the quantum prediction $\delta = 0.35$ (see above, after eq. (19)). We have shown[18] that a perfect agreement is obtained if, instead our previous estimates for signal intensity ($2h\nu$) and threshold (3/2), we use 2.204 $h\nu$ and 1.431, respectively.

The comparison with real experiments must include the real efficiencies of polarizers, represented by their measurable maximum and minimum transmission, ε_M and ε_m, for linearly polarized macroscopic light. To do that, we propose to change (23) by writing

$$Q(\varphi, \psi, \theta) = (\beta - \nu)^{-1} \int_0^{\pi/2} \sin 2\Psi_0 \, d\Psi_0 (I - \nu)_+,$$

$$I \equiv \beta (\varepsilon_M \cos^2\Psi + (1 - \varepsilon_M) \sin^2\Psi) + \varepsilon_m \sin^2\Psi_0 + \tag{24}$$

$$+ (1 - \varepsilon_m) \cos^2\Psi_0, \quad \beta = 2.204, \nu = 1.431.$$

Also, the prediction is different for different kinds of polarizer (calcite, pile of plates, etc.)[18], but we do not consider this here for the sake of simplicity. With the measured values of ε_M and ε_m in the performed experiments we have calculated the predicted value of the parameter δ for the different measured parameters in the last two Orsay experiments[19]). The results are shown in Table I, where a comparison with the quantum predictions and empirical results are also shown. A correction for depolarization is included both in the stochastic optical and quantum theoretical calculations[18]. (A review of the first four experiments is given in Ref. 16, for the Orsay ones see Ref. 19 and for the Stirling experiment see Ref. 20). It can be seen that our stochastic optics model agrees with experiments even better than quantum mechanics, becuase it explains the "Holt-Pipkin anomaly", i.e., the fact that the only experiment performed with calcite polarizers violated the quantum mechanical predictions.

Our stochastic model can be extended to experiments with additional polarizers, $\lambda/2$ and $\lambda/4$ plates, and good agreement is also obtained with the empirical results[18].

TABLE I. Predictions for the tested parameters

	(a) Stochastic Optics	(b) Quantum Optics	(c) Experiment	(a)-(c)
Freedman-Clauser	.309+.007	.301+.007	.300+.008	.009+.010
Holt-Pipkin	.242+.001	.266+.001	.216+.013	.026+.013
Clauser	.280	.284	.288+.009	-.008+.009
Fry-Thompson	.296+.007	.294+.007	.296+.014	-.000+.016
Orsay I	.313+.004	.308+.002	.307+.004	.006+.006
Stirling	.260+.008	.272+.008	.268+.010	-.008+.013
Orsay II	2.715+.040	2.70+.05	2.694+.015	.018+.043
Orsay III	.101+.010	.112+.010	.101+.020	.000+.022

6. CONCLUSIONS

Stochastic electrodynamics is classical electrodynamics with the assumption that there exists a random background radiation in the whole of space, corresponding exactly to the zeropoint field of quantum electrodynamics, but taken as real. Stochastic optics is the part of stochastic electrodynamics dealing with the interaction between light (including the zeropoint radiation) and macroscopic bodies, and completed by phenomenological postulates for the emission and absorption of light, two phenomena outside the scope of the theory

Stochastic optics offers an extremely attractive explanation of those optical phenomena thought to be "non classical". The explanation, however, is qualitative or semiquantitative for the moment, due to some unsolved difficulties. It is a promising theory still in a preliminary stage.

We acknowledge useful comments by Prof. T. Brody.
Also we acknowledge financial support from
Caja de Ahorros de Cantabria and from the
Royal Society.

REFERENCES

1. Wheeler, J.A., Law without law, in Quantum theory of measurement, eds. Wheeler, J.A. and Zurek, W.H., Priceton Univ. Press., (1983).
2. Hanbury-Brown, R. and Twiss, R.Q., Nature 177, 27 (1956).
3. Grangier, P., Roger, G. and Aspect, A., Europhys. Letters 1, 173 (1986).
4. Clauser, J.F., Phys. Rev. D. 9, 853 (1974).
5. Marshall, T.W. and Santos, E., `Stochastic optics - a classical alternative to quantum optics'. Preprint. Universidad de Cantabria, 1987.
6. Marshall, T.W. and Santos, E., Found. Phys. (in press) (1988).
7. Marshall, T.W., `Stochastic electrodynamics and the EPR argument', in Quantum mechanics versus local realism - The Einstein Podolsky and Rosen paradox. Ed. Selleri, F. Plenum, 1988.
8. Einstein. A., in The Born-Einstein letters, (Macmillan, London, 1971) (Born translates Prinzip der Nahewirkung as "Principle of contiguity". We believe "Principle of Local Action" is more accurate).
9. Pfleegor, R.L. and Mandel, L., Phys. Rev. 159, 1084 (1967).
10. Mandel, L., Prog. Opt., 13, 27-68 (1976). See specially p. 64.
11. Tersoff, J. and Bayer, D., Phys. Rev. Lett., 50, 553 (1983).
12. Pais, A., Subtle is the lord (Oxford Univ. Press, 1982) p. 430.
13. Kyprianidis, A., Sardelis, D. and Vigier, J.P., Phys. Lett., 50 A, 228 (1984); Cufaro Petroni, N., Kyprianidis, A., Maric, Z., Sardelis, D., Vigier, J.P., Phys. Lett. 51 A, 4 (1984).
14. Boyer, T.H., in Foundations of radiation theory and quantum electrodynamics, ed. A.O. Barut, (Plenum, New York, 1980)) page 49.
15. de la Peña, L., in Stochastic processes applied to physics and other related fields, eds. Gomez, B., Moore, S.M., Rodriguez-Vargas, A.M. and Rueda, A., (World Scientific, Singapore, 1983) page 428.
16. Clauser, J.F. and Shimony, A., Rep. Prog. Phys. 41, 1881 (1978).
17. Selleri, F., ed., Quantum mechanics versus local realism - The Einstein, Podolsky and Rosen paradox (Plenum, New York, 1987). In press.
18. Marshall, T.W. and Santos, E., `Stochastic optics. Analysis of the optical tests of Bell inequalities', Preprint Universidad de Cantabria (Santander, Spain, 1987).
19. Aspect, A., Grangier, P. and Roger, G., Phys. Rev. Lett. 47, 460 (1981); Phys. Rev, Lett. 49, 91 (1982); Aspect, A., Dalibard, J. and Roger, G., Phys. Rev. Lett. 49, 1804 (1982).
20. Perrie, W., Duncan, A.J., Beyer, H.J. and Kleinpoppen, H., Phys. Rev. Lett., 54 1790 (1985).

QUASIPROBABILITY DISTRIBUTIONS IN QUANTUM OPTICS

G.J. Milburn
Department of Physics and Theoretical Physics
Australian National University
Canberra, A.C.T. 2601
Australia.

D.F. Walls
Department of Physics
University of Auckland
Auckland
New Zealand

ABSTRACT. We review some generalizations of phase space probability distributions used in quantum optics. We also present an application of a particular distribution, the Q-function, to the concept of interference in phase space and the oscillations in the photon number distribution for a squeezed state. The method enables a straightforward discussion of the effect of dissipation on these oscillations.

INTRODUCTION

It has long been realised that the statistical properties of light at optical frequencies provide clear evidence of the unusual features of probability in quantum mechanics. There are a number of reasons why this should be so. Firstly thermal excitation of field modes at optical frequencies may usually be neglected leaving only intrinsic quantum fluctuations. Secondly many nonlinear optical devices generate field states which have no classical analogue such as antibunched states, squeezed states and coherent superposition states. Finally, the statistical properties of light are accessible to direct investigation through either phase - independent measurements such as photon counting, or phase dependent measurements such as homodyne detection. In this paper we will consider a number of nonclassical statistical features exhibited by light and present some techniques by which these features may be studied.

1. QUASIPROBABILITY DISTRIBUTIONS

A number of widely used methods for studying the statistical properties of light are based on quasiprobability distributions. Quasiprobability distributions are generalizations of the phase space joint probability distributions of classical mechanics. The non-commutativity of position and momentum make the definition of such distributions rather problematic in quantum mechanics, and may lead to joint distributions which become negative. Such densities cannot really be called probability distributions; thus the term quasi-probability distribution. (One quasiprobability distribution, the Q-function, does in fact turn out to be a true probability as will be shown in this section).

1.1 Glauber-Sudarshan P-representation.

The earliest quasiprobability distribution used in quantum optics has come to be known as the Glauber-Sudarshan P-representation[1-2]. Let $\hat{\rho}$ be the density operator describing the state of a single mode field. The Glauber-Sudarshan representation is then defined by

$$\hat{\rho} = \int P(\alpha) |\alpha\rangle\langle\alpha| \, d^2\alpha , \qquad (1.1)$$

where $|\alpha\rangle$ is a coherent state of the field mode. For many states of the field, such as those generated by thermal or laser sources $P(\alpha)$ is a positive well behaved function with all the properties of a true joint probability distribution.

The P-function determines normal ordered averages. Let a and $a^\dagger$ be the destruction and creation operators for the field mode. The normally ordered averages are,

$$\langle (a^\dagger)^m a^n \rangle = \int P(\alpha) (\alpha^*)^m \alpha^n \, d^2\alpha . \qquad (1.2)$$

The right hand side of equation (1.2) looks like a standard moment for a phase space density. (The phase space variables are the real and imaginary parts of α.)

Phase independent statistical features may be investigated through the second order correlation function,

$$g^{(2)}(0) = \frac{\langle (a^\dagger)^2 a^2 \rangle}{\langle a^\dagger a \rangle^2} . \qquad (1.3)$$

In terms of the Glauber-Sudarshan P-function $g^{(2)}(0)$ may be written

$$g^{(2)}(0) = 1 + \frac{\int P(\alpha)(|\alpha|^2 - \langle |\alpha|^2 \rangle)^2 \, d^2\alpha}{(\int P(\alpha)|\alpha|^2 \, d^2\alpha)^2} \qquad (1.4)$$

Clearly if $P(\alpha)$ does have the properties of a probability density the second order correlation function must be greater than or equal to unity. For example a thermally excited field mode with mean photon number $\bar{n}$ has

$$P(\alpha) = (\pi\bar{n})^{-1} e^{-\frac{|\alpha|^2}{\bar{n}}} \quad (1.5)$$

for which $g^{(2)}(0) = 2$. However there exist states for which $g^{(2)}(0) < 1$. A class of simple examples is comprised of the number states $\{|n\rangle\}$, which are eigenstates of the number operator $a^{\dagger}a$ and for which

$$g^{(2)}(0) = 1 - \frac{1}{n} . \quad (1.6)$$

States for which $g^{(2)}(0)$ is less than unity are said to exhibit antibunching. We expect that the description of antibunched states in terms of the Glauber-Sudarshan representation would not be very convenient.

Phase dependent properties of the field may be investigated through the moments of the hermitian operators $\hat{X}_1$ and $\hat{X}_2$ defined by

$$\hat{X}_1 = \frac{(a+a^{\dagger})}{2} \quad (1.7a)$$

$$\hat{X}_2 = \frac{(a-a^{\dagger})}{2i} . \quad (1.7b)$$

The operators $\hat{X}_1$ and $\hat{X}_2$ correspond to the real and imaginary parts of the complex field amplitude, and are usually referred to as quadrature phase operators. These operators do not commute,

$$[\hat{X}_1, \hat{X}_2] = i/2 , \quad (1.8)$$

and thus satisfy the uncertainty principal

$$\Delta X_1 \Delta X_2 \geq 1/4 . \quad (1.9)$$

States which realise the lower bound in this product are called minimum uncertainty states. The coherent states $\{|\alpha\rangle\}$ are minimum uncertainty states for which $\Delta X_1^2 = \Delta X_2^2 = 1/4$. Minimum uncertainty states for which $\Delta X_i^2 < 1/4$ for either $i = 1$ or 2 are known as squeezed states. Squeezed states have been the subject of intense investigation over the past decade[3] and have recently been produced in a number of nonlinear optical systems[4-7].

Using the Glauber-Sudarshan representation the quadrature phase variances may be written,

$$\Delta\hat{X}_1^2 = \tfrac{1}{4}\left(1 + \int P(\alpha)[(\alpha+\alpha^*) - (\langle\alpha\rangle - \langle\alpha^*\rangle)]^2 d^2\alpha\right) ,$$

$$(1.10)$$

$$\Delta \hat{X}_2{}^2 = \tfrac{1}{4}\left(1 - \int P(\alpha)[-i(\alpha-\alpha^*) + i(\langle\alpha\rangle-\langle\alpha^*\rangle)]^2 d^2\alpha\right).$$

(1.11)

We see that the condition for squeezing in $\hat{X}_1$ requires that $P(\alpha)$ takes on negative values. Thus just as in antibunching, squeezed states cannot be described by a positive well behaved Glauber-Sudarshan distribution.

Antibunching and squeezing by no means exhaust the possibilities for a departure of quantum from classical statistics for light. When we consider correlations between two modes of the field further possibilities arise[8]. Consider a two mode field characterised by the annihilation operators a and b. A weak inequality is imposed by quantum mechanics. Since $a^\dagger a$ and $b^\dagger b$ are positive operators we must have $(a^\dagger a + \lambda b^\dagger b)^2 \geq 0$ for all real λ. The left hand side of this inequality is quadratic in λ and the inequality means this quadratic has no real roots. Thus it must have a discriminant less than or equal to zero which gives

$$\langle a^\dagger a b^\dagger b \rangle^2 \leq \langle (a^\dagger a)^2 \rangle \langle (b^\dagger b)^2 \rangle.$$

(1.12)

If a positive Glauber-Sudarshan representation exists a stronger inequality may be derived as follows. We write the normally ordered expectation values as follows

$$\langle (a^\dagger)^2 a^2 \rangle = \int |\alpha|^4 P(\alpha,\beta) d^2\alpha d^2\beta,$$

$$\langle (b^\dagger)^2 b^2 \rangle = \int |\beta|^4 P(\alpha,\beta) d^2\alpha d^2\beta,$$

$$\langle a^\dagger b^\dagger a b \rangle = \int |\alpha\beta|^2 P(\alpha,\beta) d^2\alpha d^2\beta.$$

If $P(\alpha,\beta)$ is a true probability density then one may prove the Cauchy-Schwartz inequality,

$$\langle |\alpha\beta|^2 \rangle^2 \leq \langle |\alpha|^4 \rangle \langle |\beta|^4 \rangle,$$

(1.13)

corresponding to the operator inequality

$$\langle a^\dagger a b^\dagger b^2 \rangle \leq \langle (a^\dagger)^2 a^2 \rangle \langle (b^\dagger)^2 b^2 \rangle.$$

(1.14)

A simple example of a system which violates this inequality is the non-degenerate parametric amplifier[8].

Another unusual feature of the quantised electromagnetic field which has received a lot of attention recently is the oscillation in the tail of the photon number distribution for squeezed states found by Schleich and Wheeler[9]. The significance of this feature resides in the fact that it may be explained in terms of the more general concept of interference in phase space. This phenomenon is also connected to the nonexistence of a positive Glauber-Sudarshan representation and will be discussed in more detail in Section 2.

1.2. Drummond-Gardiner generalized P-representations.

Drummond and Gardiner generalised the Glauber-Sudarshan representation to describe nonclassical field states in terms of a distribution which gives normally ordered averages[10]. One writes

$$\hat{\rho} = \int_D d\mu(\alpha,\beta) \, P(\alpha,\beta) \, \Lambda(\alpha,\beta) , \quad (1.14)$$

where

$$\Lambda(\alpha,\beta) = \frac{|\alpha\rangle\langle\beta^*|}{\langle\beta^*|\alpha\rangle} , \quad (1.15)$$

and where $d\mu(\alpha,\beta)$ is an integration measure which may be chosen to define different classes of possible representations and D is the domain of integration. If $d\mu(\alpha,\beta) = d\alpha d\beta$ with α and β treated as independent complex variables integrated over contours c and c' the resulting distribution is known as the complex P-representation. If $d\mu(\alpha,\beta) = d^2\alpha d^2\beta$ with α and β integrated over the entire complex plane, $P(\alpha,\beta)$ may always be chosen positive and is a true probability density albeit in twice as many variables as degrees of freedom would seem to require.

1.3. The Wigner function and the Q-function.

The first quantum generalization of a phase space probability density was given by Wigner[11] and is known as the Wigner function. It may be defined in terms of the position matrix elements by

$$P^{(W)}(q,p) = \frac{1}{2\pi\hbar} \int e^{\frac{ipy}{\hbar}} \langle q - \frac{y}{2}|\hat{\rho}|q + \frac{y}{2}\rangle \, dy .$$

The Wigner function enables one to calculate symmetrically ordered averages of the annihilation and creation operators. For many states however the Wigner function is negative and thus cannot be considered a true joint phase space probability density.

The Q-function is usually included as a quasiprobability distribution. However, it is in fact a true joint probability distribution for a special class of simultaneous measurements of two canonically conjugate variables such as $\hat{X}_1$ and $\hat{X}_2$. It is defined as the matrix elements of the density operator in coherent states,

$$Q(\alpha) = \langle \alpha | \hat{\rho} | \alpha \rangle . \tag{1.16}$$

Once the Q-function is known all matrix elements of $\hat{\rho}$ may be constructed[12]. However, it directly determines antinormally ordered averages

$$\langle a^n (a^\dagger)^m \rangle = \int \frac{d^2\alpha}{\pi} \alpha^n (\alpha*)^m Q(\alpha) . \tag{1.17}$$

As $\hat{\rho}$ is a positive bounded operator one has that

$$0 \leq Q(\alpha) \leq 1 , \tag{1.18}$$

thus $Q(\alpha)$ may be considered as belonging to a sub-class of phase-space joint probability densities. The fact that $Q(\alpha)$ is bounded seems to be a reflection of the uncertainty principal and ensures that it cannot have too small an area of support in phase space.

The Q-function is a true joint phase space probability distribution for a particular class of measurements[13-14]. A simple model for such measurements was given by Arthurs and Kelly[14]. A similar explanation of the Q-function is given by Davies[13].

2. INTERFERENCE IN PHASE SPACE

A topic of current interest in quantum optics is the unexpected oscillation in the tail of the photon number distribution for squeezed light, and the interpretation by Schleich and Wheeler of this oscillation as evidence for interference in phase space[9]. In this section we will apply some of the general principals discussed in Section 1 to this topic.

The photon number distribution for a field mode state $|\psi\rangle$ is determined by

$$P(n) = |\langle n | \psi \rangle|^2 . \tag{2.1}$$

The well known correspondence between field modes and the harmonic oscillator enables one to view $P(n)$ equivalently as the probability distribution for very accurate energy measurements made on a harmonic oscillator, that is the probability to obtain the result $E_n = \hbar\omega(n+\tfrac{1}{2})$.

In classical mechanics the energy distribution for a one dimensional harmonic oscillator is determined by the phase space probability density, $P(\alpha)$, describing the classical state of the system, by

$$P(n) = \int d^2\alpha \, \delta^{(2)}\left(|\alpha|^2 - n\right) P(\alpha) , \tag{2.2}$$

where $\alpha = q + ip$ with q and p the dimensionless position and momentum phase space variables. If a field mode is in a state with a Glauber-Sudarshan representation the photon number distribution is

$$P(n) = \int d^2\alpha \, |\langle n|\alpha\rangle|^2 \, P(\alpha) \tag{2.3}$$

noting that

$$|\langle n|\alpha\rangle|^2 = \frac{|\alpha|^{2n}}{n!} e^{-|\alpha|^2} \tag{2.4}$$

which is sharply peaked at $|\alpha|^2 = n$ for n large, one sees that equation (2.3) is analogous to the classical result in equation (2.2). Once again we see that states with a Glauber-Sudarshan representation are in a sense close to "classical states".

If a Glauber-Sudarshan representation does not exist one may proceed more generally as follows. Using the completeness for coherent states,

$$\int \frac{d^2\alpha}{\pi} |\alpha\rangle\langle\alpha| = 1 , \tag{2.5}$$

the probability amplitude for photon number measurements may be written

$$\langle n|\psi\rangle = \int \frac{d^2\alpha}{\pi} \langle n|\alpha\rangle \langle\alpha|\psi\rangle . \tag{2.6}$$

Equation (2.6) has an appealing interpretation. The function $\langle\alpha|\psi\rangle$ is the probability amplitude for simultaneous measurement of position and momentum as $|\langle\alpha|\psi\rangle|^2$ is just the Q-function for the state $|\psi\rangle$, while $\langle n|\alpha\rangle$ is the "conditional amplitude" that the result of such a measurement corresponds to the energy E_n. Thus equation (2.6) is constructed in a similar manner to the classical result but at the level of amplitudes rather than probabilities. As in the two slit experiment the rules of classical probability are applied at the level of the appropriate probability amplitudes for the measurement under discussion.

From equation (2.6) the photon number becomes

$$P(n) = \int \frac{d^2\alpha}{\pi} \int \frac{d^2\beta}{\pi} \langle n|\alpha\rangle\langle\beta|n\rangle\langle\alpha|\psi\rangle\langle\psi|\beta\rangle , \tag{2.7}$$

or in terms of the density operator

$$P(n) = \int \frac{d^2\alpha}{\pi} \int \frac{d^2\alpha}{\pi} \langle n|\alpha\rangle \langle\beta|n\rangle \langle\alpha|\hat{\rho}|\beta\rangle . \tag{2.8}$$

Equation (2.8) indicates that interference effects may arise if the density operator is not diagonal in the coherent state basis.

As an example let $|\psi\rangle = |\beta_0, r\rangle$, a squeezed state with non-zero coherent amplitude β_0. The integral in equation (2.8) will only be non-zero over those regions of phase space where there is significant overlap between the two functions $\langle n|\alpha\rangle$ and $\langle\alpha|\beta_0, r\rangle$. In figure 1 we have plotted contours of the functions $|\langle n|\alpha\rangle|$ and $|\langle \alpha|\beta_0,r\rangle|$ for $\beta_0 = 7$ and $e^{2r} = 21$. Clearly there are two regions of overlap. An asymptotic analysis for large n, similar to that performed by Wheeler and Schleich, show these two regions contribute the same amplitude but opposite phases. Thus

$$\langle n|\beta_0,r\rangle = \sqrt{A_n} \left(e^{i\phi_n} + e^{-i\phi_n}\right), \qquad (2.9)$$

where A_n and ϕ_n are given in reference (9). Taking the modulus square of the result in equation (2.9) clearly indicates the source of the oscillations in P(n) for n sufficiently large. In figure 2d we have plotted P(n) which illustrates these oscillations.

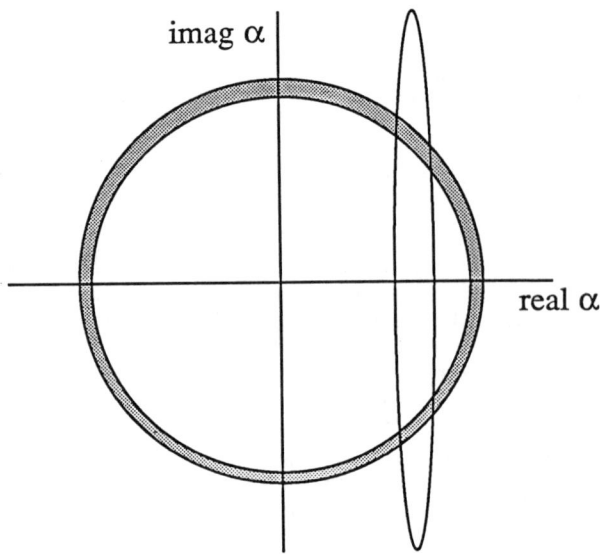

Figure 1. Schematic indication of the contours of the two functions, $\langle n|\alpha\rangle$ and $\langle \alpha|\beta_0,r\rangle$, plotted against the real and imaginary parts of α.

The probability distribution P(n) may also be represented in terms of the Wigner function[15]

$$P(n) = 2\pi \int_{-\infty}^{+\infty} dq \int_{-\infty}^{+\infty} dp \, P_n^{(W)}(q,p) \, P_{sq}^{(W)}(q,p), \qquad (2.10)$$

where $P_{sq}^{(W)}$ denotes the Wigner function for a squeezed state and $P_n^{(W)}$ is the Wigner function for a number state. An asymptotic analysis of the integral in equation (2.10) also shows the effect of interference in phase space.

A number of authors[16-18] have shown that coherent superposition states and associated interference effects are particularly sensitive to dissipation. We now show that the oscillation in the photon number distribution discussed above shows a similar dependence on dissipation[19].

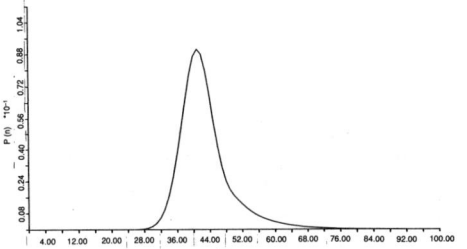
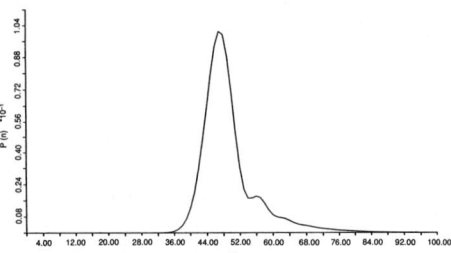
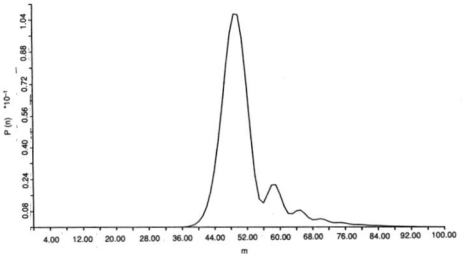
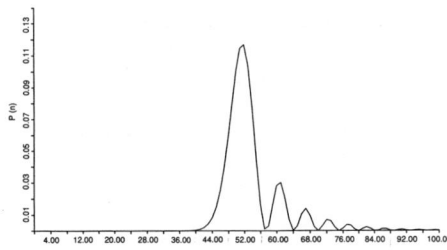

Figure 2. Plot of $P(m,\mu)$ versus m for the squeezed state $|\beta_0,r\rangle$ with $\beta_0 = 7$, $r = 1.52$ and varying values of quantum efficiency μ;
(a) $\mu = 0.8$ (b) $\mu = 0.92$ (c) $\mu = 0.96$ (d) $\mu = 1.0$.

For a single mode field weakly coupled to a zero temperature heat bath the photon number distribution at time t is [19]

$$P(m,t) = \sum_{n=m}^{\infty} P(n,0) \binom{n}{m} \mu^m (1-\mu)^{n-m}, \qquad (2.11)$$

where $\mu = e^{-\gamma t}$ and $\gamma/2$ is the decay rate of the field amplitude. Substituting equation (2.7) into equation (2.11) and evaluating the sum over m

$$P(m,t) = \int \frac{d^2\alpha}{\pi} \int \frac{d^2\beta}{\pi} \langle \alpha|\rho|\beta \rangle \langle m|\alpha\sqrt{\mu} \rangle \langle \beta\sqrt{\mu}|m \rangle \langle \beta|\alpha \rangle^{1-\mu}, \qquad (2.12)$$

The off-diagonal contribution to the integral is seen to be suppressed by the factor $\langle \beta|\alpha \rangle^{1-\mu}$. This suppression factor is identical to that obtained for the destruction of coherence in a superposition of two coherent states[18]. For short times ($\gamma t \ll 1$) the modulus of the suppression factor is

$$\exp\left[\frac{-\gamma t}{2} |\alpha - \beta|^2\right],$$

indicating that the rate of decay of the off-diagonal elements is proportional to the square of the separation of the phase space points corresponding to the superposed states. Thus interference arising from widely separated phase space points is rapidly destroyed. This indicates that the oscillations in the photon number distribution for squeezed states with large squeezing on large coherent amplitude would be particularly sensitive to dissipation, as in these cases the region of overlap of the two functions in the integrand in equation (2.6) are widely separated.

Equation (2.11) may also be interpreted as the probability to detect m photons in time t with a detector of quantum efficiency μ. The actual time dependence of μ in this case depends on the particular detection scheme [20]. This suggests that the oscillations in the photon number distribution may be difficult to observe in practice. In figures 2a-d we have plotted $P(n,\mu)$ for the squeezed state $|\beta_0,r\rangle$ with $\beta_0 = 7$ and $r = 1.52$ for various values of μ.

REFERENCES

1. R.J. Glauber, *Phys. Rev.* **131**, 2766 (1963).
2. E.C.G. Sudarshan, *Phys. Rev. Letts.*, **10**, 277 (1963).
3. For a review see, D.F. Walls *Nature* (London) **306**, 141 (1983); R. Loudon and P.L. Knight, *J. Mod. Opt.* **34**, 709 (1987).
4. R.E. Slusher, L.W. Hollbert, B. Yurke, J.C. Mertz and J.F. Valley, *Phys. Rev. Letts.*, **55**, 2409 (1985).
5. R.M. Shelby, M.O. Levenson, S.H. Perlmutter, R.G. De Voe and D.F. Walls, *Phys. Rev. Letts.*, **57**, 691 (1986).

6. L. Wu, H.J. Kimble, J.L. Hall and H. Wu; *Phys. Rev. Letts.* **57**, 2520 (1986).
7. M.W. Maeda, P. Kumar and J.H. Shapiro; *Opt. Lett.* **12**, 161 (1987).
8. M.D. Reid and D.F. Walls, *Phys. Rev. A* **34**, 4929 (1986).
9. W. Schleich and J.A. Wheeler, *Nature* (London), **326**, 574 (1987).
10. P.D. Drummond and C.W. Gardiner, *J. Phys. A* **13**, 2353 (1980).
11. M. Hillery, R.F. O'Connell, M.O. Scully and E.P. Wigner, *Phys. Rep.* **106**, 121 (1984).
12. C.L. Mehta and E.C.G. Sudarshan, *Phys. Rev.* **138**, B274 (1965).
13. E.B. Davies, 'Quantum Theory of Open Systems' (Academic, New York, 1976).
14. E. Arthurs and J.L. Kelly Jr., Bell Syst. Tech. J. **44**, 725 (1965).
15. W. Schleich, D.F. Walls and J.A. Wheeler, submitted to *Phys. Rev.A*.
16. D.F. Walls and G.J. Milburn, *Phys. Rev. A* **31**, 2403 (1985).
17. A.O. Caldeira and A.J. Leggett, *Phys. Rev. A* **31**, 1059 (1985).
18. G.J. Milburn and C.A. Holmes, *Phys. Rev. Letts.* **56**, 2237 (1986).
19. G.J. Milburn and D.F. Walls, submitted to *Phys. Rev. A*.
20. M.D. Srinivas and E.B. Davies, *Optica Acta* **28**, 981 (1981); *Optica Acta* **29**, 235 (1982). See also L. Mandel, *Optica Acta* **28**, 1447 (1981).

STOCHASTIC-DYNAMICAL APPROACH TO QUANTUM MECHANICS

Mikio Namiki
Department of Physics
Waseda University
Ohkubo, Shinjuku-ku
Tokyo 160
Japan

ABSTRACT   We mainly discuss the Parisi-Wu stochastic quantization as a possible stochastic-dynamical approach to quantum mechanics, which gives quantum mechanics by thermal equilibrium limit of a hypothetical stochastic process in a new time other than the ordinary time. First we sketch its historical background in which one of the central problems is how to replace or reformulate quantum mechanics with a classical stochastic dynamics. After a brief survey of technical merits of the method, we examine a possibility of formulating the stochastic quantization in the Minkowski space-time, and speculate possible roots of the hypothetical stochastic process in the new time.

1. INTRODUCTION

For sixty years since birth of quantum mechanics, many physicists, including both those who agree and disagree to the Copenhagen interpretetion, have confidentially been conceiving an idea of replacing or reformulating quantum mechanics with a classical stochastic dynamics. Really we have also heard many theories and opinions on it at this conference.

It is well known that we have a formal similarity between a quantum-mechanical free particle and a simple Brownian-motion particle, as suggested by their basic equations,

$$\partial\psi/\partial(it) = (\hbar/2m)\nabla^2\psi, \quad \text{(Schrödinger equation)}, \quad (1.1a)$$

$$\partial\psi/\partial t = D\nabla^2\psi \quad \text{(diffusion equation)} \quad (1.1b)$$

together with the following correspondences

$$t \leftrightarrow -it, \quad \hbar/2m \leftrightarrow D \quad (1.2)$$

where m, $\hbar$ and D are particle mass, Planck constant and diffusion constant, respectively. Using Eq.(1.2), we can mutually transform the uncertainty relations in both theories: $\Delta x \Delta p \simeq \hbar \leftrightarrow \Delta x \Delta v \simeq 2D$. The formal similarity would lead us to a possible "hidden-variable" theory, under such a naive idea that quantum fluctuations originate in classical random motions caused by interactions with unknown "ether". Indeed, we had ever seen a few preliminary attempts [1]. However, the NO-GO theorem given by J. von Neumann had strongly prevented to make a class of hidden-variable theories [2].

Breaking through the von Neumann's barrier, D. Bohm tried to make a possible hidden-variable theory [3]. Putting

$$\psi = |\psi| \exp(iS/\hbar), \quad \vec{p} = \vec{\nabla} S \tag{1.3}$$

he derived a "Newton equation" of the form

$$d\vec{p}/dt = -\vec{\nabla}(V+V_Q) ; \tag{1.4a}$$

$$V_Q = -(\hbar^2/2m)(\nabla^2|\psi|/|\psi|) \tag{1.4b}$$

from the Schrödinger equation

$$i\hbar \partial \psi / \partial t = [-(\hbar^2/2m)\nabla^2 + V]\psi \tag{1.5}$$

The "Newton equation", Eq.(1.4), contains the so-called quantum mechanical force, $-\vec{\nabla}V_Q$, to represent a classical random force rooted in a hypothetical ether. The hidden-variable theory has been developed by himself, J. P. Vigier (stochastic interpretation) [4], T. Takabayashi (hydrodynamical quantization) [5], and others.

Bohm wanted to erase quantum mechanics with a classical stochastic dynamics. It is, however, noticed that his approach can be regarded as a sort of quantization procedure, if we trace back to the Schrödinger equation, Eq.(1.5), from a classical Newton equation, by adding the above quantum mechanical force, Eq.(1.4b), to the latter. Along a similar line of thought, E. Nelson achieved to elegantly formulate his stochastic quantization [6]. Both Bohm's theory and Nelson's stochastic quantization are very interesting, from the fundamental point of view on how to give a stochastic interpretation to quantum mechanics. Nevertheless, it seems that both theories can hardly be applied to complicated systems, for example, to many-particle systems or fields.

Needless to say, a new dynamics, if proposed or constructed, must bring in not only a fundamental interpretation but also technical merits for solving important problems in frontier physics. From this point of view, we are very much interested in the Parisi-Wu stochastic quantization method (SQM) [7], which gives quantum mechanics by thermal equilibrium limit of a hypothetical stochastic process with respect to a new (fictitious) time other than the ordinary time. It seems that SQM has a big technical merits, as will be seen shortly. Remember that any kind of new dynamics can never be living in future unless it yields technical merits in frontier physics.

Both SQM and Nelson's quantization method are based on a sort of Markoffian process of the Wiener type with Gaussian white noises. One of the most important differences between them is that the former has introduced a hypothetical stochastic process in a new (fictitious) time, while the latter is formulated in terms of a random process in the ordinary time. Behind SQM, there exists a basic idea that a D-dimensional *quantum* system is equivalent to a (D+1)-dimensional *classical* system with random fluctuations. It is not quite new, because the idea has already been developed in quantum theory of spin systems [8]. As far as we are concerned with this point, the new time is only a mathematical tool for practical calculations, but it seems to have no deeper physical basis. However, one may imagine that exotic space-time worlds surrounding ours would make classical random disturbances, on our system, which we observe as quantum fluctuations. We shall come back to this topics later.

It is also noted that the usual formalism of SQM has so far been using the *imaginary* ordinary-time or the Euclidean space-time as in the above naive view, while Bohm and Nelson formulated their theories in terms of the real ordinary-time. Recently, however, a few authors have attempted to formulate Minkowski SQM, keeping the ordinary-time real [9].

We have another interesting method named "micro-canonical quantization" [10], which is based on a rather drastic idea that a D-dimensional quantum system is equivalent to a (D+1)-dimensional classical (deterministic) system with a new (fictitious) time other than the imaginary ordinary-time. Those who are using the method expect to obtain quantum fluctuations from classical chaotic behaviors of the system in the new time. Rigorously speaking, however, the method can not always quantize every dynamical system, because some systems have no chaos. Nevertheless, the method can work in its applications to numerical simulations, because successive updations of dynamical variables by a finite time step often distribute in a random manner as in SQM.

All the above quantization procedures are formulated in configuration space or in momentum space. Finally, we talk about quantum mechanics in phase space which was initiated by E. P. Wigner [11]. The Wigner function is not considered to be a probability distribution due to lack of the positive definiteness. Over forty years ago, however, K. Husimi proposed a positve-definite phase-space distribution function [12], given by averaging the Wigner function over a small cell around each point in phase space. Recently, many authors have formulated similar theories. It is certain that the Husimi function describes a sort of stochastic process, but we do not know what kind of classical random processes are working behind it. Coherent description of quantum mechanics and geometrical quantization are considered to belong to quantum mechanics in phase space. Most recently, some authors have attempted to formulate SQM in phase space [13], in which a classical Wiener-Markoffian process with Gaussian white noise in a new time is to yield quantum fluctuations.

Table I serves us for looking over all the above stochastic-dynamical approaches to quantum mechanics.

Amang many approaches, it seems that the Parisi-Wu stochastic quantization method enables us to deal with a wider class of dynamical systems, compared with the Nelson's one, and it further extends the territory of quantum mechanics beyond that of the conventional theories, *i.e.*, the canonical and path-integral quantizations. This is the reason why we mainly discuss the Parisi-Wu stochastic quantization method (SQM).

Table I. Classification of stochastic-dynamical approaches to quantum mechanics

   I-a. Quantum mechanics in configuration or momentum space.

space-time	imaginary ordinary-time	real ordinary-time	origin of fluctuation
4-dim.	Naive view Schrödinger equation ↔ diffusion equation		classical "ether"
		hidden-variable th. Bohm, Vigier, ... Takabayashi Nelson's theory	
(4+1)-dim.	Parisi-Wu SQM D-dim. quantum system ↔ (D+1)-dim. cl. system with random forces. real Langevin eq.	Minkowski SQM Hüffel-Rumpf, Gozzi Nakazato-Yamanaka compl. Langevin eq.	classical "exotic" worlds
	micro-canonical quantz. Callaway, Kogut, ... D-dim. quantum system ↔ (D+1)-dim. classical system with chaos.		

I-b. Quantum mechanics in phase space.

4-dim.		Wigner function Husimi function coherent st. descr. geometrical quantz.	?
(4+1)-dim.	SQM in phase space Kapoor, Ohba, ...		classical "exotic worlds

2. BASIC IDEA OF SQM

We first outline the basic idea of SQM in comparison with the conventional quantization methods, that is, the canonical and path-integral ones. Consider quantization of a dynamical system described by action functional, $S[q]=\int dx L$ (L being Lagrangian), depending on dynamical variables, $q(x) = \{q_i(x)\}$, in which x is the ordinary-time for particles or the 4-dimensional coordinates for fields.

The standard canonical quantization starts by putting the commutation relations, $[q_i, p_j] = i\hbar \delta_{ij}$, for canonical variables, q_i and p_j, and the Heisenberg equation of motion, $i\hbar dF/dt = [F,H]$, for an observable F, H standing for Hamiltonian. The quantum-mechanical state is introduced as a mechanism to give expectation values, $\langle F \rangle$, ... , to observables, F, ... Needless to say, the canonical quantization method requires us to have canonical variables and Hamiltonian.

The path-integral quantization starts by putting the functional integral

$$\langle f | i \rangle = \int \delta q \exp(-S[q]) \tag{2.1a}$$

with respect to the dynamical variables, q's, for the probability amplitude, or

$$\langle q(x_1) q(x_2) \ldots \rangle = \int \delta q \, q(x_1) q(x_2) \ldots \exp(-S[q]) \tag{2.1b}$$

for the Wightman function, apart from the nomalization constant. Here we have used the imaginary ordinary-time or the 4-dimensional Euclidean coordinates and the corresponding action denoted with the same letters, x and S, respectively, in order to make the functional integrals well-posed. Speaking in principle, we have only to prepare Lagrangian but not Hamiltonian for the path-integral quantization. It is also noted that this is a c-number quantization.

Our first task is to derive Eq.(2.1) by means of SQM. Let us introduce an additional dependence of q on a fictitious time, say t, other than the (imaginary) ordinary time, x, and then set a Langevin equation

$$\gamma \partial q_i(x,t)/\partial t = -\gamma^{-1} \delta S[q]/\delta q_i \big|_{q=q(x,t)} + \eta_i(x,t) \quad (2.2)$$

to govern a hypothetical stochastical process in t, where η_i's are Gaussian white noises subject to the statistical law

$$\langle \eta_i(x,t) \rangle = 0, \quad \langle \eta_i(x,t)\eta_j(x',t') \rangle = 2\alpha \delta_{ij} \delta(x-x')\delta(t-t') \quad (2.3)$$

α being the diffusion constant of the stochastic process to be equated to Planck constant $\hbar$ later. The bracket stands for an average over η's. In Eq.(2.2) we have put an arbitrary constant γ to adjust the speed of the process in t, but it never modifies thermal equilibrium distributions. Equation (2.2) together with Eq.(2.3) certainly describes a typical Wiener process.

Solving Eq.(2.2), we obtain $q(x,t)$ and then $F(q(x,t))$ as a function or functional of η's. The expectation value of F, $\langle F \rangle$, is given by averaging $(F(q(x,t))$ over η's. If we introduce the probability distribution functional $P[q;t]$ defined with

$$\langle F(q(x,t)) \rangle = \int \delta q \, F(q) P[q;t] \quad (2.4)$$

we obtain the Fokker-Planck equation

$$\gamma \partial P[q;t]/\partial t = \gamma^{-1} \alpha \int dx \sum_i \delta/\delta q_i(x) [\delta/\delta q_i(x) + \alpha^{-1} \delta S/\delta q_i(x)] P[q;t] \quad (2.5)$$

which yields

$$\lim_{t \to \infty} P[q;t] = P_{eq}[q] = N \exp(-S/\alpha) \quad (2.6)$$

at the thermal equilibrium limit, irrespective of γ, N being the normalization constant. Thus, Eq.(2.4) turns to the path-integral formula, Eq. (2.1), at the infinite time limit, provided that we put $\alpha=\hbar$. This means that we can obtain quantum mechanics through the above hypothetical stochastic process. This is the prescription of SQM.

The field-theoretical propagator, which is definfed with

$$\Delta_{ij}(x-x') = N \int \delta q \, q_i(x) q_j(x') \exp(-S) \quad (2.7)$$

in the path-integral formalism, can be reproduced by

$$\Delta_{ij}(x-x') = D(x-x',0) \quad (2.8)$$

in which D is the stationary correlation function

$$D(x-x',t-t') = \lim_{t\to\infty;\,|t-t'|=\text{fixed}} \langle q_i(x,t)q_j(x',t')\rangle \quad (2.9)$$

given by means of the prescription of SQM. Here and hereafter we use the natural unit in which $\hbar=1$.

In order to show the procedure leading us to the propagator, we apply the prescription of SQM to the free neutral scalar field case, in which the Langevin equation becomes

$$\gamma\partial\phi(x,t)/\partial t = -\gamma^{-1}(-\Box+m^2)\phi(x,t) + \eta(x,t) \quad (2.10)$$

where $\langle\phi(x,t)\rangle=0$ and $\langle\phi(x,t)\phi(x',t')\rangle=2\delta(x-x')\delta(t-t')$. From Eq.(2.10) we can easily derive the stationary correlation function

$$D_0(x-x',t-t')$$
$$= 2(2\pi)^{-5}\int d^4k \int d\omega\,[\gamma^2\omega^2+\gamma^{-2}(k^2+m^2)^2]^{-1}e^{ik(x-x')-i\omega(t-t')}$$
$$=(2\pi)^{-4}\int d^4k\,[k^2+m^2]^{-1}\exp[ik(x-x')-(k^2+m^2)|t-t'|\gamma^{-2}] \quad (2.11)$$

Hence, the above prescription surely gives the free propagator

$$D_0(x-x',0) = (2\pi)^{-4}\int d^4k\,[k^2+m^2]^{-1}e^{ik(x-x')}$$
$$= \Delta_0(x-x') \quad (2.12)$$

It is worthy to remember that we can obtain the particle mass, not only from the ordinary-space-time asymptotic behavior of $\Delta_0(x-x')$

$$\Delta_0(x-x') \xrightarrow[|x-x'|\to\infty]{} (m^2/16\pi^3)(2\pi/m|x-x'|)^{3/2}\exp[-m|x-x'|] \quad (2.13a)$$

but also from the fictitious-time asymptotic behavior [15]

$$D_0(0,t-t') \xrightarrow[|t-t'|\to\infty]{} (m^2/16\pi^2)(\gamma^2/m^2|t-t'|)^2\exp[-m^2|t-t'|\gamma^{-2}] \quad (2.13b)$$

We understand the reason why both Eqs.(2.13a) and (2.13b) inform us about the particle mass, on the common basis of the dispersion formula,

$$\gamma^2\omega^2+\gamma^{-2}(k^2+m^2)^2 = 0 \quad (2.14)$$

which gives poles of the integrand of the middle member of Eq.(2.11). Note that the dispersion formula completely determines the mass spectrum. We can proceed our work to the case of interacting fields [14], [15].

The above survey tells us that SQM can start directly from the classical equation of motion but it does not require Lagrangian or Hamiltonian in principle. This is one of the most remarkable characters of SQM, by which we have possibility of quantizing not only holonomic systems but also non-holonomic ones. We will discuss later quantization of the non-Abelian gauge field under a non-holonomic constraint. It is also noted that SQM is a sort of c-number quantization.

Many authors have developed perturbation theory within the

framework of SQM, in which they have shown its equivalence to the conventional quantization methods [16]. SQM can also be applied to fermion fields by making use of Grassmann numbers [17]. Here we do not enter into their details.

3. SOME APPLICATIONS TO FIELD THEORY

One of the most important problems in modern field theory is to quantize dynamical systems with constraints. The author and his collaborators have formulated SQM in the case with holonomic constraints, and then obtained the same results as given by the path-integral method [18]. The velocity-dependent constraints have also been discussed within the framework of the phase-space SQM [13].

Here we briefly discuss another kind of constraint problem in the case of the non-Abelian gauge field, $A_\mu^a(x)$. The conventional field theory usually requires to put the gauge-fixing term $(2\alpha)^{-1}(\partial \cdot A^a)^2$ into Lagrangian, which gives the following holonomic constraint:

$$\alpha^{-1}\partial(\partial \cdot A^a)/\partial x_\mu = 0 \tag{3.1}$$

α being the gauge parameter. It is, however, well known that the gauge-fixing procedure destroys the gauge invariance and the unitarity, so that we have to introduce the Faddeev-Popov ghost field for compensation of the defects. Here we introduce a new gauge-fixing procedure [19], which is possible in SQM but impossible in the conventional ones. Replace Eq.(3.1) with a non-holonomic constraint given by

$$\alpha^{-1}D_\mu^{ab}(\partial \cdot A^b) = 0; \tag{3.2a}$$

$$D_\mu^{ab} = \delta^{ab}\partial/\partial x_\mu + gf^{abc}A_\mu^c \tag{3.2b}$$

The corresponding Langevin equation is written as

$$\partial A_\mu^a/\partial t = -\delta S/\delta A_\mu^a + \alpha^{-1}D_\mu^{ab}(\partial \cdot A^b) + \eta_\mu^a \tag{3.3a}$$

$$\langle \eta_\mu^a(x,t) \rangle = 0, \quad \langle \eta_\mu^a(x,t)\eta_\nu^b(x',t') \rangle = 2\delta^{ab}\delta_{\mu\nu}\delta(x-x')\delta(t-t') \tag{3.3b}$$

where $S=(1/4)\int dx F_{\mu\nu}^a F^{a\mu\nu}$: $F_{\mu\nu}^a = \partial_\mu A_\nu^a - \partial_\nu A_\mu^a + gf^{abc}A_\mu^b A_\nu^c$. The Langevin equation can offer us the sound base to quantize the gauge field without help of any ghost field, within the framework of SQM. Actually, the perturbation calculations have shown that the linear part of the second term, the constraint force, on the right-hand side of Eq.(3.3a) fixes and its nonlinear part automatically produces the so-called Faddeev-Popov effects. For details, see [19]. It seems that this result is a remarkable merit of SQM. The procedure is sometimes called "covariant" or "stochastic" gauge-fixing, which is also applied to quantization of the linearized gravitational field [20].

As far as we are concerned with the gauge-fixing problem, even the "covariant" or "stochastic" gauge-fixing is not necessary within the framework of SQM. Parisi and Wu have already obtained the Landaugauge propagator of the gauge field without gauge-fixing procedure, and conjectured that the Faddeev-Popov terms could automatically be produced without help of ghost fields [7]. However, they missed the initial distribution of the longitudinal component of the gauge field (not vanishing because of the gauge invariance), which yields an arbitrary gauge parameter [21]. Their latter conjecture was confirmed in the perturbation calculations [21].

Mathematically speaking, the Langevin equation (2.2) must be replaced with the stochastic differential equation of the Ito-type:

$$dq_i = -dt\, \delta S[q]/\delta q_i + dw_i \qquad (3.4)$$

where we have put $\gamma=1$ for simplicity. It is noted that dw_i is of the order of $\sqrt{(dt)}$ because $\langle dw_i dw_j \rangle$ is proportional to dt. Hence the differential of a product of dynamical quantities, $F[q]$ and $G[q]$, becomes

$$d(F[q]G[q]) = (dF[q])G[q] + F[q](dG[q]) + (dF[q])(dG[q]) \qquad (3.5)$$

because the last term also contains contributions of the order of dt. The last term, which breaks the simple Leipnitz formula, should come from the pure quantum effect. The breaking must add an anomaly term to a conservation law as a deviation from the classical Noether theorem. Actually, the author and his collaborators have derived the chiral anomaly and the conformal anomaly, based on such formulas as Eq.(3.5) [22]. It seems that SQM offers us a useful method to analyze the anomaly problem.

The stochastic differential calculus also gives us a generalized Langevin equation of an arbitrary observable, say $F(q)$, as follows;

$$\partial F(q)/\partial t = -\delta S/\delta q_i\, \partial F/\partial q_i + \partial^2 F/\partial q_i^2 + \partial F/\partial q_i\, \eta_i \qquad (3.6)$$

The second term of the right-hand side comes from a duplicate use of the random forces. Based on Eq.(3.6), we can formulate the Langevin equation of the Wilson loop itself in the lattice gauge theory [23].

Besides the above theoretical merits, we should mention that SQM is now working as one of the most powerful tools for numerical simulations in field theory. Actually, the biggest simulations are carried out by means of SQM [24]. Needless to say, the numerical simulation is a very important non-perturbative approach to field theoretical problems.

4. FOUNDATIONAL DISCUSSIONS ON SQM

In the preceeding sections, we have seen physical and technical implications of the Parisi-Wu stochastic quantization method (SQM). We know that most physicists are interested mainly in its practical merits. Our emphasis has also been laid on them rather than foundational insight into future quantum mechanics. It is certain that SQM can enlarge the territory of quantum mechanics beyond the conventional theories, as far as it is concerned with such a technical development. However, we do not know its deeper physical roots. We should re-examine and further develop SQM from a more fundamental point of view, for example, along the line of thought of the historical ideas to replace quantum mechanics with classical stochastic-dynamics, as mentioned in Section 1.

We have two fundamental questions (i) as to whether it is inevitable or not to use the *imaginary* ordinary-time or the *Euclidean* coordinates for the formulation of SQM, and (ii) as to what kind of *physical background* is hiding behind the fictitious time.

As for the first question, we have several attempts to formulate SQM as a stochastic process with respect to a fictitious time, keeping the ordinary-time real [9]. We sometimes call it "Minkowski SQM". Formally applying the prescription of SQM to a dynamical system with a Minkowski action, $S[q]$, then we obtain a complex Langevin equation, for example,

$$\gamma \partial \phi(x,t)/\partial t = \gamma^{-1} i(-\Box - m^2)\phi(x,t) + \eta(x,t) \qquad (4.1)$$

in the case of a free neutral scalar field. Equation (4.1) should require ϕ to have a complex value even if it were originally defined to be a real quantity, and it should require us to manipulate a complex probability at first sight. Then we should wonder if we could formulate the "Minkowski SQM" in terms of a real stochastic process. Among many proposals, we introduce Nakazato and Yamanaka's theory here. They add a damping term, $-\varepsilon\phi$ (ε being a very small positive number), to the right-hand side of Eq.(4.1) and then decompose the equation into two real Langevin equations for real variables, $\phi_R = \text{Re } \phi$ and $\phi_I = \text{Im } \phi$. If we start from the real Langevin equations, we need not handle any kind of complex probability but only a real probability distribution $P[\phi_R, \phi_I; t]$. They have shown that its thermal equilibrium limit, $P_{eq}[\phi_R, \phi_I] = \lim_{t\to\infty} P[\phi_R, \phi_I; t]$, really gives the same field-theoretical propagator as in the conventional field theory. Surely, we have possibility of formulating the "Minkowski SQM" without resort to the *imaginary* ordinary-time. However, the Nakazato-Yamanaka theory is not convenient to numerical simulations because of the introduction of the very small damping term. Most recently, Tanaka proposed a new "Minkowski SQM" without help of such a damping term, putting γ (or γ^{-1})=$\sqrt{(-\Box - m^2)}$ in Eq.(4.1) [25].

Strictly speaking, as for the second question, we do not know any physical contents of the fictitious time, and we have also no reasons to call it "time". However, it might be a fascinating task to seek out its physical roots in exotic space-time worlds surrounding ours. Recent particle cosmology suggests us that the primival universe with many dimensional space-time worlds has shrinked to the present 4-dimensional world. We want to supplement the story with a new idea that the exotic worlds are still alive and fluctuating around ours, under the control of many-dimensional classical dynamics. According to the idea, we can imagine that even a smooth deterministic motion in the larger world will make a random trace on our world, and that such classical motions will yield "quantum fluctuations", through the mechanism of SQM or the micro-canonical quantization. The classical motions toward the exotic worlds are to be described in terms of the fictitious time. Of course, the idea is only a dream for the moment.

This talk is based on many collaborations with H. Nakazato, N. Nakazato, I. Ohba, K. Okano, M. Rikihisa, H. Shibata, S. Tanaka and Y. Yanamaka. The author is deeply indebted to them for their assistance.

REFERENCES

[1] For example, see; Schrödinger, E., 1931, Sitzungsb. Preuss. Akad. W. 144; Metadier, J., 1931, Comptes Rendus, <u>193</u>, 1173; Fürth, R., 1933, Zeits. f. Phys. <u>81</u>, 143.
[2] von Neumann, J., 1932, *Mathematische Grundlagen der Quanten Mechanik* (Springer, Berlin).
[3] Bohm, D., 1952, Phys. Rev. <u>85</u>, 166.
[4] For example, see; Vigier, J. P., and Roy, S., 1985, Hadronic J. Suppl., <u>1</u>, 475.
[5] For example, see; Takabayashi, T., 1984, *Proc. of the International Symposium on Foundations of Quantum Mechanics* (Phys. Soc. Japan, Tokyo), eds. S. Kamefuchi et al, p. 44.
[6] Nelson, E., 1966, Phys. Rev., <u>150</u>, 107.
[7] We have many papers after the original proposal; Parisi, G., and Wu, Yong-Shi, 1981, Sci. Sin., <u>24</u>, 483. For review articles, see; Klauder, J. R., 1983, Acta Phys. Austrica, Suppl. XXV (Springer, Berlin), p.251; Okano, K., 1984, Memoirs of the School of Sci. & Eng. Waseda Univ., <u>48</u>, 23; Sakita, B., 1985, *Quantum Theory of Many Variable Systems and Fields* (World Scientific, Singapore); Damgaard, P. H., and Hüffel, H., 1987, Phys. Reports, <u>152</u>, 277.
[8] Suzuki, M., 1976, Prog. Theor. Phys., <u>56</u>, 1454; 1976, Commun. Math. Phys. <u>51</u>, 183.
[9] Hüffel, H., and Rumpf, H., 1984, Phys. Letters, <u>148B</u>, 104; Gozzi, E., 1985, Phys. Letters, <u>150B</u>, 119; Nakazato, H., and Yamanaka, Y., 1986, Phys. Rev., <u>D34</u>, 492; Nakazato, H., 1987, Prog. Theor. Phys., <u>77</u>, 20L, 802L.
[10] For example, see: Callaway, D. J. E., and Rahman, A., 1982, Phys. Rev. Letters, <u>49</u>, 613.
[11] Wigner, E. P., 1932, Phys. Rev., <u>40</u>, 749.
[12] Husimi, K., 1940, Proc. Phys. Math. Soc. Japan, <u>22</u>, 264.
[13] Chaturvedi, S., Kapoor, A. K., and Srinivasan, V., 1985, Phys. Letters, <u>B157</u>, 400; Ohba, I., 1986, Prog. Theor. Phys., <u>77</u>, 1267.
[14] Namiki, M., and Yamanaka, Y., 1986, Prog. Theor. Phys., <u>75</u>, 1447.
[15] Nakazato, N., Namiki, M., and Shibata, H., 1986, Prog. Theor. Phys., <u>76</u>, 708.
[16] For example see: Grimus, W., and Hüffel, H., 1983, Zeits. f. Phys., 8 <u>C18</u>, 129; Nakazato, H., Namiki, M., Ohba, I. and Okano, K., 1983, Prog. Theor. Phys., <u>70</u>, 298.
[17] Fukai, T., Hakazato, H., Ohba, I., Okano, K. and Yanamaka, Y., 1983, Prog. Theor. Phys., <u>69</u>, 361; Breit, J. D., Gupta, S., and Zaks, A.,

1984, Nucl. Phys., B233, 61; Damgaard, P. H., and Tsokos, K., 1984, Nucl. Phys., B235, 75.
[18] Namiki, M., Ohba, I., and Okano, K., 1984, Prog. Theor. Phys., 72, 350.
[19] Nakagoshi, N., Namiki, M., Ohba, I., and Okano, K., 1983, Prog. Theor. Phys., 70, 326; The same procedure, based on the Fokker-Planck equation, was proposed and developed by: Zwanziger, D., 1981, Nucl. Phys., B192, 259; Baulieu, L., and Zwanziger, D., 1981, Nucl. Phys., B193, 163; Seiler, E., Stematescu, I. O., and Zwanziger, D., Nucl. Phys., B239, 177.
[20] Fukai, T., and Okano, K., 1985, Prog. Theor. Phys., 73, 790.
[21] Namiki, M., Ohba, I., Okano, K., and Yanamaka, Y., 1983, Prog. Theor. Phys., 69, 1580.
[22] Namiki, M., Ohba, I., Tanaka, S., and Yanga, D., 1987, Phys. Letters, B194, 530.
[23] Namiki, M., Ohba, I., and Tanaka, S., 1986, Phys. Letters, B182, 66.
[24] For example, see: *Proc. International Symp. on the Lattice Gauge Theory,* held in Paris in 1987, to be published.
[25] Tanaka, S., in preparation for publication.

Part 6

Epistemology, Interpretation and Conjecture

O. Costa de Beauregard

RELATIVITY AND PROBABILITY, CLASSICAL AND QUANTAL

A 'manifestly relativistic' presentation of Laplace's algebra of conditional probabilities is proposed, and its 'correspondence' with Dirac's algebra of quantal transition amplitudes is displayed. The algebraic reversibility of these is classically tantamount to time reversal, or 'T-invariance', and quantally to 'CPT-invariance'. This is closely related to the *de jure* reversibility of the = negentropy information transition, although *de facto* the upper arrow prevails aver the lower one (Second Law).

1. INTRODUCTION

How is it conceivable that the paradigm of an extended space-time geometry, where occurrences are displayed all at once (which of course does not mean "at the same time") is compatible with the very ideas of chance and becoming? There is in this a technical problem and a philosophical problem, both of which will be addressed.

The possibility of a 'manifestly covariant' probability scheme is afforded by the Feynman (1949) graphs scheme, where the Born (1926) and Jordan (1926) algebra of wavelike transition amplitudes is cast in a relativistic form. Dirac (1930) expressed the Born and Jordan algebra in the form of a 'bra' and 'ket' symbolic calculus which ran, so to speak, almost by itself.

Recently, while thinking about the famous Einstein-Podolsky-Rosen (EPR, 1935) correlations, it occurred to me (1986, 1987) that Laplace's (1774) algebra of conditional probabilities can be presented in a form exactly paralleling the Born-Jordan-Dirac-Feynman algebra. Thus a precise 'correspondence' (in Bohr's sense) can be exhibited between the two algebras, the

184

classical one adding partial, and multiplying independent probabilities; and the quantal, wavelike one, doing the same with amplitudes. Let it be emphasized at the very start that the two epithets 'conditional' and 'transitional' are synonymous, and thus exchangeable as suits the context.

Any algebra is by definition timeless, and so no temporal relation is essentially implied in the probability concept.

Algebraically speaking any transition is between two 'representations' of a system (to borrow a convenient expression from the quantal paradigm). In physics, of course, time very often enters the picture; then, one is led to consider the 'preparation representation' and the 'measurement representation' of an 'evolving system', together with (either) the classical 'transition probability', or the quantal 'transition amplitude' from the one to the other. This is equivalent to a 'conditional probability', or 'amplitude', which can be thought of either predictively or retrodictively. So a temporal aspect, when relevant, dramatizes the picture, somewhat like the Greek drama dramatized the metaphysics of Fate.

The formal parallelism, or correspondence, between the classical Laplace and the quantal Dirac schemes breaks down completely at the level of interpretation, because of the basic formula

$$(1) \qquad (A|B) = |\langle A|B \rangle|^2$$

expressing the quantal transition or conditional probability $(A|B)$ in terms of the (complex) transition or conditional amplitude $\langle A|B \rangle$. The off-diagonal terms on the right hand side (which, physically speaking, are interference or beating style terms) entail (as we shall see in more detail) the often 'paradoxical' aspects of (algebraically speaking) 'nonseparability' or (geometrically speaking) 'nonlocality'. Of course, by using an 'adapted representation', one hides these terms, in a way somewhat similar to that in which, by looking in one of three privileged directions, one sees a parallelepiped as a rectangle. But this affords no more than a deceptive semblance of the classical, Laplacean way of handling occurrences. As we shall see in more detail, while the classical summations over mutually exclusive possibilities could be thought of as being over 'real hidden states', quantal summations cannot, and are thus said to be over 'virtual states'.

2. LAPLACE'S 1774 ALGEBRA REVISITED

Expanding upon Bayes's discussion of conditional probability, Laplace wrote a series of papers, the first of which is the famous one of 1774 entitled 'Memoir on the probability of causes by the events'. There, he introduces the concept of what I will call, and denote an intrinsic, reversible conditional probability

(2) $\qquad (A|B) = (B|A)$

the probability 'of A if B' or 'of B if A'. Later, however, he discarded this concept in favor of the two converse, usually unequal, conditional probabilities which I denote |A|B), 'of A if B' and |B|A = (A|B|, 'of B if A', and which have been used ever since (see Jaynes 1983, formulas A6 and A7 p. 216). These are related to the (essentially symmetric) joint probability |A)·(B| 'of A and B' according to the formula

(3) $\quad A)·(B| \equiv |B)·(A| = |A|B)(B| = |A)(A|B|,$

which reads 'the joint probability of A and B equals the conditional probability of A if B, times the prior probability of B', or the converse. The dot I insert in |A)·(B| is by analogy with a scalar product of vectors.

It turns out that it is very useful to reintroduce Laplace's initial (A|B) concept, and thus to rewrite equations (3) in the form

(4) $\quad |A)·(B| \equiv |B)·(A| = |A)(A|B)(B| = |B)(B|A)(A|.$

Obviously, the two extrinsic conditional probabilities |A|B) and |B|A) are related to the intrinsic conditional probability (2) according to

(5) $\quad |A|B) = |A)(A|B)(B|, \quad |B|A) = |B)(B|A);$

the reason I call them "extrinsic" is that they contain the prior probabilities |A) ≡ (A| or (B| ≡ |B).

Setting B = A we rewrite formulas (4) and (5) as

(6) $\quad |A)\cdot(A| = |A)(A| = |A)(A|A)(A|,$

thereby expressing three things: the joint probability of A and A is idempotent; the intrinsic conditional probability of A if A is unity, i.e. $(A|A) = 1$; and the extrinsic conditional probability of A if A equals the prior probability of A, $|A|A) = |A)$. Thus, introducing complete sets of mutually exclusive occurrences A, ..., and (borrowing a term from quantum mechanics) calling them 'representations of a system', orthonormalization is possible in the form

(7) $\quad\quad\quad\quad\quad (A|A') = \delta(A,A').$

No temporal element has been implied up to now. For example, $|A)\cdot(B|$ can denote the joint probability that a U.S. citizen has the height A and the weight B; this is a number, devoid of any idea of succession. If the sampling is refined by considering mutually exclusive categories, such as men and women, the prior probabilities $|A)$ and $|B)$ of A and B in these categories enter the picture, and we end up with formula (4), which is preferable to formulas (3). Thus we can speak of the 'height representation' and the 'weight representation' of a U.S. citizen, and of the transition probability connecting these as being 'naked' $(A|B)$, or 'dressed' $|A)(A|B)(B|$.

'Lawlike reversibility versus factlike irreversibility' (to borrow Mehlberg's (1961) wording) has been discussed by Laplace and, later, by Boltzmann (1898) in essentially equivalent terms, as was noted by Van der Waals (1911). It may be expressed thus:

(8) $\quad |A|B) \neq |B|A)$ *iff* $|A) \neq |B).$

For example, basket ball players are usually tall and light; therefore, considering high values of height A and weight B, the two extrinsic conditional probabilities $|A|B)$ and $|B|A)$ are far from being equal for this class. This is a logical sort of irreversibility, with no time sequence implied.

Of course time enters the picture if $|A)$ denotes the 'preparation representation' and $|B)$, the 'measurement representation' of a system. In such cases both Laplace and Boltzmann assumed maximal irreversibility, in the form 'all prior probabilities (B| of the final states are equal among themselves', so that there is no need to mention them. This is an extremely radical irreversibility statement which (although quite consonant with the phenomenological claims of Carnot and Clausius) breaks the mathematical harmony - and, even worse, alters the very meaning of the probability concept, as we shall see later.

It is often useful to express a conditional or transition probability (A|C) with an intermediate summation

(9) $$(A|C) = \sum (A|B)(B|C),$$

the word "intermediate" being understood algebraically. For example C can denote the waist measurement of a U.S. citizen, while A and B retain their previous meanings. Formula (9) is known as the generating formula of Markov chains. If a temporal element is implied, it follows from the Laplacean symmetry (2) that the chain can zigzag arbitrarily throughout either space-time or the momentum-energy space (if such a picture is preferred), completely disregarding the macroscopic time or energy arrow. In that case, the word "intermediate" assumes a topological meaning.

As a first example consider the joint probability |A)·(C|, or the conditional probability (A|B), that, in a particular U.S. National Park, we will find at place A the male and at place C the female of a couple of bears. Coupling means interaction. Thus the two bears can meet at some 'real hidden place B' and a summation over B is needed. Logically speaking it makes absolutely no difference whether the AC vector is spacelike, with the meeting place B before, or after, A and C; or whether it is timelike, with, say, B after A and before C. Considered as valid in all three cases, the formula (9) is topologically invariant with respect to V, Λ, or C shapes of the ABC zigzag; and so is formula (4).

As a second example consider two Maxwellian or Boltzmannian colliding molecules. The unnormalized collision probability |A)·(C| is the product of three independent probabilities: the mutual cross section (A|C), and the two initial occupation numbers |A) and (C| or, more generally, their estimated values). This is for prediction.

Retrodictively the same formula holds, with |A) and (C|, then denoting the final occupation numbers.

Formulas (4) and (9) also hold for a C shaped ABC zigzag. As previously mentioned (A|C) then denotes the 'naked transition probability', and |A)·(C| the 'dressed transition probability, |A) the initial occupation number of the initial state and (C| the final occupation number of the final state. It so happened that neither Boltzmann, nor Laplace in similar cases, multiplied (A|C) by (C|, but both multiplied it by |A). This, as previously suggested, was their (common) way of expressing their belief in 'maximal physical irreversibility'. But this was 'intrinsically illogical' for the following reason: multiplication by |A) does imply 'statistical indistinguishability'; thus, there are (C| ways in which the transiting molecule can reach the final state; therefore, multiplication by (C| is imperative, and is a corollary to multiplication by |A).

Of course, as used in this fashion, formula (4) is none other than the formula prescribed by the quantal statistics - with |A), |C) = 0,1,2...,n for bosons, and |A), |C) = 0,1 for fermions. The implication, then, is that the internal consistency of (both) quantal statistics is greater than that of classical statistics - and that the quantal statistics entail the past-future symmetry set aside by Laplace and Boltzmann.

So, in the case of colliding molecules, formulas (4) and (9) have topological invariance with respect to V, Λ or C shapes of an ABC zigzag; the intermediate summation is over 'real hidden states'; in the case of spherical molecules these can be labeled by the line of centers when the molecules touch each other.

A complete Markov chain is expressed as

(10) $|A)\cdot(L| = \sum\sum ... |A)(A|B)(B...K)(K|L)(L|$.

The Bayesian approach to probabilities has it that the end prior probabilities |A) and (L| are shorthand notations for conditional probabilities (E|A) and (L|E') connecting the system with the environment. Thus we are left with essentially two concepts: the joint probability |A)·(B|, and the intrinsic conditional probability (A|B).

3. THE 1926 BORN AND JORDAN ALGEBRA REVISITED

'Corresponding' to Laplace's symmetry assumption (2) we have the Hermitian symmetry

(11) $\qquad \langle A|B\rangle = \langle B|A\rangle^*$

assumed for transition amplitudes, which are also conditional amplitudes. Going from the real to the complex field not only enlarges the paradigm of probabilities, but, as we shall see, also deepens it.

The joint amplitude of two occurrences A and B is expressed as

(12) $\qquad |A\rangle \cdot \langle B| = |A\rangle\langle A|B\rangle\langle B|$,

with $|A\rangle$ and $\langle B|$ (The Dirac (1930) ket and bra vectors) denoting prior amplitudes, the absolute squares of which are occupation numbers (or, more generally, the estimated values of these). Setting B = A we get

(13) $\qquad |A\rangle \cdot \langle A| = |A\rangle\langle A| = |A\rangle\langle A|A\rangle\langle A|$,

showing that $|A\rangle\langle A|$ is a projector, and that $\langle A|A\rangle = 1$; thus, 'orthonormalization' of a 'representation of a system' is allowed, according to

(14) $\qquad \langle A|A'\rangle = \delta(A,A')$.

'Corresponding' to the generating formula (9) of Markov chains we have that one of the Landé (1965) chains

(15) $\qquad \langle A|C\rangle = \sum \langle A|B\rangle\langle B|C\rangle$.

Whenever a space-time, or a momentum-energy connotation is attached to the occurrences being considered, the Landé chain can, due to the symmetry property (11), zigzag arbitrarily throughout either space-time or the momentum-energy space, disregarding the macroscopic time or energy arrow. Thus formula (15) has topological invariance vis-à-vis V, Λ or C shapes of an ABC zigzag.

An important generalization of the Landé chain is the Feynman (1949) graph, a concatenation, so to speak, where more than two links $\langle A|B\rangle$ can be attached to a vertex A. Topological invariance is a well known property of Feynman graphs.

The end prior amplitudes $|A\rangle$ or $\langle L|$ of a Landé chain

(16) $\quad |A\rangle\cdot\langle L| = \sum\sum ... |A\rangle\langle A|B\rangle\langle B...K\rangle\langle K|L\rangle\langle L|.$

or a Feynman graph are shorthand notations for conditional amplitudes $\langle E|A\rangle$ or $\langle L|E'\rangle$ linking the system to the environment. Thus Dirac (1930) and Landé (1965) write state vectors $\psi_a(x)$ and $\omega_a(k)$ in the form $\langle a|x\rangle$ and $\langle a|k\rangle$, in the spacetime and the momentum-energy pictures, respectively.

The exact formal parallelism between the 1774 Laplace and the 1926 Born and Jordan algebras, as displayed in the present and preceding section, breaks down completely, however, at the level of interpretation. This is because of the basic formula (1) expressing the quantal transition or conditional probability $(A|B)$ in terms of the transition or conditional amplitude $\langle A|B\rangle$. Let it be recalled that the Born (1926) and Jordan (1926) radically new wavelike probability algebra was adjusted so as to fit the Einstein and de Broglie wave-particle dualism: the addition of partial probabilities had to yield to the addition of partial amplitudes, due to the physical phenomenon of interference or beating.

Because there are off-diagonal terms in the right-hand side of formula (1) (even if formally concealed by use of an adapted representation), the intermediate summations in formulas (15) or (16) can no more be thought of as implying real states; thus they are said to imply 'virtual states'. And this may have very dramatic consequences.

Whenever a space-time or a momentum-energy connotation is attached to the occurrences A,B.,..., the V, Λ and C shapes of an ABC zigzag receive different physical interpretations.

A V shaped ABC zigzag describes what is called an 'Einstein-Podolsky-Rosen (EPR, 1935) correlation' proper between two distant measurements at A and C issuing from a common preparation at B. The so-called 'EPR paradox' (Einstein, 1949, p. 681) consists of the fact that in formula (15) the summation $|B\rangle\langle B|$ over the states inside the source cannot be thought of as implying 'real hidden states'. So, borrowing Miller's and Wheeler's (1983) wording, wee say that it is a 'smoky dragon'.

Accordingly, the states measured as $|A\rangle$ and $\langle C|$ do not preexist in the source. This is why quite a few authors: myself (1953, 1977, 1979, 1983, 1985, 1986, 1987), Stapp (1975), Davidon (1976), Rayski (1979), Rietdijk (1981), Cramer (1980, 1986), Sutherland (1983), Pegg (1980), and possibly others, have introduced the idea of a 'zigzagging arrowless causality'.

A ∧ shaped ABC zigzag illustrates a 'reversed EPR correlation' between two distant preparations at A and C merging into a common sink B. Of course, only those paired particles A and C having the right phase relation are absorbed in the sink B; this illustrates 'factlike irreversibility'. To understand what is meant here by 'lawlike reversibility' one must think retrodictively.

For a more detailed discussion of advanced and retarded causality in the context of EPR correlations proper and reversed, I refer to previous publications (1977, 1979, 1987a).

A C shaped ABC zigzag illustrates Miller's and Wheeler's (1983) 'smoky dragon' metaphor. In what 'state' is an 'evolving system' between its preparation as $|A\rangle$ and its measurement as $|C\rangle$? Is it in the retarded state generated by $|A\rangle$ (as classical physics had it)? Or is it in the advanced state converging into $|C\rangle$, as the Hermitian symmetry (11) allows ? It cannot be in both states if there is a transition. And then, due to the symmetry expressed by (11), why should it be in the one rather than in the other?

The truth is that the evolving system is neither in the retarded state issuing from $|A\rangle$, nor in the advanced state passing into $|C\rangle$, because it is actually transiting from $|A\rangle$ to $|C\rangle$. Borrowing an analogy from classical hydrodynamics, we say that it feels symmetrically the 'pressure from the source' $|A\rangle$, and the 'suction into the sink' $|C\rangle$. In other words, it is Miller's and Wheeler's 'smoky dragon', of which nothing definite is known, because its expression is a summation over products $|B\rangle\langle B|$ of virtual states. Only the 'tail' of the dragon held as $|A\rangle$, and its 'mouth' biting as $|C\rangle$, are, so to speak, in our world. The dragon itself lives above, in the A⊗C Hilbert space. For example, in the well known two slits thought experiment, 'the photon' neither passes through one slit or the other, nor through both. Such images, borrowed from the Laplacean paradigm, have no ready place in the Born.-Jordan-Dirac paradigm.

In the V shaped diagram previously discussed, there is at B, in the source, only a smoky dragon coiled in there, with two mouths biting at A and C; this dramatizes 'the EPR paradox'. In the Λ shaped diagram the dragon, coiled in B, has two tails held at A and C. It is in the nature of dragons to be fantastic.

4. LORENTZ INVARIANCE AND CPT INVARIANCE

That the classical Laplace algebra and the quantal, wavelike, Dirac algebra are both easily endowed with the Lorentz invariance fitting space-time descriptions must be quite obvious by now.

The topological invariance of these two algebras vis à vis the V, Λ and C shapes of an ABC zigzag, in either space-time or the momentum-energy space, leads to a consideration of invariance under the geometrical reversal of all four axes; this can be thought of either as an 'active' or as a 'passive' transformation. The natural guess then is that elementary physical laws are invariant under the geometrical reversal of all four axes.

Classically this can be called 'covariant motion reversal' PT, which (actively) exchanges emissions and absorptions or preparations and measurements, or (passively) exchanges prediction and retrodiction.

Things are more subtle in the quantal paradigm, due to the pairing between particles and antiparticles, and to the Stueckelberg and Feynman expression of their exchange by the reversal of 4-velocities in the x picture, or of energy-momentum in the $p = \hbar k$ picture.

Figures 1a, b, and c show, either in space-time or in the momentum-energy space, the three operations C (particle-antiparticle exchange), PT (covariant motion reversal), and CPT (geometrical, 'strong' space-time reflection $\Pi_\circledast$, the last expressed by Lüders (1952) as :

(17) $$\Pi_\circledast = CPT = 1.$$

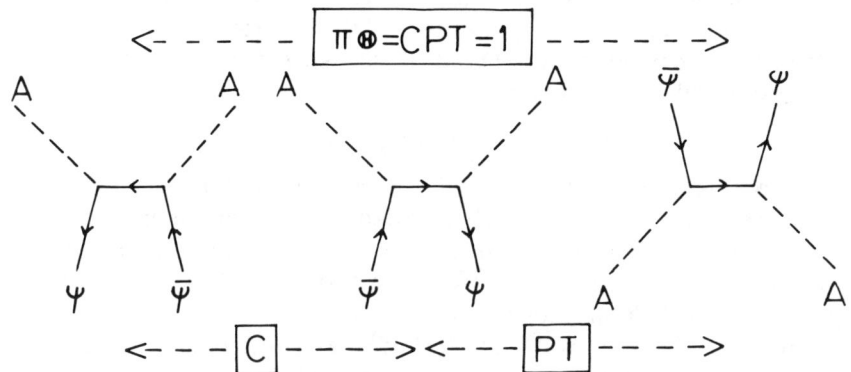

Another symbolization expressing the C, the PT, and the CPT = 1 symmetries is as follows:

(18)   C: $\langle A|B\rangle = \langle A|B\rangle^*$,

        PT: $\langle A|B\rangle = \langle B|A\rangle$,

        CPT: $\langle A|B\rangle = \langle B|A\rangle^*$.

One thus sees that CPT-invariance is built into the very scheme of Hermitian symmetry, and in the second quantization algebra that has been expanded out of it. Extending the relativistic jargon, we can say that C and PT are two 'relative images' of essentially the same operation.

CPT-invariance entails the following 'principle of detailed balance':

(19) $A + B + \ldots = C + D + \ldots$

where (beware) a bar means particle on the left hand side and antiparticle on the right side (and vice-versa). This shows that CPT invariance is the faithful quantal-and-relativistic extension of the Loschmidt (1876) T-symmetry. Therefore, *physical reversibility is expressed by the CPT- symmetry, and not by the PT-Symmetry.*

A consequence of this important remark is the following. The Tomonaga (1946)-Schwinger (1948) transition amplitude between a preparation $|\Phi\rangle$ and a measurement $|\Psi\rangle$ has the three equivalent expressions:

(20) $\qquad \langle \Psi | \cup \Phi \rangle = \langle \Psi \cup | \Phi \rangle = \langle \Psi | \cup | \Phi \rangle.$

The first one, projecting the retarded preparation upon the measurement, illustrates the so-called 'state vector collapse'. The second one, projecting the advanced measurement upon the preparation, is a 'retrocollapse'. And the third, symmetric expression, I call 'collapse-and-retrocollapse'.

Fock (1948) and Watanabe (1955) have independently explained that, in quantum mechanics, retarded waves are used in prediction and advanced waves in retrodiction; this is expressed by the first equality (19). Therefore (beware) this Fock-Watanabe symmetry between retarded and advanced waves, and between prediction and retrodiction, is equivalent to the CPT, not the PT symmetry.

Temporal succession, as we have said, provides a dramatization of both the Laplace and the Dirac algebras. In the latter, collapse of the wave function' should not be taken litterally (ipso facto causing much worry with Lorentz invariance). It is merely a computational recipe -and a much less palatable one than the Lorentz-and-CPT invariant transition amplitude concept. Borrowed from the wardrobe of macrophysics, the 'evolving state vector' concept (a cloth loosely draping Wheeler's 'smoky dragon') should be dismissed for being no less noxious to clear thinking than was the late 'luminiferous aether': see Aharonov and Albert (1980, 1981) and myself (1981, 1982).

5. CAUSALITY AS IDENTIFIED WITH CONDITIONAL PROBABILITY OR, BETTER STILL, WITH CONDITIONAL AMPLITUDE

Is there any definition of causality more operational than the following : 'If you do this, then the (predictive) probability that

RELATIVITY AND PROBABILITY

that will occur is ...'; or : If you find that, then the (retrodictive) probability that this has occurred is ...'? Is there any binding of the subjective and the objective sides of chance that is more subtle or strong than this ?

This is an intrinsically arrowless conception of causality. As previously explained, it affords a topologically invariant rendering of, for example, the classical coupled-bears parable, the colliding molecules problem, and some other cases I have discussed elsewhere (1987). It also affords a topologically invariant rendering of the quantal EPR correlations (proper or reversed), and of Wheeleer's smoky dragon metaphor - all in one stroke. In all these cases, $(A|B) = \sum \langle A|B \rangle \langle B|C \rangle$, or $\langle A|C \rangle. = \sum \langle A|B \rangle \langle B|C \rangle$ are, respectively, the conditional probability or amplitude linking two measurements, two preparations, or a preparation and a measurement (depending on the V. Λ or C shape of the zigzag). Was Laplace perhaps unwittingly anticipating this, when he spoke of the "probabilities of causes" ? In any case, this is the very essence of the S-matrix philosophy.

What we have is a stochastic game, be it classical or quantal, implying questions (the preparations) and answers (the measurements). In between there is, so to speak, the 'wiring of the space-time computer' - that is, in quantum mechanics, the Feynman graph, with Feynman propagators as transition amplitudes $\langle A|B \rangle$.

6. RELATIVITY AND STATISTICAL FREQUENCY

How did the probability concept fit the classical, deterministic paradigm? Two classical stochastic tests were said to be identical when they were the same with respect to all parameters except those assumed to be of negligible influence, the latter being neglected. For example, an aerodynamics engineer repeating a measurement will not worry about the phase of the moon, nor the stock market exchange.

The neglected parameters were allowed to vary, and assumed to do so 'according to the laws of chance'. The probabilistic scheme was thus tested, and either confirmed or 'falsified', the latter, in Laplace's terms, consisting in some 'sufficient reason' having been overlooked. Guessing what this overlooked 'sufficient reason' was amounted to learning something new-eventually something important, for example the violation of the principle of the equipartition of energy and the discovery of the quantum.

In the Bayesian treatment of probabilities this sort of game is called 'hypotheses testing', or 'extracting signal from noise'.

A transposition of these remarks into the field of relativistic probability theories reconciles the use of statistical frequencies with an extended 4-dimensional geometry. Two stochastic tests will be termed identical if they are the same on all parameters up to those considered negligible. In this sense they will be 'translationally deducible from each other'(either in a spacelike or a timelike fashion). In quantum mechanics there is always a set of parameters which are allowed to vary 'according to the laws of chance': those parameters that are conjugate with the ones actually prepared or measured. This is exactly how the Feynman graphs are used at CERN or at Lawrence-Berkeley. Incidentally, two CPT associated Feynman graphs must be thought of as framed pictures, because the environment cannot be CPT reversed, due to two very strong factlike asymmetries: the preponderance of particles over antiparticles, and of retarded over advanced waves.

7. METAPHYSICS OF A RELATIVISTIC PROBABILITY CALCULUS

As there is only one world history, and it makes no sense to think of rewriting it, strictly speaking no frequency interpretation of probability is possible inside a relativistically covariant paradigm. Thus, the only acceptable conception of probability is the so-called 'subjective' one, held by Bernouilli, Bayes, Laplace, and advocated today by Jaynes (1983). This being so, one must take care to note that *information is a two faced concept: gain in knowledge* (say, in the transition: *negentropy → information* at the reception of a message) and *organizing power* (say, in the transition: *information → negenteropy* at the emission of the message). That the one face is so trivial, and the other so

RELATIVITY AND PROBABILITY

much hidden, is a direct consequence of physical irreversibility -and truly the most profound expression of it.

The evaluation of any universal constant in 'practical units' expresses some aspects of our existential-situation-in-the world: that Joule's J 'is' not far from unity means that the First Law belongs to common place physics; but, that Einstein's c 'is' very large means that the relativistic phenomena are largely outside the domain where the meter and the second are appropriate units. So then, what is the meaning of the 'smallness' of Boltzmann's k, or, rather, of the conversion coefficient k Ln 2 between a negentropy N expressed in thermodynamical units, and a piece of information expressed in bits, according to the formula

(20) $$N = (Ln\ 2)\ k\ I?$$

Letting k go to the limit zero would summarize a theory where gain in knowledge is rigorously costless, and free action utterly impossible - the once fashionable theory of 'epiphenomenal consciousness'. Cybernetics has rediscovered the 'hidden face' of information. It requires consciousness-the-spectator to pay her ticket, at a very low price; and this alone allows consciousness-the-actor to exist, but at exorbitant fees, because the exchange rate applies in the opposite direction.

While the $N \to I_2$ 'learning transition' (with de facto if not de jure $N > I_2$) is a generalization of the Second Law, as emphasized by Brillouin (1967), and expresses the familiar 'retarded causality', the $I_1 \to N$ transition (with then in fact $I_1 > N$) expresses 'finality'. This is the very sort of anti-Second-Law phenomenon radically excluded by the irreversibility statements due to Laplace, Boltzmann, Carnot and Clausius - but also the very sort of phenomenon Bergson (1907) and other thinkers deem necessary for understanding 'Creative Evolution'. The intrinsic or lawlike symmetry in the formulas suggests that natural occurrences are, so to speak, symmetrically 'pushed into existence' by efficient causes and 'pulled into existence' by final causes, as Lewis (1930) suggests - somewhat in the way that a hydrodynamical flow is determined jointly by the pressure from the sources and the suction from the sinks.

198

Accepting this picture does not entail that 'one can kill one's grandfather in his cradle' (as some have insinuated). 'Rewriting history' makes no sense. 'Final cause' does not mean reshaping the past, but does mean shaping it; as Lamarck puts it 'the function creates the organ'.

In other words, 'factlike irreversibility' does not mean suppression, but does mean repression of advanced actions. There are two aspects in an advanced action: 'seeing into the future' (precognition) and 'acting in the past' (psychokinesis; influencing falling dice implies doing so before the outcome is displayed).

These are things that can be, and have been statistically tested in repeatable experiments (Jahn and Dunne, 1987).

The intrinsic reversibility of probabilities implies that ascribing prior probabilities to later occurrences (Laplace's effects') can make sense. The New York Museum of Natural History displays the line of evolution leading to the horse. Given the structure of the 'eohippus' can one predict the structure of the horse? Of course not. But by using Laplace's and Boltzmann's prescription in reverse, one can correctly retrodict the 'primeval molecular soup' - or an equivalent of it. Objectors will then argue that the evolution problem should be placed in context, that systems which are far from equilibrium behave as parasites on the universal negentropy cascade, borrowing negentropy from upstream (Brillouin 1967) and dumping entropy downstream (Nicolis and Prigogine, 1977). All right. Of course, looking at things differently requires reassessing the prior probabilities. But there was nothing illogical in arguing from the eohhippus per se - so little so that, as Boltzmann (1964, pp. 446-448) puts it, "this, by stimulating thought, may advance the understanding of facts".

Others will argue that, after all, the eohippus was not so improbable, as the skeleton of one is displayed in New York.. But this is 'blind statistical reasoning', and *does* imply the finality assumption.

Weighing the prior probabilities of the 'effects' may well be an affair of willing no less than of knowing awareness. Relying as it does on the N → I transition, knowing awareness naturally perceives causality, while using the I → N transition, willing

RELATIVITY AND PROBABILITY 199

awareness naturally experiences finality. Eccles (1986) argues that will actually consists of biasing the end prior probabilities inside the nervous system, so that it truly is 'internal psychokinesis'. Libet (1985) should also be read in this respect.

Can psychokinesis be observed outside the envelope of our skin ? This is what Wigner (1967, pp. 171-183) implies on the basis of physical symmetry arguments analogous to mine - and is indeed also implied in a statement found in many quantum-mechnanical textbooks: 'Due to the finiteness of Planck's h (and, I would add, also due to the finiteness of Boltzmann's k) there is an unavoidable reaction of the measuring apparatus upon the measured system'. Where should one sever the two ? Between the dial or screen and the eye; somewhere along the optical nerve; or where? So, this seemingly harmless remark entails that *a reaction of the observer* upon the observed system exists in principle.

In the Princeton University School of Engineering there is an 'Anomalies Research Laboratory' where repeatable psychokinetic experiments have been conducted under Dean Robert Jahn, using sophisticated protocols and electronic equipment. His book (1987) makes interesting reading and, in my opinion, may well herald a scientific breakthrough.

Institut Henri Poincaré (France)

BIBLIOGRAPHY

Aharonov, Y. and Albert, D.Z. (1980), 'States and Observables in Relativistic Quantum Mechanics'. Physical Review 21: 3316-3324.

Aharonov, Y. and Albert, D.Z. (1981), 'Can we make sense out of the Measurement Process in Relativistic Quantum Mechanics'. Physical Review 24: 359-370.

Bergson, H. (1907), L'Evolution créatrice. Alcan, Paris.

Boltzmann, L. (1964), Lectures on Gas Theory (transl. by S. Brush). Univ. of California Press.

Born, M. (1926), 'Quantenmechanik des Stossvorgänge'. Zeitschrift für Physik 38: 858-888.

200   O. COSTA DE BEAUREGARD

Costa de Beauregard, O.:
(1953), 'Une Réponse à l'Argument Dirigé par Einstein, Podolsky et Rosen Contre l'interprétation bohrienne des Phénomènes Quantiques'. C.R. Académies des sciences 236: 1632-1634.
(1977), 'Time Symmetry and the Einstein-Podolsky-Rosen Paradox'. Il Nuovo Cim. 42B: 41-64.
(1979), 'Time Symmetry and the Einstein-Podolsky-Rosen Paradox, II'. Il Nuovo Cimento 51B, 267-279.
(1981), 'Is the Deterministically Evolving State Vector an Unnecessary Concept?, Lettere al Nuovo Cimento 31: 43-44.
(1982), 'Is the Deterministically Evolving State Vector a Misleading Concept'. Lettere al Nuovo Cimento 36: 39-40.
(1983), 'Lorentz and CPT Invariances and the Einstein-Podolsky-Rosen Correlations'. Physical Review Letters 50: 867-869.
(1985) 'On Some Frequent but Controversial Statements concerning the Einstein-Podolsky-Rosen Correlations'. Foundations of Physics 15: 871-887.
(1986), 'Causality as Identified with Conditional Probability and the Quantal Nonseparability'. Annals of the New York Acad. of Sciences 480: 317-325.
(1987a), 'On the Zigzagging Causality EPR Model: Answer to Vigier and Coworkers and to Sutherland'. Foundations of Physics 17: 775-785.
(1987b), Time, The Physical Magnitude. Reidel: Dordrecht.
Cramer, J.G. (1980), 'Generalized Absorber Theory and the Einstein-Podolsky-Rosen Paradox'. Physical Review 22: 362-376.
Cramer, J.G. (1986), 'The Transactional Interpretation of Quantum Mechanics'. Reviews of Modern Physics 58: 847-887.
Davidon, W.C. (1976), 'Quantum Physics of Single Systems'. Il Nuovo Cimento 36: 34-40.
Dirac, P.A.M. (1930), The Principles of Quantum Mechanics. Clarendon Press: Oxford.
Eccles, J. 'Do Mental Events Cause Neural Events Analogously to the Probability Fields of Quantum Mechanics?'. Procedings of the Royal Society 22: 411-428.
Einstein, A (1949), 'Reply to Criticisms'. In Albert Einstein Philosopher Scientist, edited by Shilpp P.A.. Illinois: The Library of Living Philosophers: pp. 665-688.
Einstein, A., Podolsky B. and Rosen, N. (1935), 'Can Quantum Mechanical Description of Physical Reality be Considered Complete'. Physical Review 48: 777-780.
Feynman, R.P. (1949a), 'The Theory of Positron'. Physical Review 76: 749-759.
Feynman, R.P. (1949b), 'Space-time Approach to quantum Electrodynamics'. Physical Review 76: 769-789.

Fock, V. (1948), 'On the Interpretation of the Wave Function Directed Towards the Past'. Doklady Acad. Nauk SSSR 60: 1157-1159.

Jahn, R., and Dunne, B.J. (1987), Margins of Reality. Harcourt: Brace, Jovanovich.

Jaynes, E.T. (1983), Papers on Probability, Statistics and Statistical Physics, edited by Rosenkranz, R.D.. Dordrecht: Reidel.

Jordan, P. (1926), 'Ueber eine Begründung des Quantummechanik'. Zeitschrift für Physik 40: 809-838.

Lande, A. (1965), New Foundations of Quantum Mechanics. Cambridge: Cambridge University Press.

Laplace, P.,S. (1891), 'Mémoire sur la Probabilité des Causes par les Evénements'. In Oeuvres Complètes, Paris: Gauthier Villars, vol. 8: pp. 27-65.

Lewis, G.N. (1930), 'The Symmetry of Time in Physics'. Science 71: 570-577.

Libet, B. (1985), 'On Conscious Will in Voluntary Action'. Behavioral and Brain Sciences 8: 529-615.

Loschmidt, J. (1876), 'Ueber der Zustand des Warmgleichgewichtes eines Systems von Körpern mit Rücksicht auf die Schwerkraft'. Sitz. Adak. Wiss. in Wien 73: 139-145.

Lüders, G. (1952), 'Zur Bewegungsumkehr in Quantiesierten Feldtheorien'. Zeitschrift für Physik 133: 325-339.

Mehlberg, H. (1961), 'Physical Laws and the Time Arrow'. In Current Issues in the Philosophy of Science, edited by Feigel, H. and Maxwell D.. New York: Holt Reinhart, pp. 105-138.

Miller, W.A., and Wheeler, J.A. (1983), 'Delayed Choice Experiments and Bohr's Elementary Phenomenon'. In Proceedings of the International Symposium on the Foundations of Quantum Mechanics in the Light of New Technology, edited by Kamefuchi S. and alii. Tokyo: Phys. Society, pp. 140-152.

Nicolis, G. and Prigogine, I. (1977), Self-Organization in Non-Equilibrium Systems. New York: Wiley.

Pegg, D.T. (1980), 'Objective Reality, Causality and the Aspect Experiment'. Physics Letter 78A: 233-234.

Rayski, J. (1979), 'Controversial Problems of Measurement within Quantum Mechanics'. Found. Phys. 9: 217-236.

Rietdijk, C.W. (1981), 'Another Proof that the Future can Influence the Present'. Found. Phys. 11: 783-790.

Schwinger, J. (1948), 'Quantum Electrodynamics, I: a Covariant Formulation'. Physical Review 82: 914-927.

Stapp, H.P. (1975), 'Bell's Theorem and World Process'. Il Nuovo Cimento 29B: 270-276.

Sutherland, R.I. (1983), 'Bell's Theorem and Backwards-in-Time Causality'. Intern. J. Theor. Phys. 22: 377-384.

Tomonaga, S.I. (1946), 'On a Relativistically Invariant Formulation of the Quantum Theory of Wave Field'. Prog. Theor. Phys. 1: 27-42.

Waals, J.D. van der, 'Ueber die Erklärung des Naturgesetze auf Statistisch-Mechanischer Grundlage'. Phys. Zeitschrift 12: 547-549.

Watanabe, S. (1955), 'Symmetry of Physical Laws'. Reviews of Modern Physics 27: 26-39.

Wigner, E.P. (1967), Symmetries and Reflections. Cambridge Massachussets: M.I.T. Press.

CLASSICAL AND QUANTUM PROBABILITIES

Eftichios Bitsakis
Department of Philosophy, University of Ioannina
Department of Physics, University of Athens

ABSTRACT. The object of this paper is to bring out differences as well as similarities between classical and quantum probabilities. The first part of the paper is devoted to a brief analysis of the two main epistemological interpretations of probability — the positivist and the realist interpretation. The second part concerns the essential premises and characteristics of classical probabilities. The problem of the "nature" of quantum probabilities and of their differences from classical ones is the subject of the following part. The main thesis supported here is that classical probabilities are, in the general case, <u>actual</u>, before the measurement, while quantum probabilities are the measure of the <u>potentialities</u> of the quantum ensemble. Quantum statistical determinism is a concept transcending the classical definitions of determinism; it is the epistemological generalization of the fact that quantum probabilities are determined in a specific way, characteristic of the microphysical world.

INTRODUCTION

According to the Copenhagen Interpretation (C.I.), there is an epistemic difference between classical and quantum probabilities. Classical probabilities are — they say — of subjective character: they are the expression of incomplete knowledge. Quantum probabilities, on the contrary, are considered as objective, primordial, essential, irreducible to a dynamical form of determination. It is worthy to examine the validity of the above argument.

It must be noted from the outset, that the validity of the axiomatic probability theory in Quantum Mechanics (Q.M.) will not be put in question. It will be accepted that the formalism of quantum mechanics satisfies the axioms of probability theory[1]. As Jauch argues, in the controversies concerning this problem, the calculus as such, in its definite form given by Kolmogorov in 1930, was never put in question[2].

The above thesis concerns the axiomatic level, as well as the empirical one — the level of direct observations — where the limit frequency interpretation seems satisfactory. However, the physical background, the processes underlying classical and quantum probabilities, are not necessarily identical. The object of this paper is to bring out similarities as well as differences between classical and quantum probabilities. Consequently, epistemic differences, if any, do not concern the axiomatic foundation of probability.

1. REALISM AND POSITIVISM IN PROBABILITY THEORY

We now pose the question: Is it possible to give a definition of probability and/or of probability calculus?

According to von Mises, "probability calculus or probability theory is a mathematical theory of a specific area of phenomena, aggregate phenomena or repetitive events"[3].

It is evident that the above quotation is not a definition. For other writers, a definition of probability is simply impossible. Thus Popper writes: "Nothing is further from my mind than an attempt to solve the pseudo-problem of giving a definition of the meaning of probability. It is obvious that the word 'probability' can be used perfectly properly and legitimately in dozens of senses, many of which, incidentally, are incompatible with the formal calculus of probability"[4]. The above position is not perhaps the only legitimate. But I want to leave open the question of a definition of probability.

The classical definition: $P_i = \lim_{n \to \infty} \frac{n_i}{N}$

appears similar to that of the limit frequency interpretation. But classical theory did not arrive to bridge its theoretical foundation with the observed statistics of real events. Nevertheless, classical theory had set the goal of defining probability on the basis of the objective structure of phenomena, to explain the stability of real frequencies and to calculate frequencies on the basis of objective data[5].

It was impossible for classical theory to realize this program. However, its approach is interesting from the epistemological point of view because it presupposes two major premises of classical physics: scientific realism and determinism. More than that: this program is not obsolete today, on the condition that it is liberated from its mechanistic constraints.

The classical conception of probability was an integral part of a more general, realist determinist and mechanistic conception of the Universe. In fact the so-called laplacian conception of probability was founded on a certain number of ontological premises:

1. Objectivity of nature and — consequently — of physical systems (scientific realism).
2. Objectivity of physical interactions, propagated with infinite velocity (postulate of the action-at-a-distance).
3. Absolute Euclidean space — unlimited scene of phenomena — and absolute universal time.
4. Unique determination of the effect by its causes. (The same causes produce the same effects, according to the classical definition of Newton).

5. Objectivity of physical laws. Laws possess an ontological and not only an epistemic status.
6. An evolutionary conception of knowledge founded on the belief that the human intellect can grasp more and more profoundly the laws of nature — to approach absolute truth without arriving to an exhaustive knowledge of Nature.

Let us recall the famous thesis of Laplace concerning determinism, probability and scientific knowledge: "Une intelligence qui, pour un instant donné, connaîtrait toutes les forces dont la nature est animée et la situation respective des êtres qui la composent, si d'ailleurs elle était assez vaste pour soumettre ces données à l'analyse, embrasserait dans la même formule les mouvements des plus grands corps de l'univers et ceux du plus léger atome: rien ne serait incertain pour elle, et l'avenir comme le passé serait présent à ses yeux" [6].

We cannot accept today the mechanistic conception of the Universe — this extreme case of reductionism — which aims to reduce the infinite qualitative variety of phenomena to the laws of mechanics. We reject the reduction of the category of chance to the status of subjective category — reduction realized in the name of an absolute, fatal determinism. Mechanism is the Achilles heel of laplacian determinism. However, this is not a reason to forget its merits: realism, ontic status of states and of laws of physics, rejection of metaphysical and anthropomorphic ideas, gnoseological optimism concerning the possibilities of our intellect. More than that: Laplace connected probability with <u>totality</u>. The Universe is a machine governed by deterministic laws and every part of it is intrinsically correlated with the rest of this wholeness. We do not know the totality of the determinations of a system. Yet, this is not a reason for abandoning our quest for a more profound explanation of phenomena — of a hidden substratum determining the probabilistic distribution in a unique way. "A la limite" the whole chain of physical events should be deterministically explained. Thus, for Laplace, chance would be relegated to the status of an epistemic, subjective category.

The above conception dominated classical physics. For the materialist Laplace (1749-1827) the inexorable laws of nature are those of an enormous machine obeing its own internal determinations. For Leibniz (1646-1716), on the other hand, ultimately, every thing is necessary and the divine Intellect not only knows and can predict everything, but is in the origin of our Universe "the best of all possible worlds".

Thus, for modern materialists as well as for idealists "rien dans la nature ne peut se faire au hasard" (P. Holbach). The Democretian conception was reproduced litterally. Consequently chance was considered as an epistemic category reflecting our partial ignorance. Probability, in its turn, has been considered as the measure of the degree of our information, that is to say, as an essentially subjective category which would cease to have meaning if we could obtain precise knowledge of the phenomenon [7].

The mechanistic conception of chance (and of probability) is outdated. But even a non-mechanistic epistemology must accept that if we come to know exactly the individual history of every particle, the probabilistic distribution of the ensemble of individual processes should be the same. Consequently, the probabilistic laws are objective, in the sense that they describe the state of evolution of a great ensemble of identical systems. Thus chance possesses an ontic counterpart and Poincaré was right supporting that chance is "autre chose que le nom que nous donnons à notre ignorance".

We know today that the Universe is not a machine. So, we must seek another way out of the impass of the classical interpretation of probability.

As is well known, von Mises tried, in the twenties, to overcome this impass. According to him, "the subject of the probability theory is a long sequence of experiments or observations repeated very often and under a set of invariable conditions" [8]. Thus probability is defined as a limit: $P_i = \lim_{n \to \infty} \frac{n_i}{N}$

Von Mises tried to found probability directly on 'observable' events. This is his objectivist orientation. But the frequency interpretation is essentially positivist. The measure of a frequency is considered, in the last instance, as the only way of a positive knowledge of probability. Because the ultimate element of the analysis is the experimental, observational datum. Thus, if you put the question of the objective conditions, of the causes determining the statility of the observed frequencies, the objectivist will tell you that your question has no meaning. According to Khintchine, "La théorie fréquentielle en critiquant les insuffisances fondamentales de la théorie classique, lui oppose le refus — de caractère agnostique — de fonder les lois statistiques à partir des propriétés spécifiques objectives des phénomènes eux-mêmes". Thus, "quant à prévoir la stabilité des fréquences à partir des lois qui régissent le phénomène lui-même, ou à partir de quelque 'symétrie' objective (comme le faisait la théorie classique) à cela il est interdit de penser" [9].

The objectivism of von Mises was leading "à la limite" to positivism. Thus frequency theory has a logical-positivist and "à la limite" a subjectivist version: Probability is described as a logical relation. It is considered as a belief about the realization of an event, as a branch of psychology, and so on [10].

For the positivistic version of the frequency interpretation there is no ontological explanation of the frequency theory. The measure of a frequency is the only means of knowing probability. For scientific realism, on the contrary, it is in principle possible to explain the stability of the frequencies given by the experience, on the basis of the stability of objective conditions. Consequently, it would be, in principle, possible to explain the invariance of the frequencies and even to predict the probabilistic distribution on the bases of relevant conditions. In this way the classical ideal becomes legitimate anew but in a new epistemological context, free of its mechanistic constraints.

Classical theory tried to explain probabilities on the basis of objective conditions. Nevertheless, at that time it was in general impossible to explain the stability of the relative frequencies and to predict these frequencies on the basis of their objective determinations. Today this goal is attained at least by quantum mechanics. In fact it is well known that quantum mechanics is able to predict, on the basis of the essential parameters of the system and the conditions, the ensemble of the possible states and the probability of each one of them. Consequently, quantum theory is a <u>de facto</u> refutation of the positivist dogma.

Is it then possible to give an ontological foundation to the frequency theory? In other words: Is it possible to explain and to predict the probabilistic distribution on the basis of the objective conditions, to explain the stability of frequencies and to give an objective basis to the limit frequency interpretation? Consequently: is it possible to overcome the fatalism of the mechanistic conception and at the same time the agnostic position of the objectivists?

The law of large numbers constitutes the first step towards the solution of this problem. If a group of n castings, writes von Mises, is repeated again and again with the same die, for which q is the success probability, almost all groups will yield a rate n_i/n of successes close to q if n is large enough [11]. Consequently the stability of conditions is the ontic background of the law of large numbers. And as already noted, in some cases it is possible, on the basis of the relevant conditions, to compute beforehand the probability of an event. If, afterwards, we modify the conditions, the probability distribution will be modified. This is an indication of the objective determination of the relative frequencies. Consequently it seems legitimate to consider chance as an ontological category, and at the same time not to preclude the possibility of a more complete knowledge of the causes and thus the possibility to predict with a probability equal to unity the occurrence of an event. This means that the status of objectivity of the category of chance is the other aspect of its epistemic status.

The stability of the conditions is necessary in order to pass from the frequency to the probability. Nevertheless, in the frame of the objectivist-positivist interpretation it is impossible to affirm with certainty that this limit really exists. As Popper and Miller note "probabilistic support is not an inductive support" [12]. But our problem is not that of induction. As Hume in his time and Popper in our time have noted, induction leaves always a certain margin of uncertainty. On the contrary, if we consider the probability as the numerical measure of the <u>potentialities</u> of the statistical ensemble under invariable external conditions, then we can overcome the limits of objectivism and explain the coincidence of the empirical limit with the actual probability on the basis of the multiple potentialities of the statistical ensemble under the given conditions. In this way the law of large numbers acquires an objective foundation and ceases to be a simple empirical law.

Nevertheless, the dominant positivist conception was and continues to be an epistemolotical obstacle to the research of a realist explanation of the law of large numbers, and of the frequency theory of probability. The subjective interpretation dominated classical physics. In quantum mechanics it is its formal negation that prevailed: Classical probabilities are considered of subjective character. Quantum probabilities are objective. But as Margenau and Cohen argue, it is wrong to view the subjective thesis as <u>opposed</u> to the objective one [13]. Probabilistic laws describe objective phenomena. Thus they possess the same ontological status as dynamical laws. At the same time it is legitimate to affirm that nothing precludes the search for dynamical processes explaining the probability distribution. <u>Chance is the dialectic negation of necessity; not of causality</u>. Thus the search of hidden causal processes determining the probability distribution does not mean the restoration of mechanistic laplacian determinism [14].

2. PROBABILITIES IN CLASSICAL MECHANICS

Classical mechanics is founded on a number of ontological premises: existence of material points endowed with mass and momentum but deprived of any qualitative characteristic; interactions (forces) propagated with infinite velocity; euclidean, absolute space and absolute, universal time; instantaneous and unique determination of the effect by its causes. Let us consider, in this frame, the question of the status of classical probabilities.

1. The state of a system is defined by the value of a number of measurable physical quantities (observables). It is accepted that measurement does not disturb the system. Thus, the measurement of an observable does not modify the information concerning the other observables defining the state. The fact that all the observables are compatible implies, on the other hand, the compatibility of the ensemble of propositions concerning the classical system.

2. Thus, classical observables are represented by real functions on Γ forming a commutative algebra. Consequently, it is possible to define classical states as points $P(p_i, q_i)$ in the phase space. The states are defined with a variance equal to zero.

3. Classical propositions correspond to Lebesgue - measurable sub-ensembles of the phase space. For every state there is a class of simultaneously true propositions. The boolean structure of the propositions concerning a classical system is a consequence of the preceeding idealizations.

4. It is well known on the other hand that the boolean structure is that of the formal logic. Formal logic, in its turn, is the logic of identity. Thus, we arrive at the most important point concerning laplacean determinism and — consequently — classical probabilities: On the basis of the preceding idealizations we accept that the system concerves its identity during the movement in space, as well as during the measurement. The measurement does not result in the creation of new elements of reality (in the sense of EPR) as it is the case in a class of measurements in quantum mechanics.

5. The above idealizations are necessary for the validity of the laplacean form of determinism. Their mechanistic character is obvious and becomes more obvious because of the following condition: The boolean structure of the propositions is a necessary but not a sufficient condition for classical determinism. A boolean system is not necessarily deterministic. <u>Atomicity</u>, on the contrary, is a necessary and sufficient condition for a boolean system of finite degrees of freedom to be deterministic. Atomicity, on the other hand means conservation of the identity of the system. Thus, mechanical determinism presupposes the conservation of the identity of the system. This is not necessarily the case for quantum mechanics.

6. In classical mechanics, there are <u>statistical</u> states represented by probabilistic measures in phase space. A classical state is repre-

sented by a non-dispersive probabilistic measure. Accordingly, a dispersive state is considered as a non-complete description of the system. Nevertheless, it is accepted that it would be in principle possible to introduce a certain number of supplementary variables and to define a non-dispersive state, that is to say to achieve a complete specification of the state. Consequently, classical probabilities are considered as reducible to classical, non-dispersive states. Chance is, accordingly considered as a subjective or epistemic category. According to the CI, the contrary is valid for quantum mechanics.

7. The compatibility of the classical observables means that it is possible to measure $a \wedge b$ for every couple of classical observables. The validity of the distributive law is an implication of the compatibility of classical observables. But, as is well known, the distributive law is not valid in quantum mechanics. The validity, on the other hand, of the superposition principle in its domain made possible the fiction of the wave packet and of its "reduction" which constitutes, according to the CI, a proof of the indeterminism inherent in quantum phenomena. But as we will argue in the next section, it is possible to interpret the so-called reduction in a perfectly realist and deterministic way: This phenomenon constitutes the most fundamental difference between classical and quantum probabilities.

8. In fact, as it will be argued in the next section, the validity of the superposition principle in Q.M. is an expression of the potentialities of the quantum ensemble, potentialities realized during the measurement, or "spontaneously" in nature. The states in that case are potential states, and the initial pure ensemble is transformed during the measurement into a mixture. In classical mechanics, on the contrary, the only type of superposition is that of a mixture. As Gudder puts it, no pure classical state is a non-trivial superposition of other pure states, and the mixture is the only kind of measurable superposition in classical physics [15]. The classical states exist before the measurement: they are actual states. Thus, we find, from this point of view also, a fundamental characteristic of classical probabilities: to be real and no potential. But this characteristic is not of universal validity as we will see later.

9. The compatibility of the classical observables implies that the probability of one of them is not modified by the measurement of other observables defining the state. Consequently, joint probabilities are possible in classical physics. This is an argument for classical determinism.

Thus for two classical variables A and B we have

$$P(A \wedge B) = P(A).P(B)$$

The above formula means that the probability of A is not modified because of the measurement of B and vice-versa. We can generalize the preceding formula for $n>2$:

$$P(A_i \wedge A_j) = P(A_i).P(A_j)$$

Compatible observables in Q.M. behave like classical ones. For such variables joint probabilities are evidently possible. On the contrary, joint probabilities are considered impossible for non-commuting variables. But as is well known, many writers maintain that even non-commuting variables are measurable with arbitrary degree of accuracy. These authors accept, therefore, the existence of joint probabilities. Accordingly, they reject complementarity and the whole of the positivist interpretation [16].

The question of the existence of joint probabilities in Q.M. is open. Yet, let us accept the positivist point of view. In that case, the inexistence of joint probabilities does not mean that probability calculus is not applicable in Q.M. Because the existence of joint probabilities is not an axiom of the theory but an implication of its axioms. Let us accept now the inverse point of view: in that case another classical property will be valid in quantum mechanics.

Till now we enumerated a number of essential characteristics of classical probabilities and we noted the differences between them and the quantum case. But now we must put the following question: What is the real value of classical idealizations and their implications, if we abandon the level of the preceding abstractions and try to face more concrete and complicated — and so, more real situations? Let us try a concise answer.

1. According to F. Bopp, the antithesis between classical and quantum mechanics concerning the possibility of description in the phase space, is not absolute. Bopp affirms that even in classical mechanics it is not possible to observe exactly a particle's position in the phase space. On the other hand, it is possible to formulate quantum mechanics as a new kind of statistical mechanics describing the notion of virtual ensembles of particles or of systems of particles, and to derive the quantum equations from principles which are connected with the notion of moving ensembles. One of these principles is that there exists a phase space in which a particle has a definite position at any time and its motion is completely determined. Evidently, Bopp does not assume that we are able to know the position in phase space exactly [17]. Also, according to Max Born, because of the uncertainty of the initial conditions due to the inevitable imprecision of measurements, the predictions of classical mechanics do not concern a precise trajectory but an ensemble of trajectories characterized by a probabilistic distribution. Even in classical mechanics it is impossible, according to Born, to calculate rigorously the evolution of the system, with the exception of ideal harmonic oscillators [18]. According to the above point of view, the antithesis between classical and quantum mechanics is, evidently, epistemic and at the same time relative, not absolute.

2. Superposition in quantum mechanics is the measure of the potentiality of the statistical ensemble. A classical superposition, on the contrary, is, as already noted, a mixture of real states. Consequently we can separate the initial ensemble into a number of sub-ensembles constituting pure states. But the situation is not so clear because we neglect in that case the eventual correlations between the elements of such systems. More than that: There are classical superpositions which it is difficult to define in a unique way. A superposition of electromagnetic waves, e.g., is a superposition of real waves from the classical point of view. The same ensemble is, for quantum theory, a mixture of photons of different frequencies. But the photon, quantum "particle", interferes with itself and an ensemble of photons manifests, under certain conditions, ondulatory properties. Another case concerns the realization of

states in the classical case. In fact, it is not true that the states are actual in every classical situation. If, for example, we throw many times a non-loaded die, it will realize six potentialities with probabilities equal to 1/6. Conclusion: The concept of potentiality or propensity (Popper) characteristic of the quantum measurement is valid also for classical games. Nevertheless, throwing a die is a mechanical act and the realization of its potentialities has not the same physical content as that of the realization of quantum eigenstates. Yet, this fact is a warning against any absolutization of the differences between classical and quantum probabilities. And the above case is not unique in classical physics and in classical statistical mechanics.

3. Finally, what is the meaning of the concepts: <u>mechanical phenomenon and mechanical interaction</u>? None of the four known physical interactions is of a mechanical character. The so-called mechanical phenomena are generated by non-mechanical causes, and they are accompanied by other, non-mechanical effects.

4. The axioms and the laws of mechanics presuppose an extreme abstraction from the real properties of things. In this realm of the "pure events" it is easy to postulate an irreducible antithesis between dynamical and statistical laws. But even from the mathematical point of view, the dynamical law is the limiting case of statistical one: it is the case when the probability becomes equal to unity. From the physical point of view, it is possible for a dynamical law to cover an enormous ensemble of random processes, as is the case of electromagnetic and gravitational waves. As Terletski puts it, Relativity theory is a macroscopic theory, that is to say, a theory of statistical mean values of microscopic quantities [19]. In an opposite sense, it is possible to consider the probabilistic distribution as a result of hidden, unknown processes of dynamical character. Thus, it is possible today to understand the profound insight of Engels when he affirms that everywhere chance seems to work on the surface, it is always under the influence of hidden internal laws and that the question is to discover them [20].

5. Chance and necessity are mutually exclusive categories only from the point of view of formal logic. From the point of view of dialectical logic, chance and necessity, statistical and dynamical laws, mechanical, dynamical and quantum-statistical forms of determination are forms under which is realized the mutual determination of phenomena.

3. PROBABILITIES IN QUANTUM MECHANICS

As already noted, classical probability calculus is not — according to a certain point of view — applicable in Q.M. The alleged non-existence of joint probability distributions for non-commuting variables is one of the arguments supporting this point of view. But as Jauch states, we must distinguish between probability calculus and probability theory — the calculus as such was never put in question. Also according to Bellentine, the formalism of quantum mechanics satisfies the axioms of probability theory and this fact refutes the thesis that classical probability theory does not apply to quantum mechanics. Ballentine uses for quantum probabilities the notation $P(A/B)$ and this in order to emphasize the fact that the probability of A will happen under the <u>conditions</u> specified by the occurrence of B.

A form of axioms according to Bellentine is the following:

(1) $0 \le P(A/B) \le 1$
(2) $P(A/A) = 1$
(3) $P(\sim A/B) = 1 - P(A/B)$
(4) $P(A.B/C) = P(B/A.C) P(A.C)$

Ballentine gives a rigorous demonstration of the compatibility of the above axioms with quantum mechanics, and refutes a number of "fallacies" concerning the opposite point of view [21].

Differences between classical and quantum probabilities were analyzed by many authors, and we noted already some of them. I want to mention now only one formulated by E. Wigner. According to this authour, the laws of quantum mechanics furnish probability connexions between results of subsequent observations carried out on a system. The laws of classical mechanics can also be formulated in terms of such probability connexions. However, they can be formulated also in terms of objective reality while the laws of quantum mechanics can be expressed only in terms of probability connexions [22].

The above argument is invalid. Because the laws of quantum mechanics are the expression of a reality deeper than that of classical mechanics, and equally real and objective. On the contrary, the above thesis of Wigner results in an explicitly solipsistic interpretation of probability in Q.M.: Probability is related to the "reduction of the wave packet". But this reduction takes place "whenever the result of an observation enters the consciousness of the observer — or to be even more painfully precise, my own consciousness, since I am the only observer, all other people being only subjects of my observation" [23]. The "principle" of psychophysical parallelism of von Neumann finds in the above thesis its clearly solipsistic formulation.

The more essential difference between classical and quantum probabilities consists in the fact that classical probabilities are in most cases actual, while quantum probabilities are potential in the most important case of a superposition. The category of the <u>possible</u> is applicable in a restricted degree in classical mechanics and also for purely mechanistic phenomena. It is, on the contrary, fundamental for quantum mechanics. As Fock says, the concept of probability is a numerical measure of the potentially possible. And this probability is determined by the internal properties of the object and by the nature of external conditions [24]. Thus we must discuss this fundamental difference between classical and quantum probabilities.

Probabilistic quantum phenomena are realized in nature or during the measurement. But this is not a general case because there are measurements which do not realize new states. Thus we must make a concrete analysis of the problem.

1. There are <u>pure states</u> transformed into pure states by the measurement. It is the case of "sharp states" described by a unidimensional Hilbert space. We have in that case two possibilities:

1a) The initial state was actual before the measurement. The state, consequently, was not modified; it is only the apparatus that realized one of its "eigenstates". Thus we have:

$$\Phi_0 \Psi_i \rightarrow \Phi_i \Psi_i , \quad i=1$$

1b) The state was potential and was realized during the measurement. We have in that case:

$$\Phi_0 \Psi \rightarrow \Phi_i \Psi_i , \quad i=1$$

Conclusion: In the case of sharp states the result of the measurement is unique. Consequently no question of probabilities arises.

2. A <u>mixture</u> of states $\{\Psi_i\}$ with respective probabilities P_i remains a <u>mixture</u> after the measurement, with the same statistical weight. In that case there is no creation of states either and the relative weights of probabilities are not modified.

3. The third case, the so-called <u>superposition</u>, is characteristic of the behavior of quantum systems. In this case, the prepared state does not correspond, under the given conditions, to an eigenvector of the vector space of the apparatus. Thus we have:

$$\Psi = \sum_i C_i \Psi_i, \quad \text{and} \quad \Phi_0 \Psi \rightarrow \sum_i C_i \Phi_i \Psi_i \rightarrow \{\Psi_i\}$$

This is the only case where the question of probabilities and of determinism arises. The so-called reduction of the "wave packet" was considered by CI as a proof of indeterminism. But the "reduction of the wave packet" presupposes, in the spirit of the single system interpretation of the Copenhagen School, that the states preexist and that they are projected or reduced or analyzed by the measurement. This single system interpretation constitutes the weak point of the CI. Because the reduction is considered "instantaneous" collapse, as a psychical phenomenon, and finally is impossible in the frame of this interpretation [25].

Let us make more concrete the above subjectivist point of view. According to E. Wigner, for example, in order to obtain a mixture as the result of the interaction, the initial state must have been a mixture already. The conclusion, writes Wigner, follows from the general theorem that the characteristic values of the density matrix are constants of motion. Measurement, which leaves the system object-plus-apparatus, in one of the states with definite position of the pointer, cannot be described by the linear laws of quantum mechanics [26]. Other writers affirm explicitely that the pointer <u>has not</u> a definite position after the measurement. The final way out of the impass of the CI is the <u>deus ex machina</u> called consciousness. The above conclusion is a direct implication of the single system, non-statistical interpretation of Q.M. But this conclusion is fallacious: according to the statistical interpretation, the initial pure state is transformed into a mixture, because of the interaction of the quantum systems with the apparatus. This transformation is an objective, irreversible and dissipative phenomenon resulting, in every case, in the realization of one of the possible states. Consequent-- ly, the pointer has every time a precise position corresponding to one of the possible eigenstates. (Evidently, Wigner is right when he affirms that this phenomenon is not described by the actual laws of Q.M.).

Let us now make more concrete the above argument, using a well-known example. We consider a flux of photons lineary polarized in the direction e_p and moving in the direction o_z. We interpose an analyzer A, making an angle θ with the direction of polarization. The photons pass in the direction e_x ; they are absorbed in the direction e_y orthogonal to e_x. The state vector representing the ensemble of photons is given by:

$$e_p = e_x \cos \Theta + e_y \sin \Theta \qquad (1)$$

According to Dirac, and more generally the dominant single system interpretation, (1) represents a <u>real</u> superposition of the two orthogonal states e_x and e_y. The analyzer decomposes this superposition into its orthogonal components, one of which is transmitted while the other is absorbed.

In conformity with the statistical interpretation, on the contrary, it is possible to maintain that (1) is not a superposition of real states, but the formal expression of the two different potentialities of the ensemble of photons, under the given conditions. Because these potentialities are determined not only by the preparation of the initial state, but also by the conditions, that is to say the angle θ between e_p and e_x. Thus, changing the conditions implies a modification of the weight of the two possible states: With an angle θ = 0 N photons are transmitted (pure state). With an angle θ=45; N/2 photons are transmitted and N/2 absorbed (superposition). With an angle 90°, all the photons are absorbed. Consequently, the same ensemble appears as a sharp state or as a superposition, depending on the experimental conditions.

The above example is characteristic of the fact that the states are created during the measurement, <u>via</u> the transformation of the elements of reality of the initial state. Thus the initial pure ensemble is transformed into a mixture. Consequently, the probabilities are not real but potential probabilities, measure of the potentialities of the statistical ensemble. This is in my point of view the most essential difference between classical and quantum probabilities.

The above interpretation of probabilities conforms with the statistical interpretation of quantum mechanics. As Einstein puts it, the Schrödinger equation "determines how the probability density of a statistical ensemble of systems varies in the configuration space with the time" [27]. Quantum mechanics describes statistical ensembles. No single quantum system. Accordingly, the concept of probability has meaning only related to ensembles of systems. As Ludwig says, "I have never succeeded in understanding as to what is meant by the 'probability of one event in physics'" [28].

The "reduction of the wave packet", that is to say the transformation of quantum systems, is a non-linear transformation not described by the actual formalism of quantum mechanics. This lack of knowledge was elevated to the status of mystery by the dominant interpretation. As Schrödinger says: "The accepted outlook in quantum mechanics is based entirely on its theory of measurement. /.../. The accepted foundation of quantum mechanics claims to be intimately linked with experimental science. But actually it is based on a scheme of measurement which, because it is entirely antiquated, is hardly fit to describe any relevant experiment that is actually carried out, but a host of such as are for ever confined in the imagination of their inventors" [29].

We have stressed till now one fundamental (but not absolute) difference between classical and quantum probabilities: the fact that classical probabilities correspond to actual and quantum to potential states. Another equally intricate difference concerns the fact that classical laws are expressed in terms of probabilities, while quantum states are described in terms of wave functions (or state vectors). Thus, in classical systems we have a direct addition of probabilities, while for quantum system it is the amplitudes of probabilities that are added. The resulting interferences are one of the most intricate problems of quantum physics. Recent experiments, as is well known, indicate the existence of auto-interference of individual particles (Dirac predicted this possibility as early as 1931). These phenomena concern the more profound structures of matter and they constitute, eventually, a direct proof of the objective existence of the wave-particle duality and of empty waves. In that case also, the probabilities are potential and are realized via this, essentially unknown, process of interference.

But even these phenomena are not exclusively quantal because the classical electromagnetic theory predicts phenomena of interference. This fact is one more indication of an essential and hidden kinship between classical and quantum description of the same "mysterious" entity called electromagnetic wave or photon.

4. QUANTUM STATISTICAL DETERMINISM

The same causes produce the same effects. This is the classical definition of determinism, given by Isaak Newton. Quantum mechanics does not respect this dogma: From one initial state we have, in the case of a superposition, more than one final states. In the mechanistic spirit of the Copenhagen School, this is a manifestation of the essential indeterminism of quantum phenomena. Probabilities in quantum mechanics are primordial, essential and so on: incompatible with the principle of determination of the phenomena by their causes.

Thus von Neumann argued that quantum mechanics is in compelling logical contradiction with causality and that the "apparent causal order" of the macrocosme has no other cause than the law of large numbers" [30]. Heisenberg in his turn claimed that the law of causality is no longer applied in quantum theory [31] and according to Bohr, in the atomic phenomena we must renounce the very idea of determinism [32].

Let us examine now this argument.

Physicists, as well as philosophers, usually consider as identical the principle of causality with that of determinism. In this way, they speak of acausal phenomena, of violation of causality and so on, although the question is about determinism, not causality. According to the point of view developed here, causality means that phenomena have causes. Determinism, on the other hand, means that the causes <u>determine in a definite way the phenomena</u>.

When the indeterminist affirms that the "reduction of the wave packet" is acausal, he wants in reality to say that it is an indeterminate phenomenon. Because the quantum phenomena are causal, in the sense that they are the product of at least partially known causes. In fact, we know some of the causes operating in the quantum level. In many cases, the internal mechanisms of action of these causes are also known. There is no question of causality in quantum mechanics. The real problem is the validity of determinism in the microscopic level. Thus: what about determinism?

1. The mode of determination in quantum mechanics is not linear; it is more complicated than in classical physics. In the general case, from one initial state many final states are possible. The multiple potentialities are expressed in the statistical form of determination of the state. But what is the meaning of the concept: "statistical mode of determination"?

2. Quantum states are defined by a complete set of commuting variables. The state vector is the measure of the possible states and the corresponding probabilities. Consequently, the final states are determined by the nature of the system and the external conditions. The fact that it is possible to predict the set of the possible states on the basis of these data is a proof of the deterministic character of quantum phenomena. Thus probabilities are neither acausal non indeterminate. They are "essential", but not in the sense of the orthodox interpretation.

3. The modification of the external conditions results in the modification of the probabilistic distribution of the states. This fact is a direct proof of the deterministic character of the measurement and, more generally, of the transformations of the quantum systems. Probabilities are the result of the interplay of the internal variables of the system and the external conditions.

4. There are possible and impossible states for a quantum system. Both of them are determined by the nature of the system and the external conditions. The impossible states were never observed. The conservation laws and the corresponding selection rules have never been violated. This is another argument for the deterministic character of quantum phenomena and of quantum probabilities in particular.

5. The predictions of quantum mechanics were until now verified even in the more complicated cases. This is an argument of objectivity against the dominant conventionalism and the pragmatic philosophy which reduces quantum mechanics to the status of a simple algorithm [33].

The above peculiarities are characteristic of the quantum statistical determinism and thus of the probabilities in quantum mechanics. In the case of quantum mechanics a new form of determination is manifested. In contrast to the classical-mechanistic case, the new state emerges through a process of annihilation and creation of elements of reality. The interplay of the potential and the actual, preconized in the philosophy of Aristotle and of Hegel, characterizes the quantum statistical determinism and constitutes the hidden background of quantum probabilities [34]. The non-boolean structure of quantum propositions and the validity of the superposition principle are the formal expressions of the fact that the fundamental law of formal logic — the law of identity — is not valid in quantum processes.

The orthodox school cannot understand the fact that quantum statistical determinism *transcends* the classical forms and expresses a new, multivalent form of determination: quantum probabilities are the empirically testable result of hidden quantum processes. The orthodox school identifies determinism with its laplacean form. By so doing, this school has to confront the dilemma: determinism *or* indeterminism? *Tertium non datur*!

For the orthodox school, the simultaneous knowledge of the position and of the momentum of a particle should be a necessary and sufficient condition for the validity of determinism.

This argument is characteristic of the mechanistic spirit of this school. Because this knowledge is a sufficient condition for the description of the movement of the particle in an ideal isolation. But on the basis of these mechanical data it is impossible to predict the result of a measurement. The probabilistic distribution is determined by profound and, to a certain extent, unknown processes.

The relation between causality and probability, as Margenau says, is not one of antithesis or mutual exclusion, but of coordination. In particular, the causality principle has not been abandoned since it governs the behavior of the states. The same is valid for determinism: Quantum statistical determinism is the dialectical transcendance of the known classical forms. Let us be once more reminded of Schrödinger's words: "Though the forecast is usually not precise but of probability, there is an unambiguous representative of the *state*, the state vector or state function which is supposed to change between measurements in a precisely known fashion (if the nature of the system is known) and to determine precisely the probability to recast for any measurement and any moment" [34].

I think it is not arbitrary to believe in the final victory of the ideas of Einstein, Schrödinger and de Broglie.

References and Notes

1. L.E. Ballentine, *Am.J. Phys.*, 54, 883 (1986), This volume also
2. J.M. Jauch, *Cah. Fund. Scientiae*, 27 (1975). Nevertheless, it is well known that the above point of view is not accepted by many specialists. The non-boolean structure of quantum propositions, the alleged inexistence of joint probabilities in Q.M. and the validity of the inequalities of Heisenberg are on the basis of the arguments concerning the inadequacy of the Kolmogorov formalism of probability theory for quantum systems and the need of a more general probability theory. See, e.g.: 1) G. Birkhoff, J. von Neumann, *Ann. Math.*, 37, 823 (1946); 2) F.W. Mackey, *The Mathematical Foundations of Quantum Mechanics*, Harvard Univ. Press, 1960; 3) V.S. Varadarajan, *Commun. on Pure and Appl. Math.*, XV, 189 (1962); 4) R.P. Feynman, *Proc. Sec. Berkeley Symp. on Math. Stat. and Prob.*, Berkeley, California, 1951; 5) V. Kägs-Romano, *Jour. Phil. Logic*, 6, 455 (1977). The axioms of modern probability theory were formulated by A.M. Kolmogorov in his work *Foundations of the Theory of Probability*, Chelsea, N.Y., 1956 (German original, Berlin 1933).
3. R. von Mises, *Mathematical Theory of Probability and Statistics*, Academic Press, N.Y., 1964, p.1.
4. K. Popper, in *Quantum Theory and Reality*, M. Bunge Ed., Springer-Verlag, 1967.
5. See A.I. Khintchine, in: *Questions Scientifiques*, 4 (1954), Ed. de la Nouvelle Critique, Paris.
6. Laplace, *Oeuvres Complètes*, Gauthier-Villars, Paris 1878, v.7, p.6.
7. D. Bohm, *Causality and Chance in Modern Physics*, Routledge and Kegan Paul, London, 1959, p.26.
8. R. von Mises, op.cit., p.2.
9. A.I. Khintchine, op.cit.
10. Concerning this question, cf.: 1) A.I. Khintchine, ibid, 2) J. Bonitzer, *Philosophie du Hasard*, Ed. Sociales, Paris 1984. 3) T.A. Brody, this Meeting.
11. R. von Mises, op.cit., p.343. Gnedenko and Khintchine give the following definition of the law of large numbers: "Alors qu'une quantité aléatoire, considérée isolement, peut souvent prendre des valeurs fort éloignées de sa valeur moyenne (c'est-à-dire accuser une forte dispersion), la moyenne arithmétique d'un grand nombre de quantités aléatoires se comporte à cet égard de façon tout à fait différente, en ce qu'elle n'est sujette qu'à une très faible dispersion et qu'il existe une probabilité écrasante pour qu'elle prenne exclusivement des valeurs très voisines à sa valeur moyenne" (B.V. Gnedenko, A.Ia. Khintchine, *Introduction à la Théorie des Probabilités*, Dunod, Paris 1963, p.121).
12. K. Popper, D.W. Miller, *Phil. Trans. R.Soc. Lond.*, A321, 569 (1987)
13. H. Margenau, L. Cohen, in *Quantum Theory and Reality*, M. Bunge (Ed), Spinger-Verlag, 1967.
14. Many writers identify the classical and the laplacean (mechanistic) determinism. Thus e.g. V. Fock writes: "Classical (laplacean) determinism [...] can be defined as a point of view according to which the refinement of observation methods, together with increasing accuracy

in the formulation of the laws of nature and in mathematical deductions from these laws, permits in principle a unique prediction of the whole course of events" (v. Fock, Usp. Phys. Nauk, LXII, 461 (1957)). But the possibility of unique prediction is a characteristic of the laplacean as well as of the dynamical form of determinism (electromagnetism, relativistic theory of gravitation). Identifying the mechanistic with the dynamical form of determination, which has qualitative different physical foundation, is the first step for the rejection of determinism in the field of quantum mechanics.(Cf. E. Bitsakis, Physique et Matérialisme, Ed. Sociales, Paris, 1983.Id. Found. of Phys., 18, 331 (1988)).

15. S.P. Gudder, J. Math. Phys., 11, 1037 (1970). See also, E. Bitsakis, in: Problems in Quantum Physics, World Scientific Publisher, Singapore, 1988.
16. See,e.g., H. Margenau, L. Cohen, op.cit.,
17. F. Bopp, in Observation and Interpretation, S. Körner (Ed.), Butterworths Publ., 1957.
18. M. Born, J. Phys. et Rad., 20, 43 (1959).
19. J.P. Terletski, in Questions scientifiques, op.cit., Cf. also, J.P. Vigier, in Observation and Interpretation, S. Körner (Ed.), Butterworths Publ., 1952.
20. K. Marx-F. Engels, Etudes philosophiques, Ed. Sociales, Paris 1961, p.49.
21. L.E. Ballentine, op.cit.
22. E. Wigner, Am. J. Phys., 31, 6(1963).
23. E. Wigner, The Monist, 48, N° 2, 1964.
24. V. Fock, Usp. Fiz. Nauk, LXII, 461 (1957). Id., Sov. Phys. Usp., 66 208 (1958).
25. See the articles of de Broglie and Schrödinger, in Louis de Broglie, Physicien et Penseur, Albin Michel, Paris, 1952.
26. E. Wigner, Am. J. Phys., op.cit.
27. A. Einstein, J. Franklin Inst., 22, 349 (1936).
28. G. Ludwig, in The Physicist Conception of Nature, J. Mehra (Ed.), Reidel, 1973.
29. E. Schrödinger, Nuovo Cim, I, 5 (1955).
30. J. von Neumann, Mathematical Foundations of Quantum Mechanics, Princeton Univ. Press, 1955, pp.226-228.
31. W. Heisenberg, Physics and Philosophy, Allen and Unwin, London, 1958, p.81.
32. N. Bohr, Atomic Physics and Human Knowledge, John Wiley, N.Y. 1958, pp. 71-72.
33. Cf. E. Bitsakis, "Quantum Statistical Determinism", Found. of Physics, sics, 18, 331)1988).
34. Heisenberg and Bohr had some intuitions concerning the relations between the potential and the real in Q.M. See their books cited above. For a critical analysis, cf. E. Bitsakis in Microphysical Reality and Quantum Formalism, A. van der Merwe et al. (Eds), Kluwer Acad. Press, 1988.
35. E. Schrödinger, op.cit.

The Ensemble Interpretation of Probability

T. A. Brody
Instituto de Física, UNAM
Apdo. Postal 20-364,
01000 México, D.F.
MEXICO

ABSTRACT. Current philosophical interpretations are shown to be unsatisfactory when applied to problems of scientific research. An alternative, based on the work of Einstein and Gibbs, is proposed: probability as a scientific concept, with a theoretical and an experimental component, the former based on the ensemble and averages over it, the latter on relative frequencies in (finite) sets of experimental data. The two will agree only to the extent that the theoretical background of the ensemble is satisfactory. This interpretation extends in a natural way to the time-dependent probabilities of stochastic processes. The relevance of the concept in other areas of physics is exhibited.

I. IS ANOTHER INTERPRETATION NEEDED ?

It might be said that there already exist sufficient alternative interpretations of the probability concept to satisfy any possible need. An opinion common among physicists, perhaps induced by the present vogue of formalism, is that the mathematical theory of probability obviates any need for a further interpretation. However, Kolmogorov's axioms establish the mathematical properties of probabilities, but they do not tell us how to determine the elementary probabilities, i.e. how to measure them or derive them from a physical theory, nor do they tell us when talking of probabilities makes sense (see e.g. Suppes 1968). The problem of finding an interpretation is that of establishing the relationship between the real world in which we use probabilities and the mathematical structure which describes them. What this paper will maintain, then, is that the presently available forms of interpretation do not satisfactorily solve this problem, and that the ensemble interpretation, described below, does so.

There are two main schools of interpretation. One identifies probability with the confidence that we place in a proposition; variants range from those that treat it as a logical relation like, but less certain than, an inference (Keynes 1921; or, in another vein, Jeffreys 1948) to those who, quite openly subjective, use the individual's estimations as their starting point (Ramsey 1926; de Finetti 1970). What they have in common is that for them probability describes a feature of our knowledge of some physical system, resumed in a proposition, and not an objective property of that system; if our knowledge were certain, this sort of probability would be superfluous. These views are thus based on the "ignorance interpretation": uncertain knowledge is expressed as probability. Popper (1959a) is therefore quite right in classing all these interpretations as subjectivist.

The other kind of interpretation prides itself on being objective because it refers probability to an event - something that quite objectively either happens or does not. But its chief and almost exclusive representative, the frequency interpretation, is better called positivist. Its principal exponents, v. Mises (1919, 1931) and Reichenbach (1935), were logical positivists, and their notion of probability shows significant - and problematic - traces of this philosophy. Quite characteristically, they confound the theoretical and the experimental when they define the probability of an event as the infinite limit (a theoretical notion) of the relative frequency (derived from necessarily finite experimental data). Objectivity, in probability as elsewhere, requires the interaction of theoretical and experimental elements; conflating them in an unanalysable way does not contribute to objectivity.

These two schools of interpretation are subject to several well known philosophical criticisms; but what is relevant here is that neither of them is at all appropriate for a probability concept to be used in scientific research. Now scientific concepts have a double-sided nature, with a theoretical and an experimental aspect. A theory in which such a concept occurs will describe how it relates to other concepts; there may also be a more abstract background theory for the concept itself, particularly if the concept is common to several theories. On the experimental side there exist procedures and techniques for measuring the value taken by the concept in a concrete case; what characterises such measurements is that each is associated with certain limits on its precision. The interpretation of the concept - its physical meaning - provides for linking it to a feature of the real world, for relating its experimental and theoretical aspects; but this relation cannot be either automatic or exact. It is not automatic, for a theoretically derived value and the corresponding experimental one will agree only insofar as we have both an adequate theoretical model and a satisfactory experimental technique. It is also not exact, because besides those contemplated in the theoretical model, endless other factors will influence the experimental finding; and since these factors vary in complex and often unknown ways, measurements are repeatable only to within a certain range, technically known as the error limit. Agreement between theory and observation is then to be sought only to within these limits. Indeed, if it is notably better, it becomes suspect.

In the present case, we have a background theory of mathematical form, based on Kolmogorov's (1933) axioms; we also have specific theories for the objects of research, from which we derive values of the "elementary" probabilities to which probabilistic reasoning is then applied, to give the sought-for theoretical predictions. In the (apparently) simple case of coin tossing, we predict a probability of 1/2 for heads. If we actually toss the coin, we will mostly find something between 40 and 60 heads for every 100 tosses, while exactly 50 heads is rather improbable. With a bent coin the proportion might be quite far form this range, not because probability theory has failed but because we ought to have used a different physical model that applies to bent coins. Here then a theoretical probability will predict a relative frequency, but only with limited precision and only if the physical model is adequate. By keeping the two sides of the probability concept distinct, we can achieve a relation between them that satisfies the needs of scientific research.

For the subjectivist interpretations probability concerns a proposition, and more specifically our view of that proposition, and no experimental component is offered. The link to any observed frequency then poses an insoluble problem. To circumvent it, peculiar abstract principles, such as the principle of insufficient reason (Keynes 1921) or the stability of long-term frequencies (Hacking 1966), must be invoked. But they are apt to fail at the critical moment; insufficient reason gives rise to paradoxes, and probability can be meaningfully used even if there are no stable long-term runs. The profound divorce between theory and experiment in subjectivist conceptions appears again in their inability to incorporate the Kolmogorov axioms in a natural way: indeed, the actual use of the betting quotients that some of them take as starting point may not even conform to these axioms.

In v. Mises' conception, we have the opposite problem: the observed frequency serves actually to define the theoretical probability, and so the link between them is automatic, rigid and exact. Even though now the Kolmogorov axioms hold, essentially because they hold for relative frequencies, once again we have here a viewpoint that differs fundamentally from what is needed in the sciences. The discrepancy is shown up by the frequentist's inability to countenance probabilities for singular events, although these are constantly needed in statistical theory. Again, there is no specifically experimental component of probability.

These failures become serious in the less trivial applications of probability, e.g. when it depends on some parameter such as the time, or on whether some preceding probilistic event occurred. One's knowledge, and hence a subjectivist probability estimate, will not usually change merely because time passes; and v. Mises' basic tool, the collective, requires each event in it to be statistically independent of all the others. Hence on either view stochastic processes and ergodic theory (the theory of the connections between probabilistic and time averages) cannot be consistently developed. The recent exciting discoveries concerning chaotic behaviour in classical, deterministically described, systems are even more inexplicable: their deterministic theory means that full knowledge is available and subjective probility either 0 or 1, and no frequentist collective can be constructed since all events are wholly determined by their predecessors.

Other philosophical views of probability have been propounded, which are no more satisfactory; because of space limitations they will be ignored here (but see Rédei and Szegedi 1988).

We conclude that a new interpretation of probability, scientifically sounder, is needed.

II. THE PHYSICAL SYSTEM AND ITS SURROUNDINGS

When doing scientific research (and in everyday life) we never deal with the universe as a whole, we select from it a segment designed to contain everything relevant to our purpose and nothing irrelevant; following the physicists' custom, we shall talk about the (physical) system. Such a system is not simply given geometrically, for even of the objects involved only those aspects that interest us are taken into account: in calculating the trajectory of the inkpot that Luther threw at the devil, we do not ask what the colour of the ink was, nor do we worry whether Newton's apple tasted sweet; for calculating the tide we include the positions of both sun and moon, although they are very far yet we do not consider the boats floating on that tide.

Nor is our choice of system final; half the battle in doing research is finding a satisfactory system, for only then can we build an adequate theoretical model. It is only on the basis of an appropriate system that we can even design and carry out relevant experiments. In daily life past experience is usually a good enough guide to lead us quite rapidly to a satisfactory system; in research it may take many repeated cycles of setting up a system, developing a theoretical model, making experiments, and from the discrepancies seeing how to revise the system.

We also use several systems in parallel, either to devote special attention to certain aspects, to provide several different levels of approximation, or to study a problem with different aims in mind. Thus there must be many hundreds of different models of our earth, some treating it simply as a mass point, others as a perfect sphere, still others as a flattened one, others yet taking into account its surface iregularities, and some even treating it as an infinite flat body. And in probability theory we contemplate an infinite number of systems.

A system is specified both theoretically and experimentally. Normally we begin with the theoretical side, with the mental image, which we shall call its model. From it we derive the experimental specification of the system (through the use of suitable equipment like thermostats and so on). This is intended to isolate the system as well as possible from the remainder of the universe; nevertheless, an almost endless number of interactions of diverse nature and strength still link them. It may be possible to ignore these "outside" factors; if the model explains the system's behaviour in terms of a finite set of factors with negligible outside dependence, then we call the system closed. Some outside factors act so as to stabilise the system and so allow us to use a simpler one; thus the force of gravity reduces the billiard table to a two-dimensional system, by not allowing significant motion in the third dimension. Many small interactions may be neglected at first and later be taken into account as corrections. Finally there are almost always very many outside factors whose individual effect is negligibly small but whose joint influence may be quite large and may even change the system's character. In such cases we can neither ignore these factors, nor can we include them explicitly, for individually they are quite irrelevant and we are interested only in their joint action. Thus we need a new approach: the probabilistic one.

III. ENSEMBLES, AVERAGES AND PROBABILITIES

To each concrete and tangible physical system there corresponds essentially only one model, for if we change the model, we include new features and drop old ones, so that a different system is described; but one model may describe many systems, differing from each other in features not contained in the model. We could have built lower-level models that make these differences explicit; but instead of actually building them, we think of them as having been averaged over, and now a higher-level model built using these averages takes what above we called the irrelevant factors into account - jointly but not individually.

This procedure was first developed for the kinetic theory of gases. Traditional thermodynamics gives us a description of a gas in terms of pressures, volumes, temperatures, and so on. Yet if one could treat a gas simply as a mechanical system made up of a great many loosely interacting molecules, the same results should appear. However, they do not; instead we have endless facts about the positions and velocities of all the molecules at different moments, all of them strongly dependent on their initial positions and velocities. The problem of how to remove the unwanted information that depends on the initial conditions while keeping what does not so depend was solved by Einstein (1902, 1903, 1904) and by Gibbs (1902), using the method indicated above; they considered not one model, with the appropriate initial values for positions and velocities, but a whole collection of models with all conceivable combinations of initial values. Averaging over this set of models - technically known as an ensemble - is a trivial operation mathematically, but one that has the power of creating new concepts of quite different characteristics: for instance, ensemble averages no longer depend on initial conditions, and in statistical mechanics possess the properties needed for temperatures and so on.

Now ensembles are usually discussed in terms of an already established probability concept (e.g. Tolman 1938 or Balescu 1975); but the ensemble notion and the averaging operation can be quite straightforwardly defined without this concept, simply by clothing the preceding discussion in somewhat more formal terms. Probability is then nothing but a particular kind of average over the ensemble. Indeed, statistical mechanics can be presented in terms of no more than these two notions, ensemble and average (Fowler and Guggenheim 1939). We indicate briefly how this more formal argument might run:

Consider a set of hypothetical physical systems, all described by a common higher-level model and each by a lower-level one which is not made explicit but merely labelled by variables collectively denoted by ω. This could simply be a numbering, but in statistical mechanics the initial conditions are used. Any property f of the systems will then depend on ω, and we write it $f(\omega)$. Its ensemble average is then

$$F \equiv \langle f \rangle = \int_\Omega f(\omega)\, d\mu(\omega) \,/\, \int_\Omega d\mu(\omega) \qquad (1)$$

Here Ω is the range of possible values of ω, the angle brackets $\langle\ \rangle$ are standard notation for an ensemble average, and the function μ characterises the ensemble. If we now consider various related functions f, then eq. (1) will connect the corresponding averages F in a new theory, from which the underlying variables ω and and any information explicitly dependent on them have disappeared. In the case of statistical mechanics, the ensemble concept has thus allowed us to pass from the level of initial-condition dependence, in the mechanical model, to that of a new theory, thermodynamics, no longer so dependent.

Probability appears as a particular average. Given a property A that some systems in the ensemble possess but others not, we consider the indicator function $\chi_A(\omega)$, which is 1 if the system labelled ω has the property A and 0 if not; the probability of A is then simply the ensemble average

$$\text{Prob}(A) \equiv \langle \chi_A \rangle = \int_\Omega \chi_A(\omega)\, d\mu(\omega) \,/\, \int_\Omega d\mu(\omega) \qquad (2)$$

If Ω is a finite set of points (or systems) and μ the same for all of them, this is equivalent to a (theoretical) relative frequency; but (2) is more general, and works for an infinite set or a continuous Ω. The definition (2) is easily seen to satisfy the Kolmogorov axioms, since

$$\chi_{A \cap B}(\omega) = \chi_A(\omega)\, \chi_B(\omega)$$
$$\chi_{A \cup B}(\omega) = \chi_A(\omega) + \chi_B(\omega) - \chi_{A \cap B}(\omega) \qquad (3)$$

Therefore the probability defined by (2) is indeed that dealt with by standard probability theory. (This definition may be further generalised.)

The two fundamental quantities Ω and μ are not determined by probability theory; they must be derived from the particular theory (physical or other) that covers the application envisaged. In some cases they can be fully specified in this way; more commonly Ω is so determined, but μ must be found by research into the problem, usually in fact by trial and error. In other cases μ need not even be fully detailed: provided a central-limit theorem exists, all we need to establish is that μ belongs to the class to which it applies; the commonest such case gives rise to Gaussian distributions, and explains why this distribution occurs so frequently as to have earned the epithet "normal". In certain cases it is even sufficient to know that μ exists in principle.

It may however happen that Ω has a structure for which no μ can be found, or Ω itself may not exist. The notion of probability does not then apply. The truth of a scientific theory is a case in point. The ensemble here would consist of situations where the theory is true and others where it is not, and it is easily established that there is no consistent way to formulate such an ensemble. One cannot therefore speak meaningfully of the probability that a theory is true, and induction is better forgotten about.

The construction of a suitable ensemble - when possible - provides us with the theoretical aspect of the probability concept. At least one experimental method derives from such intuitive views as saying that if the probability of a coin falling heads is 0.5, then about one half of the coin tosses will come up heads. In other words, a relative frequency (obtained on a finite data set) is an experimental correlate of a theoretical estimate found from an ensemble. But these two will coincide only exceptionally; and the relative frequency will have error limits. For short data series these limits will be dominated by the so-called statistical errors, i.e. the errors due to the fluctuations in the numerous factors outside the model but still of some influence on the system; statistical theory describes how to estimate and treat these errors. But as the data series grows, other error sources becomes increasingly important, until at length the coin we are tossing becomes too worn to be treated as symmetrical or the two faces as distinguishable.

Determining a relative frequency is the most general method of estimating probabilities. If the successive events in the data series accumulate very rapidly, we may only be able to observe their rate of occurrence: thus the intensity of a light source is really the rate at which individual molecules emit and are deexcited. In other cases neither the relative frequency nor a corresponding intensity are accessible, but we can measure the value of a function like the F defined in eq. (1), which by using the discrete equivalent of (2) may be rewritten as

$$F = \sum_i f(\omega_i) \, \text{Prob}(\omega_i) \qquad (4)$$

We cannot of course invert (4) to find the probability; but if several different such functions have been measured we can often make a reasonable guess at the probability distribution.

Clearly all methods of measuring a probability except that of counting instances to give a relative frequency depend rather strongly on the particular application. And all of them have finite error limits, a fact which if forgotten causes numerous confusions, some of which still haunt the philosophical discussion of probability. Many of the paradoxes surrounding the "principle of indifference" spring from this source.

Having briefly indicated what the ensemble interpretation of the probability concept consists in, I should add that I am in no sense its inventor (Brody 1975). It is, if anything, part of the common culture of physicists, and I suspect of other scientists too. It does not correspond, of course, to what scientists are taught in university courses, but their daily practice forces it on them. Analogous ideas appear to have been in the minds of some very outstanding mathematicians; thus Kolmogorov (1969) comes very close to an explicit statement of the ensemble view presented here, and Kac (1959), after presenting some very illuminating examples, sums up as follows:

> At this point it should have become clear to the reader that probabilistic reasoning consists in imbedding a particular situation in an ensemble of like situations and replacing statements about individuals by statements about the ensemble.

Could the viewpoint outlined here be presented more succinctly?

IV. PROBABILITY IN STOCHASTIC AND CHAOTIC SYSTEMS

We turn now to the problem, mentioned in section I, of the time dependence of probabilities. A probability may vary in time through three mechanisms: under an external time-varying influence, because it belongs to a stochastic process and therefore depends on preceding events, or because a deterministic system has entered a region of chaotic behaviour.

A good example of the first case is found in the incidence of an epidemic that is being combatted by an effective public-health programme, so that the probability of falling a victim to the disease diminishes. Such a time dependence creates a difficulty for the frequentist, since at any given time there is only a finite sequence of events available, and lengthening the sequences by admitting different times gives series that do not converge to a well-defined limit. For the subjectivist the difficulty is rather that now probabilities alter without any further evidence being adduced. No problem arises in the ensemble view; any ensemble that represents reality adequately may be expected to evolve in time, and so also its probabilities.

In a stochastic process all probabilities are conditional on what has happened earlier to the system; their time evolution will therefore be different for different members of the ensemble, being given by a subensemble defined by the common prior history. As a result, the average along a trajectory (for a given system over its history) and over the ensemble (all systems, whatever their history, at a given time) will in general differ. But there is a special type of ensemble for which they coincide: the so-called ergodic systems. This concept (and the corresponding theory) plays an important role in much of modern physics; it is also relevant to the understanding of probability, in that non-ergodic systems (now known to be much the commonest case) will not show any long-term stability of relative frequency, such as many authors (e.g. Hacking 1966) try to use as underpinning for their probability views. The ensemble view provides a probability concept to be used even in such cases; more in general, it is the only one in which stochastic processes and ergodic theory can be given a consistent treatment.

In the last few decades it has been found that non-linear systems, even though they are described by perfectly deterministic equations of evolution, can exhibit rather startling behaviour of random character (Hao 1984; Cvitanović 1984). These phenomena are so complex (and as yet so little understood) that I only discuss one of the simplest and oldest cases, the generation of random numbers in a computer. The chief method used is the congruence method (Lehmer 1951): given suitably chosen integers a, c and m, one chooses a starting integer u_0 and calculates successively for all i

$$u_{i+1} = (a u_i + c) \bmod m \qquad (5)$$

where "$x \bmod y$" denotes the residue after division of x by y. Dividing u_i by m now gives a random number between 0 and 1. Other algorithms (e.g. Brody 1984) are often advantageous.

Now a computer is a deterministic device and no algorithm should, according to traditional views, produce a sequence of numbers with probabilistic behaviour. However, if the algorithm is well designed, the sequence has statistical properties notably closer to random than natural processes; one should not then sidestep the philosophical problem by talk of "pseudo-random" numbers, as is often done. But an ensemble can be constructed here:

the event space is, for the congruential generator (5), simply the set of integers from 0 to m; the "systems" are the different sequences generated and may be labelled by their initial value u_o, which determines them completely. A good generator can be proved to be ergodic - averages along a sequence coincide with averages over the ensemble for fixed i, which here plays the role of a time. And even if we know the values of the three parameters in (5), we cannot from a number in the sequence deduce its predecessor, since there are many different possible ones. Not that our ignorance is necessary: the computer could easily be programmed to keep a record; but we do not normally do so, because the algorithm can make no use of it. Just as in the case of statistical mechanics, in other words, we deliberately ignore part of what is knowable; and by ignoring the individual trees, here irrelevant, we are able to see the wood. To put this point more generally: the probability concept is not needed because of our ignorance; on the contrary, it permits us to eliminate unwanted information. To link probability with ignorance, as so many authors do, is thus a mistake; the paradox that we need probability to deal with an excess of information comes closer to reality.

Neither the subjectivist views nor the frequency interpretation are useful in understanding random-number generation or indeed the chaotic behaviour of deterministic systems; the best they can do is summed up in the quip that "anyone who considers arithmetical methods of producing random numbers is, of course, in a state of sin" (v. Neumann 1951).

V. PROBABILITY AND QUANTUM MECHANICS

We have discussed the interpretation of probability in statistical mechanics where it is the physical application that has helped to clarify the philosophical concept; in quantum mechanics, inversely, the philosophical confusion has become reflected in the conceptual confusion surrounding this theory. Soon after the development of matrix mechanics, its founders saw the need for an interpretation, which they based on the idea that a quantum description (in terms of a state vector, or equivalently a Schrödinger wave function) refers to a single physical system only. The square of the wave function then gives the probability of finding the system in the relevant state; but for a single system, what does such a probability signify? Each expectation value derived from it has a certain spread, and Heisenberg had shown that for a pair of conjugate variables the product of the spreads has to be greater than a certain limit; neither of them can thus be zero. Heisenberg interpreted the spreads as the theoretical minima of the experimental errors, which thus become irreducible. Does this mean that the variables are, to this extent, indeterminate? And how is it possible for the extent of the indeterminacy to depend on what else is being measured? It seemed necessary to conclude that it is the measurement procedure that determines the value of what is measured, which therefore is not really a property of the physical system. Indeed, one might say that it is the measurement which creates the quantity measured. The spiral into steadily more subjective views was impossible to stop, and even outstanding physicists fell into fully subjectivist and in the end solipsist world views (Bohr 1936, 1948; Jordan 1938; Heisenberg 1951; Pauli 1954; Houston 1966) or the acceptance of quantum mechanics as incomprehensible (Dyson 1958).

That a probability may be stipulated for a single physical system is only possible, as our previous discussion has shown, within one or another of the subjectivist views, where probabilities belong to propositions; the attempt to reach consistency then leads to these undesirable consequences. The alternative approach, that the state vector represents not one but an ensemble of systems, is conceivable for the frequency interpretation and the only possible one for the ensemble interpretation. In quantum mechanics this approach was adopted first by Slater (1929) and developed further by Einstein (1936, 1953), Blochinzew (1953), Lamb (1969, 1978), and others; but much work remains to be done on it, a fact which has given rise both to criticism of various kinds, often conceptually confused, and to the refusal of most textbooks even to mention that there are alternative interpretations. Yet the ensemble view offers many advantages among which its conceptual straightforwardness stands out; it eliminates the endless confusions, misunderstandings and paradoxes which afflict the "orthodox" view (see e.g. Ballentine 1970; Ross-Bonney 1975).

One criticism levelled at the ensemble viewpoint is that the EPR argument (Einstein, Podolsky and Rosen 1935) that quantum mechanics is incomplete would have to be accepted. Without entering into technicalities, what the argument implies is that quantum mechanics, being a theory whose predictions take the form of ensemble averages, is the "thermodynamics" to a statistical mechanics which remains still to be formulated. But is this not precisely the case? In the last few years, an increasing number of physicists have come round to the idea that work in this direction is worthwhile (Brody 1983 and references given there).

Work along such lines might have been initiated many years ago had an adequate conception of what probability is been generally accepted. As it is, except for some less significant addition and a certain amount of largely formal rewriting, non-relativistic quantum mechanics has remained unchanged for half a century, while experimental knowledge has progressed by leaps and bounds. The single-system view has even misled many people into concluding that quantum mechanics is our last fundamental theory, so that physics, apart from improvements in technical detail and new applications, has essentially come to an end. Such views arise if one argues that quantum probabilities are basic and not further to be analysed (as the ensemble view of course permits and fosters).

VI. PHILOSOPHICAL CONCLUSIONS

The survey of the preceding sections leads us to conclude that traditional views on probability have not generally been helpful in physics and have sometimes done damage; the ensemble interpretation, on the other hand, aids in untangling conceptual confusions. There still remain some points of philosophical import to be mentioned.

(i) A probability concept based on ensembles might, on a superficial view, be considered subjective, since ensembles are thought constructions. But we cannot use just any ensemble: like any other scientific concept, an ensemble must, by repeated cycles of reformulation and comparison of experimental reality, be made to describe that reality fittingly, in accord with our purposes; and like any scientific concept, it acquires objectivity in the course of reformulation and adaptation. Note that objectivity in a concept can only mean that it correctly represents a selected aspect of the world. Objectivity is, therefore, a property we confer on the concept through our work in formulating and reformulating it. Correspondingly, no concept is purely objective; it is precisely its subjective aspect that allows us to capture, understand, adapt and use it. What is wrong, then, with the subjective views of probability is not that they possess a subjective aspect; it is that they possess nothing else and are therefore subjective where they should not be. Concretely, therefore, the objectivity of each particular application of the probability notion must be achieved through research work and must be critically examined; there cannot be a global solution. There may be numerous and difficult problems in validating the particular probabilistic approach taken; but these problems are not in essence distinct from those facing any other kind of theoretical model.

There is also a much more general sense in which probability is objective: probability arises as a way of recognising the non-isolated character of the system we are dealing with, without describing the "outside" influences explicitly. Probability connects the level of full description (outside influences made explicit) with that of statistical description, of description via the elements of behaviour common to all systems; these two, as we have seen, may be of entirely different character. The two-level structure is a consequence of the fact that physical systems are only partially separable from the rest of the universe, and so is probability, the tool we use to describe it; the probability concept is objective inasmuch as it yields an account of this partial separability.

(ii) Probability is also relative, and that again in a double sense. Each probability is meaningful when referred to its ensemble; probabilities belonging to different ensembles are not comparable, and such comparisons give rise to well known absurdities. And for a given event we can find a large number of different probabilities, according to the various ensembles into which this event will fit; the choice between them is guided by our purpose, but given that purpose is essentially determined (to within indefinitions not discussed here). Probability is relative in the sense in which distances are relative: given a reference point, they are perfectly objective, without it they are meaningless. Points (i) and (ii) are adequately catered for by the ensemble conception, but at most with great difficulty by other views.

(iii) The ensemble view of probability allows the construction of an ensemble each of the members of which are ensembles in their turn; indeed, constructions of three or more levels are conceivable (though they do yet seem to have been used). This is of practical utility: in statistics, where it underlies the notion of the probability of a probability; in quantum mechanics, where it is needed for quantum statistical mechanics; and in the biological theory of evolution. In the last, the top level concerns the evolutionary process itself, and the individuals making up the ensemble are evidently the biological species, which are born, change without becoming a new species, change into a new species, divide into two or more species, or become extinct. It is important to note that this process involves a time-dependent ensemble whose (stochastic) evolution is highly non-ergodic; there is thus no equilibrium state to which most trajectories tend, and final states, in themselves of low probability, are "frozen in". Each member of this ensemble must itself be an ensemble, however, in order to be able to account for the evolution of the species in terms of individual histories. If this double structure is not recognised, it becomes impossible to see what is actually the bearer of evolution, and sterile and confusing debates have hence arisen. But the ensemble-of-ensembles construction makes it clear that it is the species that evolves, not the individual, and least of all the gene (indeed, since an individual living being has one and only one set of paired genes, it seems unlikely that a third level of ensembles could be useful).

(iv) Finally, as the probability concept developed here suggests and the study of chaotic systems demonstrates, the philosophical dichotomy between probabilistic and deterministic world views is very inadequate. Neither is applicable to the universe as a whole but only to specific systems; and in every case, given a deterministic model we can build a probabilistic one for the same system, and inversely. The two model types complement each other. It is possible that instead we face a much more fundamental distinction (so far ignored by philosophers): in a non-chaotic system, only the precision and completeness of our knowledge of the initial conditions limits the time range over which we can predict future behaviour; a chaotic system, on the other hand, possesses an inbuilt limit to predictability. In such a system, trajectories starting starting at almost coincident initial conditions will quite rapidly diverge so widely that improving our knowledge of the initial conditions will hardly help at all. A good example is that of weather forecasting; although the atmosphere is a rather well understood "deterministic" system, it is capable of very varied chaotic behaviour (Lorenz 1963, Mason 1968), so that even with ideally precise and abundant data we could not foretell the weather for more than a few days. Since this distinction between the predictable and the unpredictable characterises the physical system rather than its theoretical model, we have here an ontological difference; that between a deterministic and a probabilistic description being that between two aspects, possessed by almost every system, but brought out by different models.

REFERENCES

R. Balescu (1975), "Equilibrium and Non-Equilibrium Statistical Mechanics", J. Wiley, New York

L.E.Ballentine (1970), Revs. Mod. Phys. **42**, 358

D.I.Blochinzew (1953), "Grundlagen der Quantenmechanik", Deutscher Verlag der Wissenschaften, Berlin

N. Bohr (1936), Erkenntnis **6**, 263

N. Bohr (1948), Dialectica **2**, 312

T.A.Brody (1975), Rev. Mex. Fís. **24**, 25

T.A.Brody (1983), Rev. Mex. Fís. **29**, 461

T.A.Brody (1984), Comp. Phys. Comm. **34**, 39

P. Cvitanović (1984), "Universality in Chaos", Adam Hilger, Bristol

F.J.Dyson (1958), Scientific American **199**(9), 74

A. Einstein (1902), Ann. d. Phys., IV. Folge, **9**, 417

A. Einstein (1903), Ann. d. Phys., IV. Folge, **11**, 170

A. Einstein (1904), Ann. d. Phys., IV. Folge, **14**, 354

A. Einstein (1936), J. Franklin Inst. **221**, 349

A. Einstein (1953), in "Scientific Papers Presented to Max Born", Oliver & Boyd, Edinburgh, p. 33.

A. Einstein, B. Podolsky & N. Rosen (1935), Phys. Rev. **47**, 777

B. de Finetti (1970), "Teoria delle probabiltà, Einaudi, Torino

R.H.Fowler & E.A.Guggenheim (1939), "Statistical Thermodynamics", Cambridge University Press, Cambridge

J.W.Gibbs (1902), "Elementary Principles in Statistical Mechanics", Yale University Press, Yale, Connecticut

I.M.Hacking (1966), "The Logic of Statistical Inference", Cambridge University Press, London

Hao Bai-Lin (1984), "Chaos", World Scientific, Singapore

W. Heisenberg (1951), Naturwissenschaften **38**, 49

W.A.Houston, Amer. J. Phys. **34**, 351

H. Jeffreys (1948), "The Theory of Probability", Oxford University Press, Oxford

P. Jordan (1938), "Die Physik des zwanzigsten Jahrhunderts", Vieweg, Braunschweig

M. Kac (1959), "Probability and Related Topics in Physical Sciences", Interscience, London and New York, p. 23

J.M.Keynes (1921), "A Treatise of Probability", Macmillan, London

A.N.Kolmogorov (1933), "Grundbegriffe der Wahrscheinlichkeitsrechnung", J. Springer Verlag, Berlin

A.N.Kolmogorv (1969), Probl. Inf. Trans. **5**, No. 3, 1

W.E.Lamb, Jr. (1969), Phys. Today **22**(4), 23

W.E.Lamb, Jr. (1978), in S. Fujita (ed.), "The Ta-You Wu Festschrift", Gordon & Breach, London, p. 1

E.N.Lorenz (1963), J. Atmos. Sci. **20**, 130

B.J.Mason (1968), Contemp. Phys. **27**, 463

R. v. Mises (1919), Math. Zeits. **5**, 52

R. v. Mises (1931), "Wahrscheinlichkeitsrechnung und ihre Anwendung", F. Deuticke, Wien

W. Pauli (1954), Dialectica **8**, 112

K.R.Popper (1959), "The Logic of Scientific Discovery", Hutchinson, London

F.P.Ramsey (1926), in R.B.Braithwaite (ed.), "Truth and Probability in the Foundations of Mathematics", Routledge and Kegan Paul, London

E. Rédei & P. Szegedi (1988), these proceedings

H. Reichenbach (1935), "Wahrscheinlichkeitslehre", A.W.Sijthoff, Leiden Wissenschaften, Berlin

A.A.Ross-Bonney, Nuovo Cim. **30B**, 55

J.C.Slater (1929), J. Franklin Inst. **207**, 449

P. Suppes (1968), J. Phil. Sci. **65**, 651

R.C.Tolman (1938), "The Principles of Statistical Mechanics", Oxford University Press, Oxford

QM AXIOM REPRESENTATIONS WITH IMAGINARY & TRANSFINITE NUMBERS AND EXPONENTIALS

William M. Honig,
Curtin University (which was formerly known as the
Western Australian Institute of Technology),
Perth, Bentley, 6102, Western Australia.

ABSTRACT: A presentation is made showing how imaginary numbers, exponentials, and transfinite ordinals can be given logical meanings that are applicable to the definitions for the axioms of Quantum Mechanics (QM). This is based on a proposed logical definition for axioms which includes an axiom statement and its negation as parts of an undecidable statement which is forced to the tautological truth value: true. The logical algebraic expression for this is shown to be isomorphic to the algebraic expression defining the imaginary numbers $\pm i$ ($\sqrt{-1}$). This supports a progressive and Hegelian view of theory development. This means that thesis and antithesis axioms in the QM theory structure which should be carried along at present could later on be replaced by a synthesis to a deeper theory prompted by subsequently discovered new experimental facts and concepts. This process could repeat at a later time since the synthesis theory axioms would then be considered as a new set of thesis statements from which their paired antithesis axiom statements would be derived. The present epistemological methods of QM, therefore, are considered to be a good way of temporarily leapfrogging defects in our conceptual and experimental knowledge until a deeper determinate theory is found.

These considerations bring logical meaning to exponential forms like the Psi and wave functions. This is derives from the set theoretic meaning for simple forms like 2^A which is known to be the set of all subsets of the (discrete) set, A. The equal symbol in equations which are axioms, and all its other symbols, can be mapped to a transfinite ordinal. Imaginary exponential forms (like $e^{i\theta}$) can be shown to stand for the (continuous) set of all subsets or the set of all experimental situations (which thus includes arbitrary sets of experimental situations) which are based on the axiom, θ, a transfinite ordinal.

PREAMBLE

This preamble should make clear the point of view on which the subsequent parts of this paper are based. It is generally motivated by the idea that axioms are always tentative and will in time be progessively replaced by deeper axioms. Its main thesis is that the study of the nature of axioms should be in a field free of the Boolean strictures where such an axiomatic evolution can be considered and effected. In discussing a possibly useful redefinition of axioms, that is, replacing the canonical definition: axioms are tautologically true, it is here held that the logical meaning of an axiom redefinition need not conform with any Boolean logical definitions or operations as long as such axiom definitions are explicit and clearly stated. The only requirement is their logical definition, i.e. their description in logical terms, conforms with how we as humans might actually understand and use axioms.

It is necessary to emphasize the primacy of accurate description rather than consistency or reasonableness when one considers the logical nature of axioms. A preliminary discussion according to this viewpoint is now given, including critical remarks which have been kindly provided by a colleague.

The nature of axioms is a subject which must be estranged from that of all other theoretical fields in science because the logical status of axioms is peculiar. This is because the presently perceived nature of axioms has prevented their consideration as objects of formal analysis.

In a colloquial sense, an axiom is usually taken to be a statement from which one reasons in the usual or Boolean way. The axiom statement itself is taken as something which is a primitive statement or one with no nonequivalent precedent. Its content is then assumed or stated to be true and it is not to be analysed any further. Canonically, an axiom is a tautology (a declared truth). This is, of course, a subjective judgment, but the subsequent logical manipulations of axioms according to rules which are themselves axioms have proven their great power in almost all fields of human endeavor. This is illustrated by the case of the theorems of, say, Euclidean geometry. Inside the field of Euclidean geometry the rules of Boolean (or classical) logic operate to generate the theorems and at the borders (i.e., the foundations) are the axioms. Thus axioms are the objects from which reasoning proceeds and to which reasoning (Boolean operations) cannot be applied. To go beyond this, two fields of discussion must first be defined.

Working descriptions for the fields where discussions will proceed are:
1. The Boolean field - As in the case of Euclidean geometry above this is a local field ringed with self consistent axioms inside of which the usual canonical logical operations can occur.
2. The Non-Boolean field - This is a global field which may contain many inconsistent considerations and/or ideas. It is a covering field so that Boolean fields may lie inside it. It is the arena of the consciousness, see further remarks on Varela below.

The considerations in this paper are undertaken primarily with respect to the axioms of QM (but with some reference to STR). A presentation has been made[1,2] which tends to show that the tenets of QM and of STR each have sets of antithetical axioms. This means that each of these theories make use of global axioms which are axiom pairs consisting of an axiom, say α, and its negation $-\alpha$. Each (local) axiom of the pair is needed to deal separately with each part of the antithetical experimental situations in QM or STR in a Boolean (usual logical) way although the complete theory would lie in a Non-Boolean field. For example in QM, fundamental entities can alternately and separately be taken as waves or particles depending on the particular experiment involved. In STR, phenomena in different rest frames are each analyzed in a Boolean (i.e., self-consistent) field whereas globally this is not so. The global axioms in QM and STR as exemplified by Wave-Particle Complementarity and the Relativity Principle, respectively, are not self consistent but this is not visualisable or operationally testable. The great success of QM and STR, however, in delivering dazzling experimental predictions and accuracies is the reason for their acceptance.

These theories are in contrast with the previous Newtonian approach where a single set of self consistent axioms are always used for all situations. One may well ask of what relevance are antithetical axiom pairs with respect to such classical theories. It is here where the concept of progressions to deeper theories can also prove to be of value. In considering the ongoing history of such a reigning theory, more and more experimental facts are usually discovered which this ruling theory continues to cover. Finally experimental facts may be found which this theory is unable to explain. For example, electrons gave evidence of behaving like waves instead of particles, or Galilean rest frames showed inadequacies and relativistic rest frames were more successful in treating electromagnetic phenomena. What these examples really show is that not only are the original axioms inadequate but that the opposite of those original axioms suddenly acquire a surprising relevance and usefulness.

Thus, the first faults in the theses of the original axioms make it a necessity to consider antithesis axioms to treat the new phenomena. For example, the very first step in developing a deeper theory to the Newtonian one is to say: suppose that the Galilean rest frames are not true; or the very first step by Planck in his considerations of hot bodies was when he considered the antithetical idea that electromagnetic waves were not continuous.

Generally, at this point one expects that a Hegelian synthesis would occur and a set of deeper axioms can be found to reconcile and assimilate the disparate older and newer phenomena. If this happens the deeper theory is one that again can be set in a Boolean field with a single set of consistent axioms. Even so, the future usefulness for carrying along an additional set of conflicting axioms for further cycles of this process is evident. This suggests the mechanics of the progression to deeper theories.

The present paradigms, QM and STR, appear to be theories each of which is in the Hegelian stage of thesis-antithesis. This is a new state of affairs compared to the classical theories. One might view this situation as one in which QM and STR are thesis-antithesis theoretical constructions which are merely tentative attempts to leapfrog defects in our determinate understandings and that the future will eventually provide these understandings in the progression to deeper theories. The possibility for a future synthesis to a Boolean theory is evident in the dynamics of such a process. This attitude is the basis for the definition of the global axioms to be presented as an antithetical $(\alpha, -\alpha)$ axiom pair.

This is illustrated in Charts One and Two. Chart One shows the Hegelian progression emphasizing the renaming of the synthesis theory as a thesis theory from which antithesis statements can be easily found when further developments are necessary. Chart two gives a comparison between classical and non-classical theories. The upper portion of the chart gives synonyms for each kind of theory. The lower portion of the chart shows that the classical theory uses only one set of consistent global axioms although its paired (partial) antithesis axioms are dormant. The non-classical theory, i.e., QM or STR, has a global set of axioms containing antithetical local axiom sets which are simultaneously available. A particular local axiom set is to be used as the occasion demands, this is illustrated further in Appendix B. The above matters are also discussed further in this section.

A candidate axiom statement which is symbolized by α is considered in a Non-Boolean field where its definition as an axiom will be developed. In line with the idea of progression to deeper axioms, α can be considered as a statement whose mechanism is (as yet) unknown. It is because of this, that α cannot be true or inferred to be true, because it is not yet an axiom or cannot be deduced from a set of axioms, respectively. Neither can α be false nor inferred to be false because its unknown basis precludes such a consideration of its falsity. A human mind considering this matter would say that the truth and the falsity of α is undecidable. This description is now developed by invoking (asserting) modus ponens (The Law of the Excluded Middle) on each part of the above sentence and restating them. Since α is not false and not true then via modus ponens α is both true and false. This is only one of the two features which will be used to redefine the notion of axiom.

The simultaneous truth and falsehood of a statement is first more generally taken as a logical definition for undecidability. The existence of an undecidable question implies that more than one state is under consideration and a decision, as to which state is the case, is necessary. This is, in fact, what an undecidable question means. In the case of the question: Is α true or is α false?, the contemplater considers the case: α is true, and also considers the case: α is false. If there is insufficient evidence to decide the case the contemplater stops right there, with both cases still under consideration.

If, however, the contemplater is asked to show his state of thinking about the question: Is α true or false?, he could exhibit his state of thought on the subject (or his desktop or notebook) which would have to show the entry: α is true, and also the entry: α is false, with whatever evidence supports each statement. A description of the status of the contemplater's thinking on the question would be the exhibition of his desktop/notebook entries: α is true **and** α is false (together with the insufficient evidence for each case). It is the **and** which indicates that these two matters are still under simultaneous consideration and

HEGELIAN PROGRESSION

THESIS
⇓
THESIS-ANTITHESIS
⇓
SYNTHESIS
⇓ renamed
THESIS
⇓
THESIS-ANTITHESIS
⇓
SYNTHESIS
⇓ renamed
THESIS
⇓
AND SO ON...

CHART ONE

AXIOM COMPARISION

NEWTONIAN/CLASSICAL THEORY (i.e., mechanics)	QM/STR THEORY
BOOLEAN	NON-BOOLEAN
THESIS	THESIS-ANTITHESIS
NON-CONFLICTUAL	CONFLICTUAL
LOCALLY	LOCALLY CONSISTENT &
(& GLOBALLY CONSISTENT)	GLOBALLY INCONSISTENT

AXIOM SETS	AXIOM SETS
IN USE: [A, B, C, ...]	LOCAL SET: [$\alpha, \beta, \gamma, ...$]
DORMANT: [-A, -B, -C, ...]	LOCAL SET: [$-\alpha, -\beta, -\gamma, ...$]
	BOTH SETS TOGETHER ARE THE GLOBAL AXIOMS, SIMULTANEOUSLY AVAILABLE, BUT USED IN DIFFERENT EXPERIMENTAL SITUATIONS

CHART TWO

that, the question is undecidable, thus the definition:

$$\text{The truth and simultaneous falsity of } \alpha \equiv \text{Undecidability of } \alpha \quad (1.1)$$

is a fair description of the desk top, notebook, or mind of the contemplater.

With the above characterization, (1.1) now begins to lose its bizarre quality. This has required, however, that the mind (the consciousness) of the contemplater be introduced (See further remarks on the work of Varela). This, then, is the basis for (1.1) above, as a realistic description of that mind which is contemplating an undecidable question and where the description of this matter is taking place in a Non-Boolean field.

Critical remarks on (1.1) above have been supplied by a colleague: "This seems confused. For example, undecidability is interpreted as the "simultaneous existence" of α and not α. If existence means truth, this is inconsistency, not undecidability. If existence means well-definedness, this is a standard requirement, not undecidability. If "simultaneous existence" means conjunction then this is falsity, not undecidability. -- The related work of F. Varela should be noted."

In reply to the above I must say that I agree with all these remarks if the field of operation for the matters under discussion is in a Boolean field. The nature of the axioms to discussed here, however, is to be considered inside a non-Boolean field. From a Boolean point of view the expressions (1.1) et seq. are indeed inconsistent and false as remarked above. Well-definedness, however, is the purpose of these proposals and they seem clear even in a Non-Boolean field and thus, refer to a description of what is in the mind of the contemplater of this matter under consideration. The reference to Varela is discussed further on.

It has been shown[3] that general axioms can be reduced to a sequence of Yes-No (or binary) questions, and this is sufficiently general to cover axioms in general. Thus (1.1) is a sufficient basis for our general definition of axioms which will follow.

Via the methods of Boolean Algebra[4,5] the left side of the above expression (1.1) can be rendered in the language of sets as:

$$[\,\alpha\,] \times [\,-\alpha\,] \equiv \text{Undecidability.} \quad (1.2)$$

where the minus sign stands for negation or falsity and where the brackets now designate their contents as sets, and x is set intersection or logical multiplication meaning: **both α and -α**.

The previous discussion has proposed only one feature for the redefinition of an axiom. The final feature which would complete our definition of an axiom is, of course, is that axioms are assumed (declared) to be true. Our final definition for an axiom is: an axiom is an undecidable statement which is assumed (tautologically or categorically declared) to be true. How can an undecidable yes-no (or α and -α) statement be true? The meaning of the word, assumed (declared), is indeed what makes it possible to say this, but a clearer way to say this is to describe the situation by saying: an axiom is an undecidable statement which is "forced" to be (tautalogically) true.

The use of "forced" is related to but not identical with the meaning of this term that has been given[6]. Forcing, as given by Cohen is carefully qualified and restricted for his applications. Thus, quoting Cohen (p. 112): "It is clear that there are some properties of it [forcing] which we would like to hold. First, it should be consistent, i.e., we should not have P forces A and P forces not A......... and these properties should correspond to the usual properties of implication" This shows that forcing an undecidable statement to be true as in the previous discussion above violates the use to which Cohen has put it.

Nevertheless, according to the discussion in this paper, forcing an undecidable statement to be true means that both an axiom and its negation are simultaneously forced to be (tautalogically) true. Such a situation can be shown to be necessary in the considerations of the logical foundations of STR and QM[1,2]. In a practical sense it merely signifies that the truth of a statement or the truth of its negation may be arbitrarily selected for the appropriate experimental situation in STR or QM. As has been shown in the above references, an axiom pair symbolized by, say, α and $-\alpha$ refers to the global theory in QM or STR. As an example, for a particular experiment in QM an electron can be assumed to be either a discrete particle or a wave, each of which would then be based on the appropriate α or $-\alpha$ assumption from which Boolean reasoning can proceed. The global QM theory, however, would contain both points of view: the α and $-\alpha$ axiom pair, or in QM terms, the particle-wave pair, see reference above and subsequent discussions.

The use of an antithetical axiom pair (α, $-\alpha$) appears to be totally wrong with respect to a Newtonian type of theory. We counter this by putting such theories inside a Boolean field. Although this is merely a legalism, the failure of such theories to explain recent phenomena is the reason for the modern theories of QM and STR. Since QM and STR obey correspondence principles they incorporate the contents of Newtonian ideas and are open ended in that we believe their use of antithetical axiom pairs can provide the clarity to support the idea of progression to deeper axioms as newer inexplicable phenomena are discovered.

Therefore, taking the field of operation for axioms to be a non-Boolean field and proceeding according to the previous discussion, the final definition for a global axiom is: "An axiom is an undecidable true-false or a global α and $-\alpha$ axiom statement which is forced to be true". It is now rendered into the language of algebraic logic[4, 5]:

$$[\alpha] \times [-\alpha] \equiv 1 \qquad (1.3)$$

where $\equiv$, the usual symbol for definition stands for forcing and 1 is the usual algebraic symbol for the truth value: true. It should be noted that the three stroke definition symbol itself can indeed stand for a forcing operation; it is another way of describing the operation: definition.

Treating the symbols in (1.3) algebraically one then immediately gets as the algebraic solution of (1.3):

$$\alpha = \pm i \; (\sqrt{-1}, \text{ the imaginary number}). \qquad (1.4)$$

This, then establishes that there is an isomorphism between the algebraic meaning for i ($\sqrt{-1}$) and the meaning of an axiom as introduced above and that there is no property enjoyed by i that is not enjoyed by -i. Thus $\pm i$ ($\sqrt{-1}$) may be used to label inverse axiom statements or symbols. This paper explores what seem to be many useful interpretations for this and related ideas in the mathematical formalism of theories where imaginary numbers and/or imaginary exponentials appear.

It could be argued that such discussions and definitions as are here presented have no useful purpose, but this remains to be established. A detailed description has been given which shows that axiom pairs consisting of an axiom statement and the negation of this axiom statement are necessary to clarify issues in both Quantum Mechanics and Special Relativity[1, 2, 7]. The above considerations are also meant to utilize mathematical concepts for the clear presentation and differentiation of axiomatic matters from non-axiomatic matters (i.e., deductions, physical quantities, etc.). For example, axiomatic statements might be represented inside an imaginary number field whereas the real number field would be used as at present for the representation of the magnitudes of deductive (physical) variables.

The previous reference to Varela is apt[8, 9]. Varela has presented a fully developed thesis for the characterization of living systems, in which the relevant features to the above discussion are that living systems are self referential and unitary. His meanings for these terms are explained in great detail, and they bear on the mind-consciousness problem. They seem to imply that that which is called the non-Boolean field here is the field of action of the consciousness of living systems. This is a solipsistic view of the mind, but Varela's developments appear to have at least some physiological and philosophical usefulness. This new emphasis on usefulness gives one hope in the future non-sterility of solipsism.

What attitude should one take with respect to a non-Boolean field? It is here where the ideas of Varela are helpful. If one considers that undecidability with respect to some question exists, then obviously there are elements of ignorance present in the mind which is considering this matter. Thus if a statement is undecidable this is an explicit tentative acceptance of ignorance.

It is our human condition that permits us to see both undecidability and ignorance as commonly experienced sensations. The consideration and the use of axiomatic considerations is also a common everyday experience which cannot be denied. Although the nature of axioms as suggested here violate Boolean rules, these considerations nevertheless occupy our minds. It serves no useful purpose to banish these considerations with the epithets: irrational, mystical, etc. They should, therefore, be given and clearly identified, in such a way that they can be expressed in an explicit and separate way from those more common and more numerous Boolean ideas which do indeed occupy most of our thoughts. Possibly another way of saying this is that the Non-Boolean and Boolean fields could refer to the subjective and the objective, respectively. The Non-Boolean field has an "Alice in Wonderland" quality about it, as it should have, if it is to be the place for the representation (description) of thought processes in the consciousness of the contemplater.

From this preliminary presentation the body of this paper proceeds.

1. INTRODUCTION

The purpose of this note is to amplify previous remarks published on this subject [10, 11, 2, 7]. On the basis of the previous discussion in the preamble, Section 2 continues the discussion of the logical status of antithetical axiom pairs and of $\pm i$ ($\sqrt{-1}$). In a logical sense axioms and i both have the same meaning: undecidable; but via the recent forcing ideas of Cohen they are "forced" to function as terms which have the status: true. The mathematical definition (the logical status) of a scientific axiom is that it is an undecidable statement which is forced to be true. The isomorphism between the logical meaning for i and logical status of axioms permits that i can be used to label number symbols which designate axioms.

The set theoretic meaning for exponentials of base 2, which comes from the Power Set Theorem of Axiomatic Set Theory [12, 13] can be extended to exponentials of base e. In Section 3, a discussion is given which suggests that 2 and e in the forms 2^a and e^b can be given a meaning which denotes that a and b apply to or represent discontinuous and continuous quantities (or sets), respectively. It then follows that these exponential forms refer to the set of all subsets (discontinuous or continuous, respectively) represented by the set numbers a and b.

Section 4 suggests a meaning and use for the transfinite ordinals as number symbols designating the equal symbol in equations. If such an equation as, for instance, $E = \underline{h}\omega$, is an axiom, then a transfinite ordinal labelled with the coefficient i can stand for the complete set $[E, \omega]$ of number pairs satisfying this equation. The transfinite number, therefore, stands for the equation and thus for the axiom directly.

Section 5 suggests that the imaginary exponential form $e^{i\emptyset}$ stands for the set of all subsets of physical values (situations) to which the axiom $\emptyset$ can refer. This section concludes with a utilitarian and epistemological discussion of the Euler form:

$$e^{i\emptyset} = \cos\emptyset + i\sin\emptyset.$$

This results in a logical interpretation for the combination of the forms $e^{i\omega t}$ and $e^{iEt/\underline{h}}$ and for each of these wave functions separately. This also suggests that taking the logarithm of a probability which is of the form $e^{i\emptyset}$ and which can be recognized as the entropy or information function can be logically interpreted as the search for axioms from the set all subsets of experimental results.

Much of the above profits from the insights offered by the Non-Standard numbers [14,

15) and from Frege[16] who linked the logical meaning of a number (zero) to the meaning of paradox.

2. THE LOGICAL STATUS OF AXIOMS & THEIR SIMILARITY TO THE LOGICAL MEANING OF i ($\sqrt{-1}$).

According to the Preamble discussion of axiom statements in a Non-Boolean field, an axiom of a theory cannot be true because it cannot be deduced from any more basic statement which we already know is true. Neither can it be false because it cannot be shown to be undeducible from any such more basic statement, This falsity, in addition, cannot exist because no one would work on the physical predictions of a theory which is established from a false axiom since it would yield results conflicting with physical reality. Thus an axiom is not true and it is not false. If we apply the Law of the Excluded Middle (Modus Ponens) to both parts of the previous statement we get that an axiom is both true and false. As explained previously, the simultaneous existence of both sides of this matter can be taken as a definition for the overall state: undecidable and this quality is one component for the definition of an axiom. An axiom has the additional crucial quality that, we, the people who consider and use theories, assume, at least tentatively or even more strongly, that the axiom is true. We arrive thus at a definition for an axiom: it is a statement (mathematical or otherwise) which in a formal sense is undecidable, but which is nevertheless assumed to be true.

The most striking feature of axioms; that they are assumed to be true, appears to be a tentative and social act. It is tentative because, a la Popper, all those who deal with an axiom and its deductions will accept the truth of the axiom as long as the physical results predicted by the theory based on the axiom appears to be in congruence with measurements. It is social because it is a necessity for the dissemination of a theory that its axioms be clearly communicated to anyone wishing to use the theory. This emphasizes that axioms are arbitrary and not deducible from anything else.

Logical forms (both Boolean and Non-Boolean) may be treated in quite an algebraic manner. This is built on the work of Stone and many others[4, 5 & refs.]. Kiss has shown that such treatments can be extended to undecidable statements.

Let a local axiom (statement or equation) be symbolized by α and its negation or the assertion of its falsity by $-\alpha$. As per the preamble discussion, the simultaneous existence of both these statements can be represented by logical multiplication, which results in the algebraic expression:

$$[\alpha] \times [-\alpha]$$

which is taken to stand for undecidability.

Forcing this to be true, suggested by the approach of Cohen, as per the preamble discussion, results in the previously given expression (1.3):

$$[\alpha] \times [-\alpha] \equiv 1. \qquad \text{(1.3 from Preamble)}$$

Treating this in an algebraic manner:

$$-\alpha^2 = 1 \qquad (2.1)$$

$$\text{or} \quad \alpha^2 = -1 \qquad (2.2)$$

$$\text{or} \quad \alpha = +i \text{ and also } -i, \qquad (2.3)$$

which can also be expressed as: $\quad \alpha = i \quad \text{and} \ -\alpha = i. \qquad (2.4)$

Statements, equations, or symbols may now be identified as axioms or as something which is simultaneously undecidable and true. This is done by labelling these statements, equations or symbols with an i (as a coefficient).

The non-Boolean expressions (1.1) and (1.3) are ways of describing clearly and accurately the content of the minds considering undecidability and the meaning of axioms. The expressions in (2.4) show that this i formalism treats equally well an axiom and its negation and results in the ($\alpha, -\alpha$) axiom pair.

The three facets for the logical definition of axioms which are presented here are summarized:

1. Axioms with their negation and $\pm i$ ($\sqrt{-1}$) can (each pair) both be taken as undecidable statements which are forced to the truth value: true.
2. Just as the real number value for i is unknown so is the mechanism for an axiom unknown (See Section 4).
3. Axioms and their negations a la Hegelian thesis-antithesis should always be carried in a reigning theory.

The meaning of zero with reference to axioms deserves comment. Zero has usually be taken as denoting any and all nullities (null elements) in a set under consideration. For example, if the set of six apples has the six apples removed from the set, then that set still contains the null element or zero apples. A similar set of oranges will have the null element also, or zero oranges. Of course, one may take (a la Frege) the number zero as consisting of the set of all null elements. In the above 2 cases the single symbol 0 (zero) is clearly applied because apples and oranges enjoy the same logical status.

They are the objects of our contemplation which can be defined in terms of other qualities or relationships; they are deductive objects. They do not have the same logical status as axioms. Simple clarity and the i labelling rule above is the reason for our belief that objects of our contemplation which have a different logical status from each other should have their null elements labelled differently. Thus 0 and i0 should be the null symbols for deductive and axiomatic null elements, respectively. This is useful in what follows and in some previous work[1, 2].

3. A SET THEORETIC MEANING FOR THE FORMS, 2^a AND e^b.

The Power Set Axiom states that the set of all subsets, [PS], of a set [A] with discrete members (including the null element) which may be finite or enumerably infinite and which set [A] has the cardinality a, is the set with the cardinality 2^a. Thus, e.g., a set with 3 non-null members will have its power set, [PS], (its set of all its subsets) consist of 2^3 members.

It is also well known[12, 13] that the [PS] can be conceived of as the insertion set [p|A], where [p] is the 2 member set [0, 1] and where [A] is the discrete set with the cardinality, a, so that:

$$[PS] = [p|A] = 2^a.$$

The set [p|A] is the set of all insertions of [p] into [A] and can be conceived of for an arbitrary subset $[A_o]$ of [A] as the insertion into [A] of the set [1, 0] with 1 applying to all the elements of $[A_o]$ and 0 applying to the remainder of the set [A]. The word "applying" is taken as an ordinary multiplicative operation between one or zero on the one hand and a particular set member $[A_i]$ on the other hand. Thus 1 preserves the existence of an $[A_i]$ and 0 denies that existence. The set of all such insertions of [p] into all the subsets of [A] results in the [PS] of [A] with the cardinality 2^a. Directing ones attention to [p]; this set with members 1 and 0 can be conceived of as a definition for the quality: discreteness. This pairs the 1, 0 set members of [p] with the qualities existence and non-existence, respectively. This is a quantal idea which does not permit any way of representing continuous variations. In this way, every set member of [A] can be explicitly picked out. If, however, the set [A] were continuous then this procedure using [1, 0] would be useless, which emphasizes how the set [p] defines discreteness.

Associating the number 1 or the number 0 with each element of a set [A] is really equivalent to defining a function, f, where:
$$f(a_i) = 1 \text{ if } a_i \text{ is to be a subset of } [A]$$
$$\text{and } f(a_i) = 0 \text{ if } a_i \text{ is not to be a subset of } [A].$$

Such functions, called characteristic functions (or ch.f.), are closely connected with the operations of Boolean Algebra[12, 13, 4, 5]. The ch.f. for the example given above consists of the set [1, 0]. Two useful theorems for such ch.f. are described in logical terms:
a) The ch.f. set corresponding to the intersection of two or more general sets are the (Cartesian) product set of the ch.f. sets of each general set.
b) The ch.f. set corresponding to the union of two disjoint general sets is the sum of the ch.f. sets of each general set.

Intersection and union of sets corresponds to logical multiplication and logical addition, respectively. Thus for two sets [A] and [B], logical multiplication corresponds to those elements of [A] and of [B] that lie in both [A] and [B] simultaneously; whereas logical union corresponds to elements of [A] and of [B] which lie in [A] or in [B] or in both. The definition for discreteness which has been given, together with the Power Set Axiom lead to a conclusion about the set theoretic meaning for 2 in the cardinality number 2^a of the set of all subsets of [A]. The meaning for the cardinal number 2 must connote that it is the cardinality of the particular set [1, 0] and that the set [A] is discrete. Thus the meaning for 2 (a la Frege) stands for the set of all sets which are discrete. The quality of discreteness must be clear, operative, and especially, unique for both logical multiplication and logical addition as presented in the previous paragraph, because both of these logical operations should apply in normal procedures associated with theories and experiments. Since the complete generality of a statement will occur only after it refers to the set of all subsets, theorems a) and b) above should be used to help create a progression from sets to the cardinality number that represents the set of all subsets which represent discreteness.

These progressions will be given for logical sums and for sets and should give the same cardinality number of 2; this is given here, for logical (and algebraic) sums: (where n approaches infinity carries these matters to the general case and in the logical sense means; all). It should be noted the the comma symbol (inside the brackets which will represent sets [,]) is logically equivalent to the plus symbol (inside the parentheses which will represent sums (+)), which reconciles their interchangeable appearance in what follows. A good illustration of this is the label for points in the complex plane which can be given as (x, y) or as (x + iy); thus the set progression should be:

[Discreteness sum] $\rightarrow\rightarrow$ [Product of n discreteness sums] $\rightarrow\rightarrow 2^1 \rightarrow\rightarrow 2.$ (3.1)
For sums this is:

$$(1 + 0) \rightarrow\rightarrow (1 + 0)^n \rightarrow\rightarrow (1 + 0) \rightarrow\rightarrow [1,0] \rightarrow\rightarrow 2^1 \rightarrow\rightarrow 2 \quad (3.2)$$
which counts the null element as an element. For sets, the progression becomes:

[Single discreteness set] $\rightarrow$ [Cartesian Product of n discreteness sets] $\rightarrow$ 2 (3.3)
In set nomenclature this is:

$$[1,0] \rightarrow\rightarrow [1,0]^n \rightarrow\rightarrow [1,0] \rightarrow\rightarrow 2^1 \rightarrow\rightarrow 2. \quad (3.4)$$

These detailed explanations above appear to be trivial with respect to the result (that 2 represents the cardinality of a number that means discreteness for exponents of this number). In view, however, of the development of the Non-Standard numbers and literal infinitesimals[15], it is tempting to view continuous sets in a similar manner to the above discussion for discrete sets. Instead of the discreteness definition set [1, 0], we take the set [1, δ] where δ is a literal infinitesimal with a non-standard meaning and which is defined as:

$$\lim 1/n \text{ as } n \text{ approaches infinity} = \delta. \tag{3.5}$$

Thus: [1, δ] is the limit of [1, 1/n] as n approaches infinity
and this set has no null element, by definition, as per Robinson. This set, [1, δ], is thus meant to represent the qualities associated with continuousness. Now 1 and δ are, for example, a point (a real number) on the continuous interval between zero and one, and an infinitesimal increment beyond that point, respectively. These elements are enough to get to any point in the continuous internal between 0 and 1. In an analogous manner to that given for the discreteness sets, one gets now for n approaching infinity (or all):

In sum progressions: $(1 + 1/n) \to\to (1 + 1/n)^n \to\to e$

and in set progressions: $[1, 1/n] \to\to [1, 1/n]^n \to\to [e] \to\to e$ (3.6)

where the comma and plus symbols are understood as above and where [e] is a non standard set with the member e and no null element.

The progression from $[1, 1/n]^n$ to e thus gives the same result for the addition expressions above it for reasons similar that of the previously discussed discreteness case. The conclusion is that e is a number which has a non-standard set theoretic meaning corresponding to the quality: continuous (or continuity) to be assigned to arguments of the exponent of e. These comments identify the exponential forms of 2 and e, say 2^a and e^b, each of which is the set of all subsets whose exponents a and b refer to discrete and continuous variables, respectively. Thus e^b can also, with this meaning, represent the set of all physical situations to which an exponent number symbol which stands for an axiom (see below) can apply.

4. A MEANING AND USE FOR TRANSFINITE ORDINALS

Generally the laws of physics are written as equations in terms of symbols describing physical reality. Thus symbols are given to concepts like force, pressure, velocity, energy, and the like. Their numerical magnitudes are coupled with dimensions which define their meaning. The numerical magnitudes lie in the real number field and a normalization can be set up, of course, which maps this number range into the [0 to 1] interval, which is necessary in what follows.

The human procedure for ascertaining some details of physical reality consists of the use of both a theoretical framework of statements and an experimental setup to find, say, the magnitude of a physical variable. Suppose one is ignorant of the magnitude of a physical variable. Upon the manipulation of theory and experiment a value for the magnitude for a physical variable is found; this with a specified accuracy. Such a procedure, however, may be incomplete because the person doing it may be ignorant of significant information affecting the outcome of the experiment. Even so, the performance of the experiment a large number of times can be used to determine the magnitude of an average value of a physical variable. Thus where a particular outcome occurs for a fraction f of the total repetitions, that fraction f is defined, as by von Mises and Reichenbach, as the probability of the outcome. This gives meaning to the real number range [0 to 1] and it refers to the relative frequency of that outcome for a well defined physical variable.

Although probability in this way can is associated at present with a subjective ignorance of the parameters defining a physical variable, it should rather be called an objective ignorance and this probability should be called an objective probability. This is because all experimenters using the same set of concepts and axioms would get the same probable values for the outcome of their experiments on such well defined aspects of physical reality. This requires that all experimenters have identical sets of definitions and the same deductions from the same theoretical framework.

On the other hand, the axioms of any theory are statements that have a peculiar logical significance. Axioms must necessarily always be statements which cannot be logically deduced from other statements. If they could be so deduced then those other statements would be called the axioms. Of course the ongoing progress of our scientific theories may reduce the axioms of a previously accepted theory to deductions in a new theory. The new theory, however, would still be based on axioms (newer, deeper axioms). Thus each in the sequence of deeper theories will always be based on axioms. If one wishes to preserve a Boolean logic in the theories it is useful to consider an axiom as a statement whose "mechanism" is not yet known. The fact that the real number value for i does not exist might be identified with the fact the mechanism for an axiom is unknown. Imaginary probability may thus adhere to such ignorance.

This is, in a literal sense, a subjective ignorance because there is no agreement on the future axioms of deeper theories which could someday be used to deduce a present day axiom. This can be expressed in a mathematical formalism which is suggested in the **Appendix A**.

It remains to illustrate the logical meaning for the form $e^{i\emptyset}$ where $\emptyset$ is a local axiom (see **Appendix B**). Starting from the preceding discussion, $e^{i\emptyset}$ can be taken as the set of all subsets to which the axiom statement or axiom number symbol $\emptyset$ applies.

To apply this in Quantum Mechanics (QM) one may take the Planck energy relation:

$$E = \hbar\omega \tag{4.1}$$

to be an axiom of QM and it should be representable with $\emptyset$. Here, of course, $\hbar$ is $h/2\pi$ and ω is $2\pi f$ where h is Planck's constant and f is the frequency. An infinite set can represent the above relation which is the set of all the number pairs [E, ω] obeying (4.1). This set needs to be continuous since E and ω are continuous.

Eq.(4.1) can be converted into a pure number axiom symbol $\emptyset$, but only after the terms on both sides of the equal sign are made into pure numbers with no dimensional designation. This is necessary in order to make it follow the only logical meaning that exists for exponentials. This comes from Axiomatic Set Theory where, for example, the number 2^A is the pure number of all the subsets that can be made from the set A. This number 2^A is for A a discrete set (as discussed above). The conversion of each side of Eq.(4.1) to a pure number can be effected by multiplying both sides by $t/\hbar$ and results in the factor t as an independent time variable :

$$Et/\hbar = \omega t. \tag{4.2}$$

The above expression is not unique, however, since it might just as well be written as $E/\hbar\omega = 1$. The reason for the use of Eq.(4.2) is because all our human considerations of these scientific matters are carried on in space-time frames. It is this which makes it a requirement that independent space or time variables be inserted into these considerations in such a way that they can be indexed directed to our space-time measurements and theoretical considerations. For the case where, say, x, the distance is needed to be the independent variable, then momentum p and wave number k ($=2\pi/\lambda$) can represented instead of E and w using the De Broglie relation instead of the Planck relation above, see final section for additional remarks.

The meaning for an exponential of e, in the form $e^{i\emptyset}$, has been shown previously to designate $\emptyset$ as a continuous set. We, therefore, take the set of all continuous sets to which the Planck relation applies, in the form of the axiom (4.2), as:

$$e^{i\emptyset} = e^{i(Et/\underline{h} = \omega t)} \text{ or } e^{(iEt/\underline{h} = \omega t)}. \tag{4.3}$$

The bizarre appearance of the equal sign in the exponent can be made more palatable if it can be mapped to a number. This will be shown below. First, with one additional supposition the above expression can be written as:

$$e^{(iEt/\underline{h} + = + i\omega t)} \tag{4.4}$$

where the supposition is that + means its logical meaning which was previously introduced. Thus it can also mean: lies next to, and the plus symbols and parentheses can be dropped. This then becomes:

$$e^{iEt/\underline{h}} e^{=} e^{i\omega t}. \tag{4.5}$$

It now appears that in (4.5) the first and last factors can be the Psi function of Quantum Mechanics and the wave function of Electromagnetic theory, respectively.

Since the arguments of each of these factors are equal by (4.2), one gets:

$$e^{iEt/\underline{h}} = e^{i\omega t}. \tag{4.6}$$

If we replace the equal symbols in (4.6) and (4.5) by the symbol β and set those terms equal to each other, then we get:

$$e^{\beta} = \beta \tag{4.7}$$

where the equal symbol in (4.7) above stands also for equality; but it must be different from the previous equality signs, possibly deeper than the meaning of b; see remarks below.

There have been a number of discussions[5, 12, 13, 17, 18] of the numerical value which satisfies the relation (4.7). These discussions suggest that (4.7) can be interpreted as referring to transfinite ordinals and where the meaning of (4.7) is that it is:

$$\text{the limit of } e^{\beta} \text{ as } \beta \text{ approaches infinity is } \beta, \tag{4.8}$$

and that β is the first transfinite ordinal; which is the ordinal number for the well ordered set $(1, 2, 3, 4,\ldots\ldots\ldots)$. It can also stand for the ordinal number of a well ordered continuous sequence, see final remarks. There are minor differences between discussions of such numbers by Fraenckel and Kamke[12, 13, 18] who appear to restrict the base number, which we have as e, to integers; whereas Sierpinski[17] permits that it be any real number. The discussion given in the last section of 2^a and e^b, however, can be taken as a justification of the replacement of 2 as a base by e as a base.

Additional discussions by the above authors appear to give additional meaning to the equal sign that appears in (4.7). It is the second step in a hierarchy of the transfinite ordinal numbers which start from the well ordered integer set above.

We retain the designation β for the first transfinite ordinal, the solution of (4.7), although it is usually referred to in the mathematical texts as ω. In order, however, not to cause confusion with ω, the radian frequency in the Planck energy relation, the β is retained. The equation $E = \underline{h} \omega$ can now be written:

$$E \beta \underline{h} \omega. \tag{4.9}$$

In order to agree with the multiplicative operations which are usually defined for the transfinite ordinals and with the requirement that pure numbers be the only terms for the expression of the axiom forms like (4.9) when it is symbolized by $\emptyset$ and when it is used in the form $e^{i\emptyset}$, the above is converted to:

$$(E/\underline{h}\omega) \beta 1 \tag{4.10}$$

or versions of (4.2) which according to the manipulative rules for transfinite ordinals can be given by b alone[5, 12, 13, 17, 18], where β is the ordinal of the sequence (see final remarks):

1, 2, 3, 4, ... or of the continuous sequence in the 0 to 1 interval. (4.11)
The infinite set of pairs [E, ω] can be matched to the sequence above and thus iβ can, according to the manipulative rules for transfinite ordinals, stand for the axiomatic forms (4.4), (4.5), and (4.6).

Two comments: first, the appearance of transfinite ordinals in these matters seems useful for the expression of scientific and human matters. This is because the definition of an ordinal as a set consisting of all its predecessors[19] can make ordinals useful for the historical or sequential time development of a subject.

Finally, (4.7) is an obvious violation of the Power Set Axiom of Axiomatic Set Theory. This point is commented on principally by Pierce [20, 21] to the effect that, for example, the 0 to 1 continuum must be richer than the Aleph-one cardinal because the Cantorian Alephs are based on and developed from discrete numbers and do not carry a strong continuity implication. Thus, saying that a number field contains all powers of that number field, as (4.7) states, may be a useful way to characterize continuity. There is a hint here that the methods of Robinson[14, 15] (the literal infinitesimals and others of that family) would be more relevant than the Cantorian transfinite number sequence.

5. MEANINGS FOR IMAGINARY EXPONENTIALS

Taking the set of all subsets (or the set of all subsets of the physical situations) to which an axiom Ø applies, as $e^{i\emptyset}$, we postulate that this form should correspond to a probability amplitude. As suggested previously, the mechanism for an axiom (which must be unknown) is associated with (an imaginary) probability. Taking the logarithm of that probability should then logically correspond to the human activity (research) consisting of the recovery or discovery of the underlying axiom from the set of all subsets of physical situations to which the axiom applies. Obviously the logarithm of the above imaginary exponential would recover the axiom symbol Ø.

The only other occasions in science when the logarithms of probabilities are of use is in the Boltzmann definition of entropy and the Shannon definition of information[22, 23]. Such a procedure cannot be of any help in finding the unknown mechanism for an axiom, but rather it may help find if the specific axiomatic statements will explain new data or if an independently conceived new axiom fits or can predict data. This will not be a mechanical way of finding new axioms, but it may permit a clearer evaluation of the physical effects that such new axioms predict. Thus, according to the ideas discussed here, axioms must always be created to fit physical situations and not the reverse.

All the previous discussion leads naturally to the 2 questions:
  1. What is the logical significance of the first and last factors of Eq.(4.5),
      which appear to be the Psi and Electromagnetic wave functions, respectively?
  2. What is the logical significance of the Euler relation:

$$e^{i\emptyset} = \cos \emptyset + i \sin \emptyset \ ?$$

The first question has been discussed in some detail[1, 2, 7].

The second question must be broken into 2 parts; first for the transfinite representation of Ø, and second, for the finite representations of the wave functions.

For the transfinite representation of Ø as an axiom, the Euler relation appears to be a mapping from the infinite number range of the axiom to the edge of the unit circle in the complex plane or to the sum of a real zero-to-one (and minus one-to-zero) continuous interval and an imaginary zero-to-one (and minus one-to-zero) continuous interval. The transfinite ordinal β that has been used for the axiom symbol previously, has only been canonically defined as the ordinal of the discrete number sequence 1, 2, 3, 4,.... but it would appear to apply equally well to a continuous interval which is well ordered as indeed is the zero-to-one interval. These considerations are suggestive but not yet very enlightening. They might represent the arena for the display of (real)magnitudes for normalized physical variables and (imaginary)number magnitudes which can be associated normalized axiom numbers.

Another way in which this can be considered is as a countless periodic demonstration in time and space of the axiom $\emptyset$. In this view the periodicity of $e^{i\emptyset}$ would logically correspond to what we mean by law: each application of the law at different times and places but with the same experimental conditions should give the same experimental result (and thus the same real and imaginary values). See **Appendix B** and the discussion after Eq.(·'.2) for the association of space-time with the axiom number symbol.

The second part of the second question has the same answer as the first question and the reader is referred to the same references as in that answer. Those referenced discussions consider the wave function of both Quantum Mechanics and Electromagnetic Theory on the one hand, and similar functions in Special Relativity on the other hand, as transformations between situations or rest frames, respectively. These situations or rest frames exist at each end of a transformation which have axiom sets that are the negative or logical inverse of each other.[1, 2] An outline of these considerations is given in **Appendix B**.

APPENDIX A

The usual definitions are made for the symbols + and x where: + is the union of the sets, say, [A] and [B] that can stand on either side of + (in the logical sense it stands for an element in either [A] or [B] or both) and x is the intersection of the sets [A] and [B] standing on either side of x (logically it is an element in both [A] and [B] simultaneously). Boole adopted the designations[24, 25]:

$$1 = \text{Universe of Discourse set}$$

and: $\quad 0 = \text{The null set.}$

We adopt the restricted designations:

$\underline{1}$ = Universe of physical phenomena which are well defined variables and are represented by their symbolic designations, i.e., force, pressure, etc., and thus are based on the axioms of a theory but they are the deductive elements of the theory.

(A.1)

$\underline{0}$ = Absence of the above, or the deductive element null set.

Using these designations we set:

$$-\underline{0} = \underline{1} \quad \text{and} \quad -\underline{1} = \underline{0} \quad\quad (A.2)$$

where - is negation, and modus ponens has been used.

Now $[\underline{1} + \underline{0}]$ is the universe of well defined deductive and physical variables or their null set. This then includes all the theorems and deductions of the theory, but it does not include the axioms of the theory. The symbol $\underline{1}$ can be decomposed into the dual set, say, [A] and its complement [-A]:

$$[A] + [-A] = \underline{1}. \quad\quad (A.3)$$

In agreement with this discussion up to now we let:

$$-[\underline{1} + \underline{0}] = i[1' + 0'] = i[M' + -M' + 0'] \quad\quad (A.4)$$

where i designates the axiom(s) of the theory, the primes refer to axioms, and M' and -M' are the set of its axioms and their complement set, respectively. These sets can be combined logically in 2 different ways:

$$[M'] + [-M'] \quad \text{and} \quad [M'] \times [-M']. \tag{A.5}$$

The second expression has already been used to express undecidability and forced to the truth value true (or one) in the earlier discussion which defined the logical status of axioms. The first expression is somewhat ambiguous; it can be written two different ways and meaningfully equated to zero or to (1' + 0'), respectively:

$$M' - M' = 0 \text{ or rather } i0 \tag{A.6}$$

and
$$[M'] + [-M'] + 0' = 1' + 0' \text{ or rather } (1' + 0'). \tag{A.7}$$

Expression (A.6) is the assertion of an axiom set followed by its denial and results in the null axiom element; this is not useful. Expression (A.7) taking the value as given in (A.4) does, on the other hand, appear to useful because of its logical meaning. It means: either an axiom or its inverse, or both. This corresponds quite well to the previous discussions given here on these matters because axioms, their inverses or both are indeed the way they are considered in a Non Boolean field. This is where the Hegelian progressions of axiom thesis to antithesis to synthesis etc. goes on.

APPENDIX B

According to the foregoing, the set of all subsets to which the local axiom a applies is $e^{i\alpha}$ and the set of all subsets to which the local axiom $-\alpha$ applies is $e^{-i\alpha}$, where (-) is negation. This is first illustrated with the global axioms of the Special Theory of Relativity (STR) and then with QM.

In STR, the 2 basic global principles are:
1. The principle of relativity.
2. The constancy of the velocity of light.

Starting with 2 inertial frames, A and B, a set of statements can be made by observers A and B, the inhabitants of these frames, respectively. These statements are an example of each of the principles 1 and 2, above:

Illustrating Principle 1 (The Principle of Relativity):
Observer A says: Rest Frame A is at rest. Rest Frame B moves at a speed v.   (B.1)
Observer B says: Rest Frame B is at rest. Rest Frame A moves at a speed v.   (B.2)

It can be shown[1,2] that the statements made by Observer A in (B.1) can be mapped to an axiom symbol, say, α; whereas the statements made by Observer B in (B.2) can be mapped to the negation of this, $-\alpha$. Principle 1, the principle of relativity, is thus exemplified by the **simultaneous existence of the axiom set α and $-\alpha$ or both (B.1) and (B.2) simultaneously.**

Illustrating Principle 2 (The Constancy of the Velocity of Light):
Observer A says: The velocity of B is v. The velocity of light is c. The velocity of light with respect to B is $c \pm v$.   (B.3)
Observer B says: The velocity of A is v. The velocity of light is c. The velocity of light with respect to A is $c \pm v$.   (B.4)

It can likewise be shown that the statements made by Observer A in (B.3) can be mapped to a local axiom symbol, say, γ, whereas the statements made by Observer B in (B.4) can be mapped to the negation of this, $-\gamma$. Thus principle 2, The Constancy of the Velocity of Light is also exemplified by the **simultaneous existence of the axiom set γ and $-\gamma$ or both (B.3) and (B.4) simultaneously.** The local axioms α, γ and their negations, respectively, can be combined and replaced by θ and $-\theta$.

Let the rest frame A and its physical contents be designated with a general symbol x, and let the rest frame B and its physical contents be designated by a general symbol x'. One wishes to transform from x to x', this means that one wishes to transfer from x to x', to examine and treat physical phenomena in x', and also take along all laws that are completely internal to x which do not change for the x to x' transformation. In addition, however, one must take along the set of all subsets which will be unique to x', to which the combined axiom -θ, applies, thus:

$$x' = x\, e^{-i\theta} \tag{B.5}$$

If one then goes in the reverse direction from x' to x:

$$x = x'\, e^{i\theta} = x\, e^{-i\theta}\, e^{i\theta} \tag{B.6}$$

which becomes x (as it should).

If x, x' are the coordinates of the rest frames, then (B.5) and (B.6) are the Lorentz transformations. The imaginary exponentials above may be considered as the well known Lorentz transformation constants whose product must be equal to one. Now these above expressions carry a meaning corresponding to the different axiomatic bases for the respective rest frames.

For QM one examines the Psi functions and the electromagnetic wave functions, $e^{\pm Et/\hbar}$ and $e^{\pm i\omega t}$, respectively. The exponents are connected via the Planck relation, $E = \hbar\omega$. The exponents have t (time) as the independent variable. (For the alternate case where, say, x, the distance is the independent variable, then momentum p and wave number k (= $2\pi/\lambda$) can be represented instead of E and ω.)

It is evident that forms such as the final terms on the right of (B.6) do indeed occur in QM. The Born probabilistic interpretation of the Ψ function requires that Ψ Ψ* be used when the probability is normalized:

$$\int dP = \int \Psi(u)\, \Psi^*(u)\, du = 1 \tag{B.7}$$

here P is probability of existence of, say, the electron and u is a one dimensional spatial variable representation. Further, the QM canonical probable value of a physical variable, say, **A**, which can be a differential operator is:

$$\langle A \rangle = \int \Psi(u)\, \mathbf{A}\, \Psi^*(u)\, du . \tag{B.8}$$

Now Ψ must be of the form $e^{i\theta}$ in order to be a solution of the Schrodinger equation which does carry physical information on the electron. The motivation for (B.7, 8) is the need to have real values for probability and for physical variables. On the other hand from the above discussion of Eqs. (B.1) to (B.6) the meaning for the occurrence of an imaginary exponential and its conjugate is that it represents a transformation from a frame x to a frame x' and then back to x. In QM x and x' can represent the laboratory rest frame and the electromagnetic (photon or photex) rest frame which have been discussed elsewhere. Since measurements can only be made in laboratory frames this transformation sequence is always needed. [1, 2, 26]

REFERENCES

1. Honig, W.M., 'Godel Axiom Mappings in Special Relativity and QM-Electromagnetic Theory', Foundations of Physics, **6**, 37-57 (1976).
2. Honig, W.M., The Quantum and Beyond, (1986) Chapters Seven and Six, Philosophical Library, 200 W.57 St., N.Y., N.Y.,10019.
3. Piron, C., Prosperi, G.M., & Jauch, J.M., Foundations of Quantum Mechanics, Proceedings of the International School of Physics-Enrico Fermi, Academic Press, N.Y. (1971).

4. Kiss, S.A., Introduction to Algebraic Logic, Westport, Conn. USA (1961).
5. Sikorski, R., Boolean Algebras, Springer Verlag (1969).
6. Cohen, P.J., Set Theory & the Continuum Hypothesis, pp. 107-127, W.A. Benjamin (1966).
7. Honig, W.M., 'Logical Meanings in Quantum Mechanics for Axioms and for Imaginary and Transfinite Numbers and Exponentials', Proceedings of 1986 NATO Conference: Quantum Violations, Recent & Future Experiments and Interpretations, to be published as Quantum Uncertainties by Plenum Press, New York (1987).
8. Varela, F.J., Principles of Biological Autonomy, Elsevier-North Holland, N.Y. (1979).
9. Maturana, H.R., & Varela, F.J., Autopoesis and Cognition, D. Reidel Publishing, Holland (1980).
10. Honig, W.M., 'Transfinite Ordinals as Axiom Number Symbols for Unification of Quantum and Electromagnetic Wave Functions', Int. Jour. Theor. Phys., **15**, 87-90 (1977).
11. Honig, W.M., 'On the Logical Status of Axioms - As Applied in QM and STR', Bull. A.P.S., **31**, 844, Washington Meeting (1986).
12. Fraenckel, A., Abstract Set Theory, Ch. II and III & pp.202-209, North-Holland (1961).
13. Fraenckel, A., Abstract Set Theory, Ch. II and III & pp. 158-159, North-Holland (1966).
14. Davis, J. & Hersh, A., The Mathematical Experience, 45-50, 237-250, Penguin, London (1983).
15. Robinson, A., Non-Standard Analysis, Ch. 1, 2, 3, North Holland (1966).
16. Frege, G., Foundations of Arithmetic, translation and original reprint pp.30-60, Oxford, Blackwell (1950).
17. Sierpinski, W., Cardinal and Ordinal Numbers, Warsaw (1958).
18. Kamke, E., Theory of Sets, Ch. III, IV, Dover (1950).
19. Gonshor, H., An Introduction to the Theory of Surreal Numbers, p. 3, Cambridge University Press, 1986.
20. Dauber, J. W., 'C. S. Pierce's Philosophy of Infinite Sets', 233-247 and its references, in Vol. II of Mathematics, People, Problems, Results, by D. Campbell & J. Higgins, Wadsworth, 1984.
21. Buchler, J., The Philosophical Writings of Pierce, Dover, 1955.
22. Kennard, E., Kinetic Theory of Gases, pp. 367-372, McGraw-Hill (1949).
23. Shannon, C., Mathematical Theory of Communication, University of Illinois Press (1949).
24. Boole, G., The Laws of Thought, Dover, New York (Reprint, 1946).
25. Boole, G., The Mathematical Analysis of Logic, Philosophical Library, Inc., New York (Reprint, 1948).
26. Honig, W.M., 'Physical Models for Hidden Variables, Non-Local Particles and All That', in the Proceedings of the 1987 Gdansk Conference "Problems in Quantum Physics", to issue in 1988.

A CHANGE IN PARADIGM: A REALISTIC COPENHAGEN INTERPRETATION
(REALISM WITHOUT HIDDEN VARIABLES)

J. Horváth and M. Zágoni
Loránd Eötvös University
Department of Natural Philosophy

Quantum mechanics has basically two types of interpretation: an epistemological and an ontological one (1). The epistemological interpretation holds that the statements of quantum mechanics are related to our knowledge, henceforth the probabilistic character, the uncertainty relations and the reduction of the wave packet talk about the boundaries and incompleteness of our theory. On the other hand, the ontological interpretation says that the predicates of quantum mechanics are predicates about reality: uncertainty relations express the joint nonexistence of some physical quantities, quantum jumps show the real discontinuity of certain physical processes, and so on.

Consequently, according to the epistemological interpretation, quantum mechanics does not serve us a new ontology since the ontology of the microworld is perfectly given in the statistical mechanics of classical pointlike particles, while the ontological interpretation holds that the microentities have no particle-like hidden states (i.e., they are de facto not particles), so they obviously need a new metaphysics.

The attraction of the first view follows the fact that it treats the microworld in accordance with our traditional conception of reality (2). But just this is its main burden: in defending everyday realism it clashes with scientific realism, the main thesis of which is that in questions of existence and nonexistence it is science which has to decide (3).

I think that nowadays the choice between the two is not a question of taste but it is a question of experiment. It seems to me that from the EPR-Aspect experiments (4) one must conclude that the assumption of local hidden variables can be held only if there is some error either in the theoretical or in the practical base of the experiment. Until such an error is not pointed out I think the best what we can do is to keep ourselves to the experimental results.

But these results have some inevitable consequences:
A. There are no local hidden variables, neither contextual nor noncontextual (5). The two photons have no polarization before the measurement and they have only one component even at the moment of measurement — photons are not classical particles with rigid predetermination.
B. The two photons must be informed by each other at the moment of measurement, so this informational process cannot be anything else than an "action-at-an-instance" — that is, an "interaction-at-a-distance".

The nonexistence of hidden variables could seem paradoxical. For, if there are no hidden particle-like states (with definite position and momentum) then the microobjects are not pointlike; if they are not pointlike then they are extended (not only in the position-space but also in spaces of other observables); if they are extended then there has to be an underlying (subquantum) level where this extendedness appears: if there is such a lower level then there must be describing parameters belonging to it — parameters that are hidden from the quantum level. Contradiction.

The solution of the contradiction lies in the fact that the hidden variables of this second kind have quite a different function in the theory from the hidden variables of the first kind. Their task is not to determine the particle-like state of the microobject (sharp values for the conjugate observable-pairs), but
a. to determine the time-evolution of the extended (non-measured) state;
b. to determine the behaviour of the microentity in the process of measuring interaction.

With other words: these "hidden variables of the second kind" do not belong to the microobjects themselves, but to their nonlocal environment — to the whole Universe. These "hidden variables" cannot be "parameters" for any physical theory, since to know them (that is, to know the result of a forthcoming measurement) would mean to know everything in the world instantaneously — for these "variables" are not only sub-quantummechanical but also sub-relativistic: they work under the level of relativistic space-time (6). These nonlocal "hidden variables" mean nonlocal connections, contextual interactions of a new type.

At the moment when a measurement is made, the whole pyramid of determinacy is turning. The sub-quantummechanical determination of the microentity comes to be accompanied by a new, higher-level determination, coming from the measuring interaction. These two components of determination only together give the full (total) determination of the microentity; and only this total level of determination leads to the emergence of those new properties that are observed in the measurements. Without the measuring interaction the "degree of determinatedness" does not allow the microobject to have definite values of the physical quantities; the level of determinatedness in this case is enough only to determine "potentialities" or "possibilities" or "propensities" to have exact values in a would-be totally determined state - a state , which has a determination completed by the measuring interaction.

So we can say that the measuring interactions in general do not only <u>read</u> the physical quantity but they take part in the <u>creation</u> of it. At the moment of measurement there is a determinational "short-circuit" or "feed-back": while the <u>acting</u> side of the interaction <u>makes</u> (objectifies) the actual value of the physical quantity, the <u>reacting</u> side <u>reads</u> it - shows or manifests it for the environment, for anything or even anyone, who is sitting at the "other end" of the measuring interaction, does not matter whether it is an observer, a measuring device or an other microobject. The actual values of observable physical quantities are born together with the highest degree of determinatedness - neither the local nor the nonlocal interactions are able to guarrantee this maximum degree alone. Without a measuring interaction the microsystems are "open" systems from the point of view of determination: only the nonlocal and the measuring effects <u>together</u> are able to make them "closed". They have no hidden closed (totally determined) states (with determined physical qualities and quantities) which would be independent of contextual affections.

SUMMING UP:

1. There is a subquantum level, containing nonlocal interactions. But they do not <u>predetermine</u> the values of physical quantities - the microobjects have no hidden particle-like states with prefixed observables.
2. The EPR-paradox can be solved only by the assumption of nonseparability; but this does not mean , that there is a unity of the microobject and the measuring device, it means that there is a unity of the microobject and its whole environment. In this interpretation "measurement" does not mean any any "intrusion of subjectivism into physics", rather an objective part of the determining physical process. This is a fully realistic conception without inner contradiction.
3. The interpretation of de Broglie and others, which holds as basic postulates <u>causality</u> (the classical, internally predetermined kind) and locality (both relativistic **and** generalized) sharply contradicts the predictions of quantum mechanics. The assumption that EPR-correlations come from the previous collision of the two particles seems to be untenable (7). This interpretation has nothing to do with the Bell-inequalities.

4. The scheme of Bohm and Hiley concedes nonseparability and causality with nonlocal interactions, but as they assume predeterminatedness only by these nonlocal hidden variables, they do not treat the role of measuring interaction in a tenable way since they do not admit its function in the objectification (actualization) and determination of certain physical properties and quantities. Otherwise, subrelativistic (nonlocal) interactions - that neither carry nor vary energy and momentum - are allowed to be instantaneous by special relativity (8,9,10).

5. Velocities larger than that of light but finite contradict the Aspect-experiment and quantum mechanics. We can say that while the two particles are separated from each other on the level of <u>local</u> interactions, they form one and the same system (and keep unity and totality) on the lower, <u>nonlocal</u> interactional level. Below (or behind, or under) the level of local interactions the physical world really is an "<u>undivided whole</u>", an "unbroken totality"; but here is no <u>predeterminatedness</u> for the quantities of a future measurement, since these nonlocal interactions take into account everything in the determination of a microobject, except one: the actual value of a physical quantity in a would-be <u>local</u> (measuring) interaction.

REFERENCES

1. See for example Werner Heisenberg: "Quite generally there is no way of describing what happens between two consecutive observations. ... We leave it open, whether this warning is a statement about the way in which we should talk about atomic events or a statement about the events themselves, whether it refers to epistemology or to ontology."
    <u>Physics and Philosophy</u>, Harper & Row 1958, p. 48.

2. For a deterministic version of this view see e.g.: E.Bitsakis: <u>Physique et Materialisme</u>, Appendix, and for an indeterministic one: K.Popper: <u>Quantum Theory and the Schism in Physics</u>.

3. About the topic of scientific realism see G.Gutting's <u>Scientific Realism</u>, <u>in</u>: The Philosophy of Wilfrid Sellars, ed.: J.C.Pitt. 1976, D.Reidel.

4. A.Aspect, D.Dalibard and G.Roger: Phys.Rev.Lett.<u>49</u>, 1982, p. 1804.

5. A. Shimony: Contextual hidden variables theories and Bell's inequalities. British Journal for the Philosophy of Science 35, 1984, p. 25.

6. As Carl F.v.Weizsacker says: "Space-time is not the background but a surface aspect of reality." Quantum Theory and Space-time, in: Symposium on the Foundations of Modern Physics: 50 years of the EPR-Gedankenexperiment. World Scientific, 1985, p. 223.

7. Sharp and Shanks: Fine's prism model for quantum correlation statistics. Philosophy of Science 52, 1985, p. 538.

8. The quantum-logician Peter Mittelstaedt writes: "Quantum mechanics is compatible with Einstein's reality principle ... . Quantum mechanics contradicts the strong locality principle, which must be replaced by the weak principle. The weak principle of locality allows for an objectification at a distance and is thus seemingly in contradiction with special relativity. However, it follows that this objectification at a distance cannot be used for the transmission of superluminal signals, which would violate Einstein causality."
    EPR-paradox, quantum logic and relativity. In: Symposium on the Foundations of Modern Physics: 50 years of the EPR-Gedankenexperiment. World Scientific, 1985, p. 171.

9. Brent Mundy showed that: "... contrary to common belief, there is no incompatibility between special relativity and spacelike (faster-than-light) causation."
    Special relativity and quantum measurement. British Journal for the Philosophy of Science 37, 1986, p.207.

10. An argumentation in the same spirit:
    G. Nerlich: Special relativity is not based on causality. Brit.Journ. for the Phil. of Sci., 33, 1982, p.361.

PYTHIA AND TYCHE: AN ETERNAL GOLDEN BRAID

Lefteris M. Kirousis[*] and Paul Spirakis[+]
(*) Department of Mathematics, University of Patras, Greece
(+) Computer Technology Institute, Greece, and
    Courant Inst. of Mathematical Sciences, U.S.A.

ABSTRACT. There are two fundamental ways to introduce non-determinism in an algorithm. One is to provide an "oracle" that the algorithm can consult at certain stages of its execution in order to get answers for questions of a prespecified type. The second is to introduce randomness, i.e., to allow certain steps of the algorithm depend on the outcome of a random experiment. Both methods are essential in developing a theory of complexity for algorithms. In other words, both methods help in understanding the inherent difficulties associated with various types of problems and consequently, sometimes, they provide the insight necessary to obtain better deterministic algorithms (as we shall explain, random algorithms - as opposed to oracle computations - are on their own important in practice). In this paper, we explain the impact of these two forms of non-determinism on Complexity Theory and we attempt to highlight the intriguing relation between them.

Already in the 17th century, Leibniz envisaged a universal formal language that would be powerful enough to express anything in any branch of science and for which there would exist a decision algorithm, i.e., an algorithm that could distinguish the true from the false statements and would supply a proof for the true ones. Fortunately enough, Gödel [2] in 1931 showed that such a decision procedure does not exist even for languages whose expressive power would suffice to describe the properties of natural numbers. Thus, the danger of a Big Prover was shown to be nonexistent.

Since then, a big collection of well defined mathematical problems were proved to be undecidable. The most notorious of them is the Halting Problem. The input to this problem is a computer program together with its input data. The **problem is to decide** whether the program will eventually halt.

Again, as Turing [8] showed in 1936, there is no algorithm capable of solving all the instances of this problem.

However, besides the question of the decidability of a language (or problem) there is the question of how difficult is to solve a problem that we know to be decidable. The latter question has led to a fruitful theory -the Complexity Theory- with many important applications. Complexity Theory classifies languages according to the difficulty of their decision procedures. If the decision procedure of a language is simple enough so that we can decide whether a given sentence is true in a number of steps that is a polynomial expression of the length of the sentence, then the language is called efficiently decidable.

Just outside the borders of efficiently decidable languages lies the language of propositional calculus. This language cannot express statements referring to all or to a number of unspecified elements of a set (such a sentence is e.g., "All even natural numbers different from 2 can be written as a sum of two primes"). But apart from that, this language is powerful enough to express negations, conjunctions and disjunctions of statements already known to be expressible in the language. Formally, propositional calculus contains sentences that either belong to an arbitrary collection of atomic sentences, which are assumed to have no semantic content other than that they can be either true or false, or they are constructed from these atomic sentences by repeated applications of negation, conjunction and disjunction. The satisfiability problem (SAT for short) is to check whether a given sentence of propositional calculus can be made true by a suitable assignment of the values "true" or "false" to the atomic sentences.

It turns out that SAT has a considerable expressive power. For example, the statement that a given computer program on a given input will stop in a given number of steps can be expressed as an instance of the SAT problem. Moreover, there is an obvious algorithm that answers any instant of the SAT problem: simply examine all possible assignments of truth values to the atomic sentences of the given sentence. But as the old king who thought that he can afford to put two grains of wheat on the first square of a chess board, four on the second, eight on the third, etc, soon discovered, so it can be soon seen that such exhaustive search algorithms are essentially non-algorithms. Moreover, all algorithms that have been so far devised for the SAT problem require a "combinatorially explosive" number of steps. Consequently, people tend to believe that the SAT problem, like many others, is inherently difficult.

However, there is something algorithmically important about the SAT problem. Suppose that an oracle supplies us with a truth assignment of the atomic sentences, which is supposed to make the given sentence true. Then, quite unlike the oracles of Pythia, we can easily check if the case is as claimed to be by the oracle: All we have to do is carry out the computation on the oracle-supplied truth values that is dictated by the structure of the given sentence. This computation can be no longer than the length of the given sentence. Problems like that are called NP (for non-deterministic, polynomial) in the jargon of Theoretical Computer Science. Loosely speaking, an NP problem is a yes or no problem such that a yes instance of it can be efficiently (i.e. in a polynomial number of steps) verified to be such by an algorithm that is allowed to consult an oracle. Essentially, the algorithm verifies the validity of the oracle's claim. As the saying goes, the ability to solve a problem is one thing and the ability to check a proposed solution is another!

The satisfiability problem not only is in NP, but, more important, is as hard to solve as any other problem in that class. In other words, if an algorithm could be devised that would efficiently solve the SAT problem without oracle consultation, then all problems in NP would be efficiently solved (without oracle consultation). There is a large number of problems in NP other than SAT that share with SAT the property of being as hard as possible [1]. A famous one (known as the Travelling Salesman Problem) is given a number of cities and a natural number, decide whether there is a way to visit all cities exactly once, so that the total intercity mileage covered is less than the given number. These problems are called NP-complete. A central question in Theoretical Computer Science is whether an NP-complete problem can be efficiently solved without oracle consultation. The consensus among the researchers seems to be that the answer is no, but there are strong indications that a proof of this is either impossible or entails a revolutionary breakthrough in the related theory.

The NP-Completeness results proved in the early 70's showed that (unless such problems can indeed be efficiently solved, against all intuition) the great majority of the problems of combinatorial optimization that arise in commerce, science and engineering, are intractable: No method for their solution can completely escape the combinatorial explosion phenomenon. How, then, are we to cope with such problems in practice? Approximations and heuristics is an (often not satisfactory) answer. It seems that we must understand what (if anything) stands on the way from efficiently solvable things to intractable.

However there are problems in NP which most likely are not NP-complete. Such one is the problem of deciding whether a given natural number is composite (non-prime). The problem is obviously in NP, since an oracle can some up with two numbers which can be efficiently shown to constitute a factorization of the given number. With some number theory it can be shown that the complement of this problem (i.e., deciding whether a given number is prime) is also in NP. So, the primality problem can be efficiently solved, in both directions, but, alas, with the aid of Pythia. The remarkable thing though is that it can be efficiently solved with the aid of Tyche as well, who, after all, is more tangible.

But before explaining how probability theory can help in solving the primality problem, let us stress that this problem is not only theoretically interesting but has some very important practical applications as well. For example, the Rivest-Shamir-Adleman [7] Public Key Cryptosystem is based on the difficulty of factoring a number which is obtained by multiplying two very large primes ("Public Key" means that the code of encryptic a message in public). So, obtaining large primes with relative ease has become a central problem in Cryptography.

It turns out that the primality problem can be efficiently solved by a random algorithm. A random algorithm is one where at certain stages, the device performing the algorithm executes a random experiment (like tossing a coin) and decides its next step by looking at the outcome of the experiment. At first sight it may seem puzzling how that the element of randomness introduced may help. An intuitive account is given by Karp [4]: An algorithm is correct if it gives the right answer to any instance of a problem; if we think now of the instances as being chosen by an all-knowing adversary who tries to fool the performer of the algorithm, then it is reasonable to assume that a certain amount of unpredictability on the part of the performer of the algorithm would render the adversary incapable of choosing the right instance. Not only that, but even the effort of convinging somebody about the correctness of the algorithm may now become easier, since it can be the case that complex situations arise only with small probability.

The cornerstone of all random primality tests is the so called Fermat's little theorem: For any b, if b^p-b is not divisible by p, then p is not prime. Of course, Fermat's little theorem cannot constitute a primality test by itself. It is true that if given p we come up with a b such that b^p-b is not divisible by p, then p is composite. Nevertheless, it may be the case that for a certain b (in the range 1 to p) b^p-b is divisible by p, yet p is composite. Fortunately, such b's are in a certain sense a minority. Actually, for hundreds of years it was thought that if 2^p-2 is divisible by p, then p is prime. It was as late as in the 19th century that the first counterexample was found (the number $2^{341}-2$ is a multiple of 341=11x31). The fact now that such b's falsely verifying that p is a prime are a minority means that if we choose at random a b in the range 1 through p and b^p-b turns out to be a multiple of p, then with probability greater than 1/2, p is a prime. As computer scientists say, any composite number has many witnesses to its compositeness.

The essence now of the famous Solovay-Strassen [6] primality test is the following: Choose a large number of random numbers b in the interval 1 through p. If for all of them, b^p-b is a multiple of p, then with overwhelming probability p is prime. If on the other hand, for at least one such b, b^p-b is not divisible by p, then p is certainly composite. The fact that in the first case we know the primality of p almost certainly, follows from that for each choice of b we have less than 1/2 probability to falsely verify that p is prime. So, altogether, the probability to falsely verify that p is prime is less than $1/2^M$, where M is the number of random choices. Note that M can be very small, compared to p (if we had the patience to try all numbers between 1 and p then Tyche would not be needed).

Why not think about a number as a bag full of many balls? A prime bag has only white balls. A composite bag has white and red balls (the witnesses) in an even proportion. The only way to totally convince ourselves about a bag being prime is to look at half the number of balls - yet, a small number of randomly chosen balls would reveal the compositeness with high probability.

Observe that in random algorithms not only some steps depend on the outcome of a random experiment, but also the final answer of the algorithm, at least in one direction, is not absolutely certain. Of course it can be easily argued that certainty with a margin of mistake of less than $1/2^{100}$ is absolute certainly in any practical sense. (The probability of error due to hardware unreliability is bigger than $1/2^{100}$. Errors in human proofs may have much bigger chance of happening).

Observe also that randomized algorithms differ substantially from the so-called algorithms with random inputs. The latter are just deterministic techniques in which the inputs are assumed to come from some reasonable probability distribution, attempting neither to foil nor to help the algorithm. This approach was used to understand why heuristics work so well in practice (i.e. in an "average" input). For example, if the cities in a travelling salesman problem are drawn independently from the uniform distribution over the unit square and if the number of the cities is very large, then a very efficient technique by Karp almost surely produces a tour whose length is very close to the length of the optimal tour.

Of course, such results are meaningful only if the assumed probability distribution of problem instances bears some resemblance to the set of instances arising in "practice" or if the analysis works equally well for a wide range of distributions. In such algorithms, the essential characteristic of random choice as an algorithm step is missing.

To return now to NP-completeness, we have seen that between the two extremes of computationally easy problems and problems that are NP-complete lies another important class of problems which are probabilistically tractable. Actually, an oracle consulting algorithm can also be thought as a random algorithm: Simply, let the oracle be a (not necessarily evenly) random experiment. Unfortunately however, in that interpretation we do not get the correct answer with an overwhelming probability. We only know that we have a non-zero chance of getting the correct answer, but the odds may be overwhelmingly bad.

Another promising line of research is to show that a problem is complete in a complexity class like NP not with deterministic reductions, but with random ones. Loosely speaking, a problem in a certain complexity class, like NP, is called complete for this class if it is as hard as any problem in this class. Usually, the first problems that are shown to be complete in a class are archetypal or generic problems that express the computation of the type of machines which are characteristic of the complexity class. Then a non-generic problem can be shown to be complete by reducing the generic to it, i.e. by giving an efficient algorithm that transforms an instance of the generic problem to an instance of the problem under study. Then, any efficient solution for the non-generic problem would be translated to an efficient procedure for the archetypal problem. Once a non-generic problem is established to be complete, it can be utilize in place of the generic one in showing that other problems are complete as well. Showing that a problem is complete has as a corollary that this problem does not belong to any class which seems to be strictly lower (in the complexity hierarchy) than the class of the problem. So, in some sense, completeness results are negative results. U. Vazirani and V. Vazirani [9] have shown certain problems to be NP-complete by utilizing transformation algorithms that are probabilistic rather than deterministic. Of course, this is a somewhat weaker notion of completeness, but still is powerful enough to show that such a problem cannot belong to an easier complexity class. We [5] have lately shown that these ideas can be utilized in proving that certain problems are most likely non-parallelizable, i.e. although they can be efficiently solved by the classical, sequential von Neumann type of computers, no substantial savings in computation time is obtained by trying to solve them on machines which utilize a number of processors in parallel.

Finally, randomness shows up in another fascinating aspect of computation theory: showing that a theorem is correct without revealing anything about its proof. By theorem we mean any computational assertion, such as that a given sentence of propositional calculus is satisfiable or that for a given set of cities we can find a tour that would not exceed a given bound on intercity mileage.

Such proofs are interactive procedures between a prover (Alice) and a verifier (Bob). Alice is assumed to have unbounded computing power, while Bob can only perform efficient algorithms. Alice must convince Bob that she can solve any instance of a given "hard" problem without revealing to him anything about the solution. For an example (due to Goldreich, Micali and Wigderson [3]), consider the NP problem of deciding whether two graphs are isomorphic. Let G_1 and G_2 be two graphs that both Alice and Bob can see. Alice can find out whether G_1 and G_2 are isomorphic. If they are, Alice can produce an isomorphism and she must convince Bob, beyond the slightest doubt, that this is indeed the case, without revealing to him anything about the isomorphism. If on the other hand G_1 and G_2 are not isomorphic, then, even if Alice tries to cheat Bob by trying to make him believe that she knows an isomorphism, almost certainly Bob will conclude that G_1 and G_2 are not isomorphic.

The game is started by Alice who sends to Bob a graph H obtained by randomly rearranging the names of the nodes of G_1. Bob reacts by randomly choosing a graph G_i from G_1 and G_2 and sending it to Alice. Alice must now reply by sending to Bob a mapping that Bob must verify to be an isomorphism between H and G_i. That is an easy job for Alice if G_1 and G_2 are indeed isomorphic. If they are not, sooner or later, G_1 and G_i will turn out to be different and so Alice will not be able to send to Bob an isomorphism between H and G_i. Thus, Bob will conclude that G_1 is not isomorphic to G_2. If on the other hand Alice passes the test a large number of times by sending to Bob mappings that are isomorphisms between H and G_i, Bob can safely conclude that G_1 and G_2 are isomorphic.

Notice that Bob gets no information about the isomorphism of G_1 and G_2. At each iteration, he learns an isomorphism between G_i and a random "encryption" of G_1. This information reveals nothing about the isomorphism of G_1 and G_2 even if G_i is G_2. Notice also that the number of iterations of the game need not be very large compared with the combinatorially explosive number of rearranging the names of the nodes of a graph.

It seems therefore that Pythia, by exploiting the blindness of Tyche (and, of course, the capabilities of a computer), can convince anyone that she is good at her job without having to reveal any of her knowledge of the future.

This intriguing relation between the two forms of nondeterminism (oracle consultation and randomness) does not only have theoretical interest. Being able to testify (using a random algorithm) the truth of a statement for which the only known way to provide a proof is to consult an oracle that is only available to another party (an adversary) has some important practical consequences. For example, it may be used in situations where it is important to check whether an encrypted message is known to somebody, without having him reveal the message. It is also plausible that these ideas may be used in designing a verification procedure for disarmament which would not reveal vital information about each contry's defense to its adversary.

The work of the second author supported in part by the NSF Contract DCR 8503497 and by the Ministry of Industry, Energy and Technology of Greece.

REFERENCES

[1] M.R. Garey and D.S. Johnson: Computers and Intractability: A Guide to the Theory of NP-Completeness (Freeman, San Francisco, 1979).

[2] K. Gödel: 'Uber formal unentscheidbare Sätze der Principia Mathematica und verwandter Systeme I', Monatshefte für Mathematic und Physik, 38 (1931) 173-198. English translation in: J. van Heijenoort: From Frege to Gödel - a Sourcebook in Mathematical Logic 1879-1931 (Cambridge, Mass., 1967).

[3] O. Goldreich, S. Micali and A. Wigderson: "Proofs that yield nothing but their validity and a methodology of cryptographic protocol design (extended abstract)", Proc. 27th IEEE Symp. on Foundations of Computer Science (1986) 174-187.

[4] R.M. Karp: "Combinatorics, complexity and randomness", Turing Award Lecture, Communications of the Association for Computing Machinery 29 (1986) 98-111.

[5] L.M. Kirousis and P. Spirakis: "Probabilistic log-space reductions and problems probabilistically hard for P", Technical Report, Computer Technology Institute (1987, Patras).

[6] R. Solovay and V. Strassen: "A fast Monte Carlo test for primality", SIAM J. Comp. 6 (1977), 84-85.

[7] R. Rivest, A. Shamir and L. Adleman: "A method for obtaining digital signatures and public key cryptosystems", Communications of the Association for Computing Machinery 21 (1978), 120-126.

[8] A.M. Turing: "On computable numbers, with an application to the Entscheidungsproblem", Proc. London Math. Soc. ser. 2, vol. 42, (1936), 230-265.

[9] U.V. Vazirani and V.V. Vazirani: "A natural encoding scheme proved probabilistic polynomial complete", <u>Proc. 23rd IEEE Symp. on Foundations of Computer Science</u> (1982) 40-44.

QUANTA OF ACTION AND PROBABILITY

Ludwik KOSTRO
Institute of Experimental Physics
University of Gdańsk
Wita Stwosza 57
80-952 Gdańsk, Poland

ABSTRACT. It will be shown that apart from Planck's elementary quantum of action there are four additional types of quanta of action which are closely tied to the four fundamental interactions and it will be indicated that the, resented until now in Quantum Mechanics, necessity of only probabilistic predictions has no origin in the quanta of action as such but in the scattering processes typical of acts of measurement.

1. PLANCK'S ELEMENTARY QUANTUM OF ACTION AND PROBABILISTIC PREDICTIONS

The real physical world is quantitatively determined and therefore physics constructing the models of the physical phenomena is able to introduce physical quantities which can be measured. Among all physical quantities introduced by physics a particular part is played by the quantity called action which unifies in itself the dynamical quantities of physical phenomena with their spatio-temporal ones

Action = force x path x time
Action = energy x time
Action = momentum x path
Action = angular momentum x angular displacement

Several physical quantities are related to each other Some relations between them are always constant. In the models of the physical phenomena constructed by physics a particular part is played by the so-called universal constants. Planck's constant h constitutes such a constant. Planck discovered, in 1899, that, in the microcosm, the physical quantity called action is quantized. Planck's constant h represents the elementary quantum of action which became the basis of quantisation of other quantities and thus the foundation of Quantum Mechanics.

The existence of Planck's elementary quantum of action manifests itself, for example, in the acts of measurement in which the position of a particle is measured with the help of a photon scattered on the particle. During such a scattering process a portion of action equal to h is always performed

$$F \Delta l \Delta t = h$$

where F is the force with which the photon acts on the particle, Δl is the path along which it acts and Δt is the duration of its action. As we can see the influence of the photon on the particle is not a point event. It is performed always along a certain path and during a certain time interval. Since the product $F \Delta l \Delta t$ is always constant therefore the less the force acting on the particle the longer the space-time interval ($\Delta l \Delta t$) of its action and conversely the greater the acting force the shorter the space-time interval.

Note that in this scattering process not only the dynamical quantities (energy and momentum) of the particle are perturbed but also its spatio-temporal ones (i.e. its position). Let us consider, for example, a particle which before the influence of the photon was at rest (i.e. its kinetic energy and momentum were equal to zero). Although, before the act of measurement, we did not know where the particle was, we knew, however, that it did not change its position. Under the influence of the photon the particle begins not only to acquire energy and momentum but also to change its position because it is accelerated. The less the accelerating force the longer the space-time interval of acceleration and therefore the greater is the perturbation of the position under consideration.

After the act of measurement we do not know either the exact spatio-temporal position or the axact value of energy and momentum of the particle. Our knowledge of the particle's mechanical state is endowed with an uncertainty. The greater the uncertainty of our knowledge of the dynamical quantities of the particle the less the uncertainty of our knowledge of its spatio-temporal position and vice versa.

Since in the considered case a total absorption of the photon is impossible and since the photon is characterized by the constant relations $ET = p\lambda = h$ therefore it is clear that the energy and momentum acquired by the particle cannot be either equal to or greater than the energy and momentum of the photon.

$$\Delta E < E_q = h/T$$
$$\Delta p < p_q = h/\lambda$$

Analogically it is clear that $\Delta l > \lambda$ and $\Delta t > T$

A simple mathematical operation leads us to the famous Heisenberg uncertainty relations for "free" particles 1).

$$\Delta p \Delta l \geqslant h$$

$$\Delta E \Delta t \geqslant h$$

(Note that there are also uncertainty relations for "bounded up" particles 1) (e.g. for the electrons bounded up with an atom)

$$\Delta p \Delta l \geqslant h/2\pi = \hbar$$

$$\Delta E \Delta t \geqslant h/2\pi = \hbar$$

and

$$\Delta p \Delta l \geqslant h/4\pi = (1/2)\hbar$$

$$\Delta E \Delta t \geqslant h/4\pi = (1/2)\hbar$$

when we take into consideration the "average uncertainties" (see also ref. 1)).

Since our knowledge concerning the particle's mechanical state is endowed with uncertainty we must limit ourselves, if we want to follow the motion of the particle, only to probabilistic predictions. It is often maintained that the main reason of the probabilistic nature of predictions in Quantum Mechanics is the existence of the elementary quantum of action h. It must be, however, emphasized that the elementary quantum of action as such cannot be the reason of probabilistic predictions because the elementary quantum of action as such expresses constant experimentally precisely determined relations between dynamical quantities and spatio-temporal ones of certain physical phenomena in microcosm. Therefore also, first and foremost, we are able to get to know the exact value of h. There are several experimental methods, used in physics, of settling the exact value of h.

The main reason of the probabilistic nature of predictions in Quantum Mechanics has to be seen in the scatter effects connected with the acts of measurement in which the constant h is only involved making the scatter effects, in certain sense, inevitable. Heisenberg's uncertainty relations are scatter relations as it has been maintained with reason by K. Popper since 1934.[2] In the considered above case we are dealing with an immediate action of a particle (photon) on another particle. Let us consider now the action in the case of two distant interacting particles.

2. QUANTA OF ACTION OF THE FOUR FUNDAMENTAL INTERACTIONS

In the mathematical formalism of the elementary particle

physics there are present and working in a latent form four types of constants which have the dimensions of action and are closely tied to the four fundamental interactions.

(1) Elementary quantum of action of the electromagnetic interactions (EMIs). Let us begin with the introduction of the constant whose existence has been already indicated by A. Einstein 3), E. Schrödinger 4) and A. Eddington 5). Here it is

$$h_e = Ke^2/c$$

where $K = 1/4\pi\varepsilon_0$ is the constant of the Coulomb Law, e - the constant elementary charge of the EMIs, c - the constant speed of light in vacuum.

(2) Quanta of action connected with the gravitational interactions (GIs)

- for two interacting particles of the same mass m_a

$$h_{G_a} = Gm_a^2/c$$

- for two particles of different masses m_a and m_b

$$h_{G_{ab}} = Gm_a m_b/c = (h_{G_a} h_{G_b})^{1/2}$$

where G is the gravitational constant and a and b are general symbols which serve to indicate the type of a particle. Note that the quanta of action connected with the GIs are as numerous as the types of elementary particles.

(3) Quanta of action connected with the strong interactions (SIs)

$$h_{s_{AB_a}} = g_{s_{AB_a}}^2/c = g_{s_{A_a}} g_{s_{B_a}}/c$$

where $g_{s_{A_a}}$ and $g_{s_{B_a}}$ are the charges of the strongly interacting particles, A and B general symbols denoting the type of the baryons and a is the symbol denoting the type of the virtual meson carring the strong interaction.

(4) Quanta of action connected with the weak interactions

$$h_{w_{AB_a}} = g_{w_{AB_a}}^2/c = g_{w_{A_a}} g_{w_{B_a}}/c$$

where $g_{w_{A_a}}$ and $g_{w_{B_a}}$ are the charges of the weakly interacting particles, A and B general symbols denoting the type of the particles interacting weakly and a is the symbol denoting the type of the virtual boson carring the weak interactions (WIs).

2.1. Coupling Constants of the Four Fundamental Interactions and the Constants h_e, h_G, h_s, h_w

Two interacting particles are coupled by their interaction. The coupling power (i.e. the intensity of their interaction) is expressed quantitatively, in the present-day elementary particle physics, by dimensionless coupling constants. In these coupling constants there are present and working as main components the four types of constants h_e, h_G, h_s, h_w.

Therefore <u>in the case of the EMIs</u>

$$\alpha_e = Ke^2/\hbar c = (Ke^2/c)/\hbar = (h_e/\hbar)$$

<u>in the case of the GIs</u>
- for two interacting particles of the same mass m_a

$$\alpha_G = Gm_a^2/\hbar c = (Gm_a^2/c)/\hbar = (h_{G_a}/\hbar)$$

- for two interacting particles of different mass

$$\alpha_{G_{ab}} = Gm_a m_b/\hbar c = (Gm_a m_b/c)/\hbar = (h_{G_{ab}}/\hbar) = (h_{G_a} h_{G_b})^{1/2}/\hbar$$

<u>in the case of the SIs</u>

$$\alpha_{s_{AB_a}} = g^2_{s_{AB_a}}/\hbar c = (g^2_{s_{AB_a}}/c)/\hbar = (h_{s_{AB_a}}/\hbar)$$

<u>in the case of the WIs</u>

$$\alpha_{w_{AB_a}} = g^2_{w_{AB_a}}/\hbar c = (g^2_{w_{AB_a}}/c)/\hbar = (h_{w_{AB_a}}/\hbar)$$

As we could see the dimensionless coupling constants are not elementary but rather are ratios of two more elementary constants. Since the denominator of these ratios is the same for all the four types of coupling constants therefore it is clear that the coupling power of the four fundamental interactions depends only on the introduced constants. We can even say that the four constants express quantitatively the coupling power of the four types of interactions.

2.2. The Physical Meaning of the Constant h_e

Let us consider the physical meaning of the constant h_e first in the frame-work of the classical theories.

(1) In order to do this we examine the Coulomb part of the EMIs between two charged particles (e.g. between a proton and an electron). The Coulomb part manifests itself particularly in the limiting case when the two particles are at rest with respect to each other i.e. in the

electrostatic case. We can use then the Coulomb law
$$F_e = Ke_1e_2/R^2 \tag{1}$$
If we take into account now the finite velocity c of propagation of the interaction then
$$R = cT \tag{2}$$
where T is the time needed to transmit the interaction from one particle to the other one. Introducing (2) into (1) and making a simple transformation we obtain
$$F_e RT = Ke_1e_2/c = Ke^2/c = h_e$$
As we see the product $F_e RT$ is always constant and equal to h_e.

(2) Let us now consider the static case of interaction between an electron and a nucleus containing more than one proton. In such a case the product $F_e RT$ is equal to an integer multiple of the constant h_e.
$$F_e RT = KZe_1e_2/c = KZe^2/c = Zh_e$$
where $Z = 1, 2, 3...$ is the number of protons in the nucleus.

(3) If we consider the case of interaction between two nuclei deprived of electrons then the product $F_e RT$ remains also equal to an integer multiple of the constant h_e
$$F_e RT = KZ_1e_1Z_2e_2/c = Z_1Z_2h_e \tag{3}$$
Since the energy of interaction is given by
$$E_e = \pm KZ_1e_1Z_2e_2/R = F_e R$$
it follows that the relation (3) can be written also as:
$$E_e T = Z_1Z_2 h_e$$
As we can see, the product of the energy of interaction E_e and of the time T needed to transmit the interaction is always constant and equal to h_e or its integer multiple. This fact has been already indicated by A. Eddington [5].

Note that the coupling power of the EMIs is quantized and equal to the ratio $(h_e/\hbar)$ or to its integer multiple.
$$\alpha_{e_{Z_1Z_2}} = (F_e RT/\hbar) = (E_e T/\hbar) = Z_1Z_2(h_e/\hbar)$$
It was sufficient to study the Coulomb part of the

EMIs to understand how the coupling power of the EMIs between two charged microobjects depends on the constant h_e. Since the denominator of the ratio ($h_e/ℏ$) is constant and the same for all kinds of interaction therefore the coupling power of the EMIs depends quantitatively only on the constant h_e or on its integer multiple.

Since in the EMIs we are dealing with an action between two distant charged microobjects and since the constant h_e has the dimensions of action I propose to call it "elementary quantum of electromagnetic action at a distance". Where action at a distance is conceived as action between two distant charged microobjects through a field with finite speed of propagation of this action. Hunter and Wadlinger consider also the constant h_e as quantum (or wad) of action at a distance [6].

2.2.1. Virtual photons carring the EMIs and the constant h_e

In the present-day elementary particle physics the EMIs are explained by means of an exchange of virtual photons (VPhs) carring the Coulomb force. The expression "electric charge of a particle" means there the ability of a particle to emit and absorb VPhs. The VPhs are photons bounded up with the particles. They are not free photons having an energy equal to $E = h\nu$. The VPhs are characterized by the same physical quantities as the free photons except the energy. A particle charged with VPhs is able to emit and reabsorb them (self-interaction) and to exchange them with another charged particle (interaction between particles).

Let me now assume consistent with my 1978 proposal [7-9] of a three-wave model of the elementary particle that the corpuscular aspect of a particle (real or virtual) is identified with the spherical L. Mackinnon soliton [10] whose diameter is equal to the particle's Compton wavelength in the case of a rest mass particle or to one wavelength in the case of a particle of zero rest mass. According to the proposed model, an electromagnetically charged particle is charged with virtual spherical photonic solitons the diameter of which are equal to all possible wavelengths. This assumption serves to explain the long-range nature of the electromagnetic interactions.

In the proposed model it is assumed also that an EMI can be performed only by an exchange of a VPh whose diameter (one wavelength) is equal to the distance between the two interacting particles and whose period is equal to the time needed to transmit the interaction. The greater the distance between the interacting particles the weaker the interaction because the longer the wavelength of the exchanged VPh, the lower its frequency. Note that the energy of interaction is given then by

$$E_e = \pm h_e \nu$$

because if $R = \lambda_C = cT_C = c/\nu$ then

$$E_e = \pm Ke_1e_2/R = \pm Ke^2/\lambda_C = \pm Ke^2/cT_C = \pm (Ke^2/c)/T_C = \pm h_e\nu$$

Note also that the electromagnetic charge is given by

$$g_e = (K)^{1/2}e = (h_e\lambda\nu)^{1/2} = (h_e c)^{1/2}$$

We are dealing here with a long-range charge because the electromagnetically charged particle is charged with VPhs of all possible wavelengths and frequencies $(h_e\lambda\nu)^{1/2}$.

2.3. The Physical Meaning of the constants h_{G_a}

Looking for the physical meaning of the constants hG_a (first in the frame-work of the classical theories) let us use the Newton Law of gravitation. The Newton Law is here a good approximation because the GIs between the elementary particles are very weak. Let us, however, assume that the GIs propagate at velocity c as it is assumed in General Relativity.

(1) The Newton Law in the case of two particles of the same mass m_a is given by

$$F_G = Gm_a^2/R^2$$

Since we have assumed that R=cT therefore we obtain finally

$$F_G RT = Gm_a^2/c = h_{G_a} \qquad (4)$$

The product $F_G RT$ is constant for all possible distances and equal to hG_a. Since $E_G = -Gm_a^2/R = F_G R$ therefore (4) can be written

$$E_G T = h_{G_a}$$

As we see the product of the energy of a GI and of the time needed to transmit it is constant and equal to hG_a. Note that taking into account the relation (4) the mass of a particle is given by

$$m_a = (h_{G_a} c/G)^{1/2}$$

Since, in the considered case, one is dealing with an interaction between two distant particles and since the constant h_G has the dimensions of action I propose to call it "elementary quantum of gravitational action at a distance".

(2) In the case of a GI between two particles having different masses one obtain

$$F_G RT = Gm_a m_b/c = h_{G_{ab}}$$

Since $m_a = (h_{G_a} c/G)^{1/2}$ and $m_b = (h_{G_b} c/G)^{1/2}$ therefore

$$F_G RT = Gm_a m_b/c = (G/c)\left[(h_{G_a} c/G)(h_{G_b} c/G)\right]^{1/2} = (h_{G_a} h_{G_b})^{1/2}$$

As we can see $h_{G_{ab}} = (h_{G_a} h_{G_b})^{1/2}$

2.3.1. Virtual gravitons carring the GIs and the quanta h_{G_a}

According to the model proposed in this paper, every particle is charged with virtual spherical gravitonic solitons (virtual gravitons VGs) whose diameters are equal to all possible wavelengths of the rest massless gravitons. In such a way the long range nature of the GIs can be explained.

According to the proposed model, the GI between two particles will be carried only by a VG whose wavelength (diameter of the spherical soliton) is equal to the distance between the interacting particles and whose period is equal to the time needed to transmit the interaction. The greater the distance between the interacting particles the weaker the GI between them because the longer the wavelength, the lower the frequency of the VG transmiting the interaction. The energy of the GI carried by the VG will be given then

$$E_G = -h_{G_a} \nu$$

Where ν is the frequency of the VG able to transmit the interaction. Note that the gravitational charge is given then by

$$g_G = (G)^{1/2} m_a = (h_{G_a} \lambda \nu)^{1/2} = (h_{G_a} c)^{1/2}$$

We are dealing here with a long-range charge because the gravitationally charged particle is charged with VGs of all possible wavelengths and frequencies $(h_{G_a} \lambda \nu)^{1/2}$.

2.4. The Physical Meaning of the Constants $h_{S_{AB_a}}$

As is well known the hadrons divided into mesons and baryons interact strongly. In the present-day elementary particle physics the SIs between baryons which are fermions are interpreted as transmited by virtual mesons VMs which are rest mass bosons. The strongly interacting baryons are charged with VMs. There are several types of mesons transmiting the SIs (e.g. pions $\pi\pm, \pi^0$, mesons ρ, ω etc.). This fact causes there to be several types of quanta h_S. They depend on the type of the interacting particles and on the type of the intermediating mesons. Thus e.g. we have

$$h_{S_{NN\pi}} = g^2_{S_{NN\pi}}/c$$

when the interacting baryons are nucleons and the intermediating VM is the pion π.

According to the proposed model, the baryons are charged with virtual spherical mesonic solitons the diameters of which are equal to their Compton wavelengths which depend on the proper mass of the mesons. This fact explains the short range nature of the SIs. We are dealing with a SI between two baryons when the distance between them is of the range of the Compton wavelength of the intermediating VM. Note that the distance between two baryons is equal to the Compton wavelength of the intermediating meson when the spherical solitonic VMs with which the interacting baryons are charged "touch each other". In such a case the two baryons interact and the energy of interaction is given by

$$E_s = \pm h_{s_{AB_a}} \nu_{C_a}$$

Note also that the charge of the strongly interacting particles is given by

$$g_{s_{A_a}} = (h_{s_{A_a}} \lambda_{C_a} \nu_{C_a})^{1/2} = (h_{s_{A_a}} c)^{1/2}$$

where λ_{C_a} is the Compton wavelength and ν_{C_a} is the Compton frequency of the intermediating VM. We are dealing here with a short-range charge.

2.5. The Physical Meaning of the Constants $h_{w_{AB_a}}$

The recently discovered (first theoretically and subsequently experimentally) bosons W^+, W^- and Z^0 are the carriers of the WIs. The WIs are responsible for numerous decays and also for several scattering phenomena in which two particles interact e.g. for the experimentally checked scattering process $\nu_\mu + e \rightarrow \nu_\mu + e$ in which the neutral boson Z^0 is the carrier of the WI. The constants h_W depend on the type of the interacting particles and on the type of the bosons which are the carriers of the WIs. And therefore the constant h_W which intervenes in the mentioned scattering process is given by

$$h_{w_{\nu_\mu e_{Z^0}}} = g^2_{w_{\nu_\mu e_{Z^0}}} /c$$

According to the proposed model, the weakly interacting particles are charged with virtual spherical solitonic bosons $W^\pm$ and Z^0 the diameters of which are equal to their Compton wavelengths. Since the proper masses of the bosons ($W^\pm$, Z^0) are remarkably greater than those of the particles charged with them therefore the Compton wavelengths of the bosons ($W^\pm$, Z^0) are remarkably shorter than those of the particles charged with them. This fact explain on the one hand several decays of unstable particles and on the other hand the very short range of the WIs. We are dealing with a WI when the distance between the interacting particles is of the range of the Compton wavelength of the intermediating boson. The energy of interaction is given then by

$$E_w = h_{wAB_a} \nu_{C_a}$$

Note that the short range charge of the WIs is given then by

$$g_{w_{A_a}} = (h_{w_{A_a}} \lambda_C \nu_C)^{1/2} = (h_{w_{A_a}} c)^{1/2}$$

where λ_C is the Compton wavelength and ν_C the Compton frequency of the intermediating boson of the WIs.

3. THE FOUR ADDITIONAL QUANTA OF ACTION AND PROBABILITY

Concluding this paper it must be emphasized that the quanta of action connected with the four fundamental interactions cannot be the reason for the use of the probability calculus because these quanta as such express constant precisely determined relations e.g. in the case of the EMIs

$$F_e \lambda T = E_e T = p_e \lambda = h_e$$

where F_e, E_e, p_e are respectively the carried by the VPh Coulomb force, energy and momentum and λ and T are respectively the wavelength and the period of this VPh. The constants h_e, h_G, h_s, h_w can be, however, involved in several processes in the description of which the probability calculus has to be used.

REFERENCES

1) W. Heisenberg, Die Physikalischen Prinzipien der Quantentheorie, Bibliographisches Institut Mannheim 1958.

2) K.R. Popper, 'Realism and Quantum Theory' in: Determinism in Physics, Gutenberg Publ. Co. Athens 1985.

3) A. Einstein, 'Entwicklung unserer Anschauungen über das Wesen und die Konstitution der Strahlung', Phys. Zeitsch., 10 (1909) 817.

4) E. Schrödinger, 'Eine bemerkenswerte Eigenschaft der Quantenbahnen eines einzelnen Elektrons', Z. Phys., 12 (1923) 13.

5) A. Eddington, New Pathways in Science, Cambridge Univ. Press 1935 p.235.

6) G. Hunter and R.L. Wadlinger,'Quantized Action-at-a Distance, An Action smaller then h,'(1987) unpublished paper.

7) L. Kostro, 'A Three-wave Model of the Elementary Particle', Phys. Lett.,107 A (1985) 429

8) L. Kostro, 'Planck's constant and the three waves of Einstein's covariant ether,' Phys. Lett., 112 A (1985) 283.

9) L. Kostro,'Einstein's conception of the ether and its up-to-date applications in the relativistic wave mechanics', in: Quantum Uncertaintes, Plenum Press, New York 1987 pp. 435-449.

10) L. Mackinnon,'A nondispersive de Broglie Wave Packet', Found. Phys., 8 (1978) 157.

COMMENTS ON POPPER'S INTERPRETATIONS OF PROBABILITY

M.Rédei and P.Szegedi
Loránd Eötvös University
H-1088 Budapest, Rákóczi út 5.
Hungary

ABSTRACT. The aim of this talk is to comment, mainly from the point of view of the two typical probabilistic physical theories, classical statistical mechanics /CSM/ and quantum mechanics /QM/, on Popper's interpretations of probability. Based on recalling some recent developments both in CSM and in QM we will try to support our main these which is the following: Neither the frequency nor the propensity interpretation of probability alone seem to be able to reflect satisfactorily the various contents of probabilities in CSM and QM. In particular arguments can be given against the adequacy either of the frequency or of the propensity interpretation in CSM, and elaboration of some probability interpretation which is able to describe the hierarchy of probabilities in QM seems to be necessary.

1. CLASSIFICATION FOR INTERPRETATIONS OF PROBABILITY

According to Popper, the aim of a probability interpretation is to interpret probability statements such as "the /relative/ probability of a given b is equal to the real number r", formally, $p(a,b) = r$. That the aim of probability interpretations is to interpret statements of this form offers a natural classification of the interpretations. They can be grouped according to the types of the entities a is assumed in the interpretation to refer to, and they can also be classified based on what is supposed to determine the validity /truth/ of the statement "$p(a,b) = r$". So it is meaningful to speak of psychological interpretations, in which a refers to a mental or emotional state /belief, expectation/ or to any human state characterizable in psychological terms; the interpretation in which a is considered to be knowledge will be called epistemological interpretation; another possibility is that a stands for /logically treated/ propositions, in which case the interpretation may properly be called logical. An important type is where a refers to objects /not necessarily things, but events for instance/ existing independently of any human state, efforts etc.. We call such an interpretation object interpretation. In the other direction of classification one can sharply distinguish two types of interpretations: Those in which the validity of probability statements is made dependent on the subject asserting the statement are called subjective interpretations, whereas those in which the correctness or incorrectness of $p(a) = r$ is considered to be independent of anyone's knowledge, belief etc. about the correctness will be called objective interpretations.

The two "variables" in terms of which Popper also distinguishes between the probability interpretations are the same /namely the types of the entities a refers to and the truth conditions of the probability statements/ as those leading to the above classification. Somewhat surprisingly, however, the classification we have given does not appear in Popper's works. Popper's texts seem to give an impression that psychological interpretations are always subjective and that subjective interpretations are always psychological, which is not necessarily the case, furthermore, as it is well known, Popper identifies the frequency /and the propensity/ interpretations with the objective interpretation, which is theoretically unjustified and which also seems to be very questionable historically in the light of the well known positivistic attitude of Founding Fathers /von Mises, Reichenbach/. Undoubtedly, both the frequency and the propensity interpretations, as understood by Popper, are objective object interpretations, but, obviously, neither an objective nor an object interpretation has to be frequency or propensity interpretation. All this is not merely a terminological problem: later on our arguing in favor of an interpretation of CSMcal probabilities, which is different from either the frequency or of the propensity interpretation is not to be understood as arguing against the objective interpretations in favor of a subjective interpretation.

2. THE FREQUENCY INTERPRETATION

Popper based his frequency interpretation on Mises's theory, which was a theory of such chance-like /virtually infinite/ sequences S of occurances that satisfy two axioms:1) Axiom of convergence /limit axiom/: the frequency of any event in the n-long subsequence formed by the first n elements of S approaches to a definite number as n tends to infinity; 2) Axiom of randomness /principle of ecxluded gambling system/: the limit ensured by the limit axiom is the same within any /infinite/ subsequence of S. For von Mises "probability" is but another term for this "limit of relative frequencies in a given S". The limit axiom ensures the meaningfulness of this notion of probability but Popper considered desirable to prove the property expressed by the limit axiom for certain sequences rather than to postulate it. An assesment of his success in doing that by constructing special mathematical chance-like sequences in which the limit axiom holds by construction [Popper 1959a], can not be attempted here. We only wish to point out the following basic problem arising in Popper's philosophy concerning the probability statements in physics: if, according to the ferquency interpretation, probability statements in physics are to be tested by relative frequencies computed from an empirical, thus always finite, sequence F then physical probability stetements are not falsifiable, simply because for any probability statement "$p(a)=r$" one can give an infinite series S of occurances such that S contains as its first elements the sequence F and the relative frequency of a in S approaches r [Popper 1959a]chap.65-66. The falsifiability being for Popper the main criteria for a statement to be scientific one must either conclude that physics is not a science or it is unclear how to understand the probability statements in physics via the frequency interpretation. Popper tries to solve this problem /[Popper 1959a] chap.68/ but, in our opinion, his solution is very problematic: he admits that in strict logical sense probability statements are not falsifiable in physics but he considers the physicist's practice in "falsifying" certain probability statements justified by the following reasoning: suppose the physicist produces an empirical sequence of physical events S in order to check the correctness of a probability statement "$p(a)=r$" and he finds the relative frequency of a in S to deviate significantly from the value r. Despite this deviation the physicist does not regard "$p(a)=r$" incorrect, says Popper, unless he can produce other empirical sequences of physical events S',S'',S''' etc., in which he finds similarly significant deviations, and, in this case, the physicist rightly considers "$p(a)=r$" falsified. On the other hand, if the physicist finds the relative frequency of a in S', S'' etc. nearly r then, for him, the occurances forming S "...would not be physical effects, because, on account of their immense improbability, they are not reproducible at will."[Popper 1959a]p.203. In other words "The rule that extreme improbabilities have to be neglected agrees with the demand for scientific objectivity."[Popper 1959a]p.202. The problem with this "solution" is first that it is clearly inconsistent: S can not be declared unphysical for it consists of genuine physical effects by definition. Secondly, improbable events are not necessarily unreproducable: think of the diffraction pattern on a photographic plate in the quantum two slit experiment: the dark lines represent areas where the arrival of a particle is extremely improbable. This example also shows that extreme improbabilities can not be neglected in general, which also implies that neglecting extreme improbabilites has nothing to do with scientific objectivity. In short, we think that Popper did not solve the above problem.

In 1938 Popper made two changes in his theory of probablity: he introduced a formal axiomatic calculus for probablity and simplified his frequency theory by analyzing the randomness in finite sequences and postponing the reference to infinite ones as long as possible./[Popper 1959a], starred footnotes and appendices./

3. THE PROPENSITY INTERPRETATION.

In 1953 Popper gave up the frequency interpretation since he found it unsuitable for resolving the interpretational problems of QM. The propensity interpretation emerged through a critique of both the subjective and the frequency interpretation and was published in several forms, which differ somewhat in what properties of the propensity interpretation are emphasized [Popper 1957a,b;1959b;1982a,b,c]. The frequency and the propensity interpretations are closely related. In a sense the propensity interpretation itself is a frequency interpretation if the latter one is taken together with the explicit stating and acceptance of a hidden assumption inherent in it. The hidden assumption of the frequency interpretation is that only those reference classes are admissible that are results of trials that can be considered as repetitions. Popper concludes that the probability of an event is rather a property of the whole generating conditions than that of the event itself, and thereby it is also meaningful to speak of the probability of a singular event. The probability of a singular event thus measures the strength of the tendency or disposition of these generating conditions to realize the event as something which is possible with respect to these generating conditions. Thus the propensity interpretation considers the probability as a numerical expression of strength of the tendency to realize a particular possibility. This numerical value can be tested by computing the relative frequency of the occurance of the event in question in repeated experiments. In Popper's view the propensity is a kind of indeterministic force that has the ability to cause random series of events, so he identifies the propensity interpretation with indeterminism.[Popper 1982 II]. This is very problematic since, as we recalled above, testing a probability statement in the propensity interpretation requires the reproduction of the same generating conditions. Obviously, however, no condition can be reproduced without some modification for if it could it would not be distinguishable from its previous existence i.e. it would not be a repetition. This means that the generating conditions are not the same in strict ontic sense and so to infer indeterminism from the propensity contradicts the very formulation of the propensity interpretation.

4. POPPER'S ANALYSIS OF CLASSICAL STATISTICAL MECHANICAL AND QUANTUM MECHANICAL PROBABILITIES.

Popper's main goal with his interpreting the probabilities in CSM and QM via the frequency and propensity interpretation is to criticize the subjectivism, which in his opinion invaded physics through the subjective interpretation of probability, according to which "$p(a,b)=r$" in physics are to be interpreted as "measure of our lack of knowledge given the knowledge or information b" or "measure of our ignorance".

4.1. Interpretations of classical statistical mechanical probabilities

Popper's main arguments against this interpretation of CSMcal probabilities seem to be the followings: A) The subjective interpretation believes itself being supported by the law of large numbers, which seems to serve as a bridge leading from subjectively interpreted probabilities to objective frequences. But this belief is unjustified: there is no way from subjective probabilities to objective frequences. The usual "derivation" is a grave logical blunder with a conceptual shift at some stage or other in the derivation, when the non-statistical meaning of the symbols is dropped and tacitly replaced by a statistical one. For instance a probability approaching 1 is interpreted as "almost certain" in the sense of "almost always to happen", instead of the sense "very strongly believed in". [Popper 1982 III]p.66-67. B) Unless physics is treated in an extereme positivist manner, the subjective interpretation is absurd: with its help one can only explain the state of our knowledge but not the state of real physicsal objects. C) A non-absurd, objective, i.e. ferquency or propensity interpretation of CSMcal probabilities is possible.

Now we comment on B) and C). Popper himself admits that the absurd views are "...implicite /our emphasis/ in the theories of some famous physicists who have made very important contributions to quantum theory."[Popper 1982 III]p.109, that is, that the physicists'interpretation of probabilities in CSM is a subjective interpretation /in Popper's sense/ is an interpretation of their views. But, because of the trivial absurdity of such a subjective interpretation, the question raises, whether Popper's interpretation of physicists' interpretation is correct. Undoubtedly one can - as can Popper too - find texts supporting Poppers interpretation but we think that what Popper attributes to physicists is not quite typical. We claim that more typical is the following
Uncritical Textbook Explanation/UTE//see eg. [Andrews 1975]/: A CSMcal system consists of a large number of particles interacting and moving according to the laws of classical mechanics and therefore the point representing the whole system in the phase space also moves deterministically. To describe the motion of the phase point would require both the exact knowledge of initial states of all the particles and the ability to solve a large number of differential equations. But one is unable to solve so many equations of motions and the initial conditions are not known either. For this reason we describe the system by probabilities.

If we look at this explanation we find that the probabilities mentioned in it are implicitely interpreted not as subjective psychological or epistemological probabilities, but as objective object probabilities for they are considered to be descriptions of CMcally determined /macroscopic/ physical states. This is so even if we reject - like Popper /rightly/ does - this explanation for this rejection does not imply that those accepting UTE do necessarily interpret probability statements in CSM subjectivistically in either Popper's sense or in the sense of our classification. We think therefore that, although Popper's critique of subjectivism in CSM based on his critique of the subjective interpretation of probability is justified and touches upon really existing and refutable subjectivist views, his analysis yet misses a point which is essential in understanding the probability statements in CSM. To make the point in question clear we recall another explanation /see eg. [Huang 1963]/:
Improved Textbook Explanation/ITE/: The UTE is untenable for even if we knew both the initial data and the exact solutions of equations of motions we could not get rid of the probabilities as long as we want to be able to derive macroscopic quantities from /micro/mechanical properties. The probabilities enter CSM not because we do not know but because we definitely do not want to know the precise initial conditions and the exact solutions of equations of motion. What we are interested in is the value of a small number of physical quantities characterizing the bulk properties of the system.

To accept ITE fully is, however not as easy as it might seem and not every textbook that deviates from UTE in its explanation of necessity of probabilities in CSM does actually arrive at ITE. /For instance [Klimontovich 1986], [Landau 1958] are somewhere in between UTE and ITE whereas there are also textbooks that manage to avoid making clear statements about this point eg. [Kubo 1965]./ The difficulty with the acceptance of ITE is that in the light of this explanation the "extremely large number" of particles does not seem to be an important factor in connection with the necessity of presence of probabilities in CSM. But, even if it is true that CSM makes only sense for systems viewed consisting of a large number of particles, there is no apriori logical reason to held only this property of the systems responsible for the appearance of probabilities/probabilistic methods. Indeed: even more has turned out to be true: "From here and many other examples, as well as the formal development of probability and statistics we have come to associate the appearance of [probability] densities with the description of large systems containing inherent elements of uncertainty. Viewed from this perspective one might find it surprising to pose the question: "What is the smallest number of elements that a system must have, and how much uncertainty must exist before the description in terms of densities becomes useful and/or necessary?" The answer is surprising and runs counter to the intuition of many. A one-dimensional system containing only one object whose dynamics are completely deterministic (no uncertainty) can generate a density of states! This fact has only become appearent in the past half century due to the pioneering work of Borel, Rnyi and Ulam and von Neumann. These results, however, are not generally known outside that small group of mathematicians working in ergodic theory."[Lasota and Mackey 1985] p.IX.

Typical examples for one dimensional dynamic systems which generate probability densities are the iterated maps on the interval [0,1] /see eg. [Lasota and Mackey]/, and the mentioned ergodic type theorems suggest the following interpretation of the generated probability measure:
A/ Measure theoretic methods, and probability measures in particular, appear as means characterizing long time, asymptotic behaviour of all trajectories of dynamic systems regardless of whether the systems are/can or are not/can not be viewed as systems consisting a large number of subsystems.
B/Concentrating on the asymptotic behaviour of the trajectories rather than being interested in the precise description of each individual trajectory means quite a radical change of the point of view of investigation, which can be characterized as a kind of shift from the whole of details to the details of the whole i.e. means aiming at the description of global qualitative properties of the dynamic system as a whole and a deliberate turning away from the goal to describe the motion of each point no matter whether such a description is possible or not.
C/ Both A/ and B/ are made possible by the special properties of the dynamic system under consideration and have nothing to do with lack of knowledge. On the contrary, to decide whether a particular measure describes the asymptotic behaviour of trajectories requires a mathematical proof, which is in most cases quite sophisticated.

Now if we knew that CSMcal probabilities arise in the same way as the measures characterizing the asymptotic behaviour of trajectories of dynamic systems determined by certain iterated maps, i.e. via ergodic theorems, then the above interpretation could be immediately carried over to the CSMcal probabilities. Unfortunately, "The flows defined by most of the models of CSM ... are almost certainly not ergodic.... Almost certainly means that there is no published proof but most of the experts regard the statement as likely to be correct."[Wightman 1985]p.19. Nevertheless an interpretation of probability in CSM in the manner given by A/-C/ can reasonably be upheld for 1) though most of the CSMcal systems are likely to be non-ergodic, some may have even stronger stochasticity implying ergodicity; "It may be that N hard spheres in a box provide one such although a complete proof for general N has not been published."[Wightman 1985]p.19. 2) there is no reason to expect that ergodicity is the only way through which probabilities as means characterizing asymptotic behaviour of trajectories emerge.

Popper's interpretation of CSMcal probabilities is, however, different. He considers the frequency /or propensity/ interpretation as the correct objective interpretation of CSMcal probabilities. In his attempt to show how the phaenomena that are subjects of CSM /eq. irreversibility/ can be accounted for by an objective theory of probability one has to differentiate the arguments in favour 1/ of the necessity of the objectivity of the probability interpretation in question 2/ of the frequency/propensity/ interpretation. As mentioned, for Popper this does not make a difference. In case of irreversibility Popper emphasizes that if the irreversibility is an objective fact, then the probabilities with the help of which the irreversibility is explained must also be objective ones. So far agreed. But the necessity of objectivity of probabilities in question is only a necessary /and not sufficient/ condition for an explanation by these probabilities to be satisfactory. And Popper's explanation that "...we explain the experimental fact that the process is irreversible by the extreme improbability..." of a spontaneous process running in the direction contradicting irreversibility, explains only, at best, the extreme large probability of irreversibility. That is irreversibility can only be given a statistical meaning within statistical mechanics, which became clear after the "Wiederkehr"- and "Umkehreinwand" had been directed against the original /non-statistical/ H-theorem of Boltzmann. Surprisingly, Popper does not mention these arguments in his analysis of irreversebility [Popper 1982 III] [1]. But this is important because the statistical explanation of irreversibility does have an important preassumption, the "equal probability of those microstates" that are possible if some macroscopic observables are kept fixed. But this equiprobability assumption is difficult to justify on the basis of either the frequency or of the propensity interpretation of probability as the following argument, due to [Krylov 1979], shows: According to the frequency interpretation the equiprobability assumption of microstates within a given phase space region R would be justified if, preparing the system many times in the same /macroscopic/ state R and computing the relative frequency $r(s)$ of any state s in R, one would find that for any s, $r(s)$ tends to the same /equi/probability as the number of preparations tends to infinity. But this is not possible since it is possible in CM to realize /as many times as we wish/ the same condition /R/ together with some further selection of microstates within R so that any /and, therefore, also the uniform/ distribution of microstates in R is violated by the experimental distribution /=limit of relative frequencies/ obtained on this further selected system. In Krylov's words: "...it is impossible to define, in terms of CM, those physical conditions in which that law of distribution [i.e.equiprobability] will manifest itself in an experiment. In other words, in CM it is impossible to define a category of tests that would suite the given notion [i.e. the frequency interpretation] of probability..."[Krylov 1979]p.64. Clearly, Krylov's argument rests on the same basic property of the frequency interpretation that is called by Popper the hidden assumption of the frequency interpretation and therefore this argument also applies to the propensity interpretation.

4.2. The propensity interpretation and QM.

Let us first formulate the propensity interpretation of a typical QMcal probability statement "The probability $W(a)$ of the event a equals to r", $W(a)=r$, where W is a QMcal state, a is an event such as $a=$"The observable A has its value in the interval $[b,c]$". The propensity interpretation interprets this statement as "The quantitative measure of the tendency of the whole experimental situation to realize the possible event a is equal to r.", where "the whole experimental situation" is to be understood as those physical conditions that determine the physical state W in the sense of QM. We wish to emphasize this latter point. For if "the whole experimental situation" contains more than what determines the physical state then the propensity interpretation would assume the existence of some additional parameters playing a role in determining the outcome of the event a, some kind of - even if unspecified - "hidden parameters", which - in our understanding of the propensity interpretation - is not intended in the propensity interpretation; and if "the whole experimental situation" is meant to include less than what does determine W in physics then it would be physically impossible to ensure the repeatability condition required by the propensity interpretation in testing the probability statement $W(a)=r$.

Before recalling the rather simple answers of Popper to the much analized old "paradoxes" of QM we recall two important results, which in Popper's opinion are consequences of the propensity interpretation: QM is a particle theory,"... the probabilities which the theory determines are always the propensities of particles..."[Popper 1982 III]p.126. "There is thus no symmetry or duality between particles and waves: the waves describe dispositional properties of the particles."[Popper 1982 III]p.127. "Thirdly - and this is the result of the propensity interpretation - quantum theory is not a theory which describes dynamical processes in time, but it is a probabilistic propensity theory which gives weight to various possibilities."[Popper 1982 III]p.137.

Popper interprets the well known Heisenberg uncertainty relation "...as a singular probability statement, and therefore as determinig the propensity of a single particle to "scatter": it predicts that the actual statistical scattering will be observed if we repeat the experiment in question many times,

COMMENTS ON POPPER'S INTERPRETATIONS OF PROBABILITY

each time with a single particle."[Popper 1982 III]p.144. The relation is considered by him testable, which implies that one must be able to have retrodictive knowledge concerning the position and momenta, which is more precise than what would be allowed by the subjective interpretation of the uncertainty relation. Consequently, the subjective interpretation is untenable. Popper's recalling the essence of the EPR paradox shows the idea of EPR very simple: "We may let two particles A and B collide. A flies away. B can be observed and measured. We may choose arbitrarily whether to measure the position of B or its momentum. If we decide to measure the position of B we can then calculate the position of the far away A. If we decide to measure the momentum of B, we can calculate the momentum of the far away A. Thus A must have both position and momentum even though the theory does not enable us to calculate both predictively at the same time."[Popper 1982 III]p.148. In the two slit experiment, as it is well known, the problem is how can a particle, which, as a particle, passes through either of the slits only, be influenced by the fact that the other slit is open or closed. Popper's answer is:"...it is the whole experimental arrangement which determines the propensities." /Popper's emphasis/[Popper 1982III]p.152. "Thus the particle will pass through only one of the slits, and in a certain sense will remain uninfluenced by the other slit. What the other slit influences are the popensities of the particle relative to the entire experimental arrangement..."[Popper 1982 III]p.153.

We now comment on these applications of the propensity interpretation in QM. Our first remark is that it is hard to see what Popper's explanation of the two slit experiment explains at all. It is obvious that it is the whole experimental situation that determines the probabilities. The question is: how? and this question seems to remain unanswered. Furthermore, Popper's "explanation" gives the impression that fixing the whole experimental arrangement, including which of the slits are open, does uniquely determine the propensities of the particles in the sense that these propensities can not change in time, which is in accordance with Popper's view that QM is unable to describe dynamical processes. But this contradicts how the two slit experiment is treated in physics and, therefore, how the propensity interpretation was formulated: The initial state of the particles /"monochromatic wave"/ is determined by the source producing them irrespective of which of the slits is open or closed and their subsequent states, their states producing the interference pattern on the photographic plate in particular, is computed in accordance with the dynamical laws of wave mechanics, i.e. the propensities associated with the particles must also change in time if we maintain that the conditions determining the propensity are identical with those determinig the physical state. /For a short analysis of the two slit experiment in terms of propensities see [Milne 1985]./[2] More generally, and this is our second remark, Popper's view that QM can not describe dynamical processes contradicts both his analysis of the QMcal interpretational problems and the recent developments in QM: both in his proof of necessity and possibility to test the uncertainty principle and in his recalling the EPR paradox dynamical concepts as meaningful in QM are involved. Concerning the recent developmnets in QM mentioned, let's first note that besides the basic probability statements "the probability $W(a)$ of the event a in the quantum state W equals to r" there are other types of probabilities too. The formula $<w,v>$ yielding the probability of the event represented by the one dimensional projector to the subspace spanned by the vector v in the state given by the state vector w has another meaning as well: it gives the transition probability between the states w and v. This latter - we may say:dynamical - meaning is physically only well defined if considered together with a /generally time dependent/ Hamilton operator and then the transition probability formula describes the probability of finding the system's state after an /infinitely/ long time to be v if its initial state was w. Probabilities of other type arise furthermore if questions concerning the stochasticity of the whole dynamic itself are posed. Since the dynamic is unambiguously given by the Hamilton operator H, the stochasticity of the dynamic is determined by the properties of the spectra of H and the probabilities characterizing the stochasticity appear as probability measures describing the irregularity of the spectra. An example for such a probability measure is the spacing distribution $p(E,s)$ which is defined by the condition that $p(E,s)ds$ gives the probability that a spacing between two consecutive energy levels around the energy E lies between s

and $s+ds$ [Casati 1985]. Numerical calculations of the spectra both of those Hamilton operators that describe Q dynamical systems whose classical counterpart exhibits chaotic dynamical evolution and of those whose classical counterpart is integrable have been carried out and the comparison of the corresponding $p(E,s)$ shows significant differences i.e. $p(E,s)$ does indeed reflect the randomness of the dynamic [Bohigas 1986]. Now the question is how to interpret these semantically different probabilities in such a way that their interrelation, which roots in the fact that they reflect various features of the same quantum dynamical system, be also clarified. It seems highly improbable that one single interpretation is enuogh in this situation, be it a frequency or the propensity interpretation. To our best knowledge there has not been made any attempt yet aiming at the elaboration of a probability interpretation /or interpretations/ that takes into acount these relatively new developments in QM.

Summing up: Despite his success in criticizing subjectivistic tendencies in interpretations of probabilities in physics, Popper's own interpretations of physical probabilities seem to have severe weaknesses. The very notions of "subjective" and "objective" in connection with probability interpretations are insufficiently defined by Popper, in addition, his concept of the frequency interpretation seems to be incompatible with his theory of demarcation between science and nonscience via the falsifiability.

More specifically, Popper's interpretation of CSMcal probabilities as relative frequencies /or as propensities/ does not suit all types of probabilities occuring in CSM: these interpretations do not make physical sense for probabilities defining the statistical ensembles, which, as we argued, are to be interpreted via a "dynamical interpretation". Also, Popper's propensity interpretation is problematic, or, at least, is "not enough" in QM: besides showing some inconsistencies in Popper's application of the propensity interpretation to the interpretational problems of QM, we pointed out that there are probability statements in QM that characterize the stochasticity of quantum dynamic /transition probability, distribution of spacings in the spectra of the Hamiltonian/. Thus one has to interpret at least three conceptually different probabilities in QM and the propensity interpretation alone does not seem to be able to reflect sufficiently these differences. In this direction further work is necessary.

5. NOTES

[1] In the discussion after this talk Professor Angelidis referred to the following works of Popper: Nature **177** (1956) 538; **178** (1956) 382; **179** (1957) 1297; **181** (1958) 402; **207** (1965) 233; **213** (1967) 320; **214** (1967) 322; where Popper was said to deal with the problem of irreversibility. We are thankful for calling our attention to these works, however, having read these papers, the only idea relevant for the relation between irreversibility and probabilities in statistical mechanics we have been able to find is the assertion that "As it is [that is as it can be found in Boltzmann's work], a statistical theory of the arrow of time seems to me unacceptable." Nature **181** (1958) 403. This assertion seems to contradict the explanation of irreversibility in [Popper III].

[2] In the discussion after this talk Professor Selleri referred to Popper's lecture in "Open questions in Quantum Physics" /ed. by G.Tarozzi and A.van der Merwe, Reidel, 1985], where Popper accepted the "empty wave" theory. For the first view Popper's identification of "empty waves" with "propensity waves" does not seem to be an organic continuation of his previous conception but this point deserves a more detailed analysis.

6. REFERENCES

[Andrews 1975]:F.C.Andrews,Equilibrium Statistical Mechanics /John Wiley, New York,1975/

[Bohigas 1986]:O.Bohigas,M.J.Giannoni,C.Schmit,'Spectral fluctuations of classicaly chaotic quantum systems', in:Lecture Notes in Physics No.263 /1986/

[Casati 1985]:G.Casati,'Limitation of chaotic motion in quantum mechanics', in:Lecture Notes in Mathematics No.1136 /1985/

[Klymontovich 1986]:Yu.L.Klimontovich,Statistical Physics /Harwood Academic Publishers, New York,1986/

[Krylov 1979]:N.S.Krylov,Works on Foundations of Statistical Physics /Princeton University Press, Princeton,1979/

[Kubo 1965]:R.Kubo,Statistical Mechanics, an Advanced Course with Problems and Solutions /Interscience, New York,1965/

[Huang 1963]:K.Huang,Statistical Mechanics /Wiley, New York,1963/

[Landau 1958]:L.D.Landau,E.M.Lifschitz,Statistical Physics /Pergamon Press, London,1958/

[Lasota and Mackey 1985]:A.Lasota,M.C.Mackey,Probabilistic Properties of Deterministic systems /Cambridge University Press, Cambridge,1985/

[Milne 1985]:P.Milne,'Note on Popper, Propensities and the Two-Slit Experiment' British J.Phil.Sci. 36

[Popper 1957a]:K.Popper,'Philosophy of Science. A personal report.'British Philosophy in the mid century, ed. by C.A.Mace /London,1957/

[Popper 1957b]:K.Popper, 'The Propensity Interpretation of the Calculus of Probability and the Quantum Theory',in:Observation and Interpretation ed. by S.Krner /Butterworth, London,1957/

[Popper 1959a]:K.Popper,The Logic of Scientific Discovery /Hutchinson, London,1959/

[Popper 1959b]:K.Popper, 'The Propensity interpretation of Probability', British J.Phil.Sci. 10

[Popper 1982 I]:K.Popper,Postscript to the logic of Scientific Discovery,Vol.I Realism and the Aim of Science /Hutchinson, London,1982/

[Popper 1982 II]:K.Popper, Postscript to the Logic of Scientific Discovery, Vol.II, The Open Universe /Hutchinson, London,1982/

[Popper 1982 III]:K.Popper,Postscript to the Logic of Scientific Discovery, Vol.III, Quantum Theory and the Schisms in Physics /Hutchinson, London,1982/

[Wightman 1985]:A.S.Wightman,'Regular and chaotic motions in dynamic systems', in:Regular and chaotic motion in dynamic systems ed. by G.Velo and A.S.Wightman /Plenum Press, New York,1985/

VAN FRAASSEN'S CONSTRUCTIVE EMPIRICISM AND THE CONCEPT OF PROBABILITY

G.Roussopoulos
11a Ephroniou St.
Athens 116 34
Greece

ABSTRACT.This paper contrasts van Fraassen's conception of probability and his general philosophy of science to realism.Van Fraassen provides his antirealistic philosophy of science with a modal frequency interpretation of probability.The objective of the present paper then is to show that the modal frequency interpretation of probability can well be embedded within the realist framework of science:we could still hold on to the view that a realism about entities is a viable option for a philosophy of science.This realism is grounded on our ability to causally interact with,or be causally affected by,the world.

'It is not thinking about the world but changing it that in the end must make us scientific realists.'[1]

1.INTRODUCTION

The dividing issues between realism and antirealism have come once again to the fore.Van Fraassen's book,The Scientific Image (SI thereafter)[2],puts forward a consistent and convincing picture of what constitutes scientific activity and of how scientists relate to it.His attempt is the most promising antirealist program ever proposed to deal with scientific theories,models and phenomena.His antirealism is embedded within a robust empiricism,charging against current versions of realism.[3]

Van Fraassen believes that his reworking of empiricism can overcome the shortcomings manifested by older versions of empiricism and positivism.In contrast to realist approaches,probability now is seen as the new modality of science,manifesting itself as 'a kind of graded possibility'.This account involves an explanation of how modality (possibility,probability) enters the scientific enterprise via language.Realists employ modalities in science but they attribute to them real,concrete status.Van Fraassen,on the contrary,develops his conception of probability within nominalist lines and in accordance with the desiderata of his empiricism.

Van Fraassen is involved in a double enterprise:in one way,he attempts to overcome the difficulties faced by logical positivism-- in particular,its insistence on construing problems of science as problems of language and logic.In another way,he desires to reinstitute language in its proper place within the philosophy of science.

In what follows,I am going to deal with van Fraassen's conception of probability.Since van Fraassen's conept of probability is embedded within his constructive empiricism,a major part of a critique of his concept of probability concerns a critique of his antirealistic view of science (theoretical entities,unobservables,etc.).

1.SCIENTIFIC REALISM AND CONSTRUCTIVE EMPIRICISM

According to realism,the theoretical terms that make up our scientific theories refer to the world,and they refer independently of us.[4] Theories formulated within the scientific enterprise are said to aim at the truth.The most natural way to establish some intimate connection between the world and our theories about it is by means of a correspondence theory of truth.Natural as this notion of correspondence may seem at first,it gradually proves to be of an elusive nature.[5] The difficulties we face do not have to be rehearsed here again.

As a result of these difficulties,many people interested in defending realism have turned to more indirect ways.Such defenses although not unique in direction,they focus on the notion of explanatory power and the alleged practical success of our scientific theories.When faced in our everyday life with so vast a success of the physical sciences,one naturally turns to scientific realism as an obvious means to understanding this phenomenon of success.Of course,everybody recognizes that the explanatory power of a theory is an important virtue of a theory and relevant as a criterion for theory-choice.What is ambiguous however is how to understand the claim that a theory explains.In particular,given the prominent role that the notion of truth is expected to play in the realist enterprise,the question is whether one has to move from mere pragmatic success of a scientific theory to the claim that it is indeed a true theory.Wouldn't it be sufficient for our needs,as van Fraassen asks,if we interpreted the practical and instrumental success of our theories as empirical adequacy?Why do we have to proceed beyond mere acceptance of a theory's empirical adequacy to belief in its truth?[6]

For van Fraassen,to accept a theory is to believe that it is empirically adequate,that what the theory says about the observables is true.[7] Van Fraassen's constructive empiricism is distinguished from previous forms of antirealism (positivism,fictionalism,instrumentalism,etc.) in that his constructive empiricism embraces an antireductive attitude toward the language in which our theories are formulated.His constructive empiricism (as much as scientific realism) does not propose to replace the physical objects in its ontology by phenomenal,sensory or mental constructions.Nor,does it reduce the properties of the objects and their physical systems to measurements, indications or readings of instruments, or even regularities in our experience. The language in which our scientific theories are formulated are not 'as if' fictions; and the logical relations derived within the context of a theory are literal and binding. Theories, when interpreted, are interpreted literally, as true theories. Scientific theories thus are not similes or metaphors; they are literal constructions.

Van Fraassen acknowledges that most of the arguments that realism produced against positivism were, by and large, successful: the positivist picture of science is no longer tenable. Thus, van Fraassen's picture attempts to take into account previous defects in order to escape the pitfalls of positivism, in particular, the latter's talk about language and its commitment to it. Van Fraassen distinguishes between the syntactic and the semantic ingredients of a theory. He claims that one of the faulting points of positivism has been its insistence on the syntactic component of a theory; on the other hand, his constructive empiricism relies on the semantic component. In the first case, a theory is identified with the particular structures of the sentences that can be validly derived within the language chosen for the expression of the theory. This, undoubtedly gives a theory a linguistic character. This picture then is contrasted to a semantic approach, where a theory is presented not via a particular language but via its class of corresponding structures, its models. Now, the language used to express the content of a theory is neither basic nor unique: the same class of models could be described in different ways. What is essential now, is the models themselves, not the sentences employed in one particular mode of expression. In this model-theoretic approach, such concepts as truth, possibility, probability, etc. become features of the particular concrete models--shifting thus away the emphasis from language and language-dependence characterizing more traditional approaches.

This way of putting the matter allows van Fraassen to state the difference between realism and antirealism, while at the same time exposing the benefits of the antirealist picture of science. When a scientist advances a theory, the realist sees him as advancing a true theory, a theory that the scientist asserts as true. The antirealist claims, in a more modest fashion, that what the scientist does is displaying a theory, 'holding it up to view, as it were, and claiming certain virtues for it'.(SI:57) Now, a theory generally speaking draws a picture of the world but, at the same time, it specifies a family of structures, its models. Certain parts of these models are candidates for the direct representation of observable phenomena. Thus, a theory is empirically adequate when it has some model that fits the observable phenomena. In this way, all we need to do is assert the empirical adequacy of a theory with respect to observables. The difference then between the realist and the antirealist attitude toward science is evident: although in both cases 'we stick our necks out', the assertion of empirical adequacy is a great deal weaker than the assertion of truth, and the restraint to acceptance, instead of belief, delivers from metaphysics.

2. PROBABILITY AS MODALITY

The model-theoretic account of the mediation between theories we formulate and the phenomena we have to account for allows one to talk about the concept of probability.The interesting part of van Fraassen's conception of probability is the general way he proposes to embed probability and modality,more generally,in the empiricist framework of science.

He views probability as 'a kind of graded possibility'(SI:198). But treating probability as modality and,genarally,admitting modality as full-fledged notion in constructive empiricism may seem as contradictory.Scientific realists employ modalities (e.g. possibility) by reifying certain scientific entities.This is evident in the case of the philosophy of space and time,where possible light ray paths of moving bodies play central role,and it has been claimed by M.Friedman that space-time is a real,concrete entity.[8] One could also be reminded of Popper's interpretation of probability as propensity of the experimental apparatus,where he seems to classify probability as a physical magnitude that could not be eliminated through reference to actual occurrences.[9] Now,van Fraassen claims that,although a frequency interpretation of probability faces hard questions,constructive empiricism can offer a conception of probability which is in general agreement with the antirealist project.Constructive empiricism does not identify probability with frequencies but it still construes probability in terms of frequencies.It amounts to a modal account,in the sense that it takes into consideration not only what is actually the case but also what could or would be the case.[10]

One should be reminded here of the distinction between the syntactic and the semantic component of a theory that was mentioned before. It is a central feature of the constructive empiricism that all that is both actual and observable should find a place within a model of the theory.Truth of a theory is to be identified with exact correspondence between phenomena (or parts of a certain physical system) and parts of one of the models of the theory under discussion.But constructive empiricism does not require that there exist a one to one correspondence between all aspects of reality of a certain phenomenon and some parts of the model of the theory.In a similar vein,all significant aspects of the model of a theory do not have corresponding counterparts in the reality we are studying.Thus in order to study the physical system of a pendulum bob,let us say,we construct a mathematical phase-space in order to discuss the behavior of the pendulum bob.The mathematical construct allows us to describe adequately the entity's trajectories in the appropriate space.But many points in it do not correspond to positions the entity actually attained at any time in its actual past or future history.And,of course,many trajectories in that space ,allowed by the physics,bear no correspondence to the entity's actual history.Such a space is only a mathematical entity.This example shows vividly that 'the locus of possibility is the model,not a reality behind the phenomena'.(SI:202)

Van Fraassen's conception of probability starts from the idea that a probability space is a model of a repeatable experiment.And

the probability function in that space,if defined properly,must be
linked with the frequencies of occurrences of outcomes.For this,we
need certain idealizations,that the experiment is an ideal experiment
performed infinitely many times under identical conditions.The actual
experiment is thought about in terms of its possible extensions to
ideal experiments.Hence,what we compare is an actual experiment with
a conceptual model,which consists of a family of ideal repeated expe-
riments.In the model,there should be an intimate relation between
frequencies and probability.If we begin with an outcome-space K and
a Borel field of events F on K and describe one ideal experiment by
means of a countable partition $\{A_n\} \subseteq F$ and a countable sequence
$s=(s_1,s_2,...)$ of members of K,then we are concerned with a certain
class of ideal experiments ('good family') as a pair $Q=(K,E)$,where
K is the set of possible outcomes and E is the set of possible experiments (partitions A_n and sequences s_n).On Q we can then define a
probability function of an event as relative frequency with which
it would occur.[11]

With respect to modality and probability,van Fraassen claims
that these should be conceived as relating to language and not to
the structure of the world.Modality enters the scientific picture
of the world in a historical context:in accepting a certain scientific theory a certain form of language is naturally employed,in particular,modal language.Thus,we do not say that burning of copper atoms
at room temperature and pressure has no counterpart in any model of
our physics;instead,we simply say that it is impossible to burn copper
in room temperature and pressure.The language we speak in such context
and with respect to such possible phenomena has a structure that derives from the physical theories we accept.

But do these views about language and probability make one an
antirealist?In particular,does one have to go by van Fraassen's restrictive ontology of observables only?I am going to argue next that we
do not have to.Van Fraassen opposes a realism as a 'grand theory'
that accounts for all scientific activity.However,a realism about
entities is still possible,provided we work with entities (theoretical,
unobservables) to which we can interact causally.

3.REALISM ABOUT ENTITIES

To facilitate the discussion,it would first be useful to introduce
some interesting and important distinctions.According to scientific
realism entities,states and processes described by theories really
do exist.'Photons,fields,black-holes are as real as toe-nails,turbines
eddies in the stream and volcanoes.The weak interactions of small
particle physics are as real as falling in love',as Ian Hacking puts
it.[12] Scientific theories do not get things right all the time,but
the realist holds that we often get close to the truth.Our theories
aim at discovering the inner costitution of things.An indication that
we proceed the right way is the fact that we have found out so much
so far.[13]

In the above formulations,realism is very often employed in at

least two different ways:realism is opposed to idealism.Realism and
idealism are opposite dogmas concerning the mode of existence of things
in the world.Second,realism is opposed to nominalism;realism and nominalism are conflicting dogmas not about the existence of things but
rather about the way things get to be classified.Both nominalism and
realism in this sense allow that the material world is 'out there',
as an immediate and primary existence,independent of us and our consti-
tution.But they differ as far as explaining how the world is to be
'taken in':do things come already classified as 'natural kinds',or
does the mind contribute to their particular organization?Are classification systems nature's way or man's?

When analyzing van Fraassen's antirealism,most people agree that
his version shares with historical forms of positivism basic features
while,at the same time,it differs from logical positivism of the thir-
ties in that theories are to be taken literally.[14] At this point,it
would be easier to bring in the distinction of realism about entities
and realism about theories.The question about theoretical entities is
whether they exist,whereas the question about theories is whether
they are true or candidates for truth,or aim at the truth.One can very
well a realist about entities but antirealist anout theories;or,a realist about theories and antirealist about entities.[15] Now,antirealism
about entities denies that theoretical entities exist;instead,it
accepts that they are convenient fictions,logical constructions useful
to reason about the world.Or less dogmatically,it may claim that
we do not have good reasons to assume that they exist.Van Fraassen
then is a case of an antirealist about entities.

Newton-Smith has offered a further classification schema that
goes beyond the things said above about realism about entities and
realism about theories.He distinguishes

(1) an ontological/semantical ingredient:scientific theories
are either true or false,and which they are depends on the way the
world is.

(2) A causal ingredient:if a theory is true,the theoretical
terms of the theory denote theoretical entities which are causally
responsible for the observable phenomena.And,

(3) an epistemological ingredient:we can have warranted belief
in theories or in entities.[16]
One may deny the ontological ingredient,that theories are to be taken
literally as true or false;they could be taken instead as tools for
predicting phenomena.But van Fraassen does not quite deny (1),since
he admits that theories are to be taken as literal constructions:they
are true or false,and which they are depends on the way the world
is.What he denies is the epistemological/pragmatic ingredient in
so far as he thinks that we do not have any warrants concerning the
unobservables:we can make sense of science without committing ourselves to belief either in unobservables or in the truth of the scientific theories.

Thus,the notion of unobservables is quite essential to van Fraassen's constructive empiricism.In particular,his ontology depends
on the observable entities.Meanwhile,has completely overlooked the
causal ingredient (3).In fact,there are criteria that allow us to

go beyond the observables. These go back to Popper's criterion of
reality which connects the practicing subject and the world via causal interactions.[17] We shall count as real whatever we can use to
intervene in the world to affect something else, or what the world
can use to affect us. This means that reality has to do with causal
interaction and our notion of reality is formed from our ability
to change the world.

But it is quite evident that van Fraassen's notion of reality
is mostly related to our ability to represent the world through framing theories about it and without caring to intervene and change
it. The notion of reality qua intervention goes beyond the criteria
of reality given by Einstein and others.[18] We could use the typical
example of electrons as theoretical entities (unobservables) to see
how the criteria apply. It is not experimenting with electrons, in
the sense of testing for their existence, that matters but interacting
with electrons: the more we understand their causal interactions, the
more we can build ingenious devices that achieve well understood
effects. Thus, it is manipulating the alleged entities through understanding their causal interactions that can convince someone of their
reality. The entities then cease to be mere theoretical constructs;
they become tools. In this way, we derive a criterion of reality about
entities which is not susceptible to van Fraassen's scepticism about
unobservables.[19]

This view makes us hard-core realists about entities but anti-realists about theories.[20] We can hold on to the reality of theoretical entities under the conditions described, and, at the same time,
we can hold on to the conception of probability as a non-naturalized
concept nowhere to be found in nature. The modal frequency interpretation of probability accords well with the realist framework (realism
about entities). Van Fraassen takes a scepticist and restrictive view
concerning unobservables because he conceives of scientific activity
--as many realists also do--as a restricted procedure, primarily invested in representing and issuing in representations of the world. To
get rid of this contemplative conception of scientific practice
one would have to view scientific practice as intervention in the
world, manipulating and changing the world, thereby changing our representations of it.

NOTES
1. Ian Hacking, Representing and Interveving, Cambridge UP, Cambridge 1983
(RI thereafter), p.XIV.
2. Bas van Fraassen, The Scientific Image, Clarendon Press, Oxford 1980.
3. Cf. E.Hooker and P.Churchland (eds), Images of Science, University of
Chicago Press 1985.
4. A.Fine 'Unnatural Attitudes: Realist and Instrumentalist Attachments
to Science', Mind XCV (1986), pp.149-74 and bibliography given there.
5. Cf. T.S.Kuhn, The Structure of Scientific Revolutions, Chicago UP
1970, p.206:
    There is I think no theory-independent way to reconstruct phrases
    like 'really true'; the notion of a match between the ontology

of a theory and its 'real' counterpart in nature now seems to me illusive in principle.
6. Major theme in van Fraassen's critique of realism. But cf. also N.Cartwright,How the Laws of Physics lie,Clarendon Press,Oxford 1983 (HLPL thereafter).
7. A definition of what is meant by the vague term 'observable' is the following:
> X is observable if there are circumstances which are such that if X is present to us under those circumstances,then we observe it. (SI:16)

A central claim for constructive empiricism with respect to what is observable is that science naturally discloses what is observable:
> I regard what is observable as a theory-independent question.It is a function of facts about us qua organisms in the world.(SI:57)

Thus,with respect to observables,a theory is empirically adequate if and only if everything it says about the observables is correct. And,with respect to the unobservables,our theory should not commit itself to anything:its empirical adequacy underdetermines its truth. As far as the use of the term 'constructive empiricism' by van Fraassen,the following reference is relevant:
> I use the adjective 'constructive' to indicate my view that scientific activity is one of construction rather than one of discovery:constructions of models that must be adequate to the phenomena,and not discovery of truth concerning the unobservable.

8. M.Friedman,Foundations of Space-Time Theories,Princeton UP 1983.
9. K.Popper,'Propensities,Probabilities and the Quantum Theory' (1957) in D.Miller (ed),Popper,Fontana Pocket Readers,pp.199-209.
10. Cf.SI:198:
> So if anyone asks:'What more is there to look at in science besides the models,the actual phenomena,and the relationships between them?' we can answer:'The structure of the language used in a context where a scientific theory has been accepted.'

11. SI:158-203.
12. Cf.Hacking RI,p.21.
13. H.Putnam,Mathematics,Matter,and Method,Volume I,p.69 and R.Boyd, 'The current Status of Scientific Realism' in D.Leplin (ed),Scientific Realism,University of California Press 1984,pp.41-82.
14. SI:12:
> Science aims to give us theories which are empirically adequate; and acceptance of a theory involves as belief only that it is empirically adequate.This is the statement of the antirealist position I advocate.

15. Thus Russell could be viewed as a realist about theories and antirealist about entities,since he tried to rewrite entities in terms of logical theories.The Holy Fathers,on the other hand,could be viewed as realists about entities but as antirealists about theories (no theories could ever capture God,etc.).Cf.Hacking RI,p.27.
16. Newton-Smith,W.,'The under-determination of theory by data',Proc. Arist. Society,Suppl.Volume 52 (1978),p.72.
17. K.Popper and J.Eccles,The Self and its Brain,Springer Verlag,Berlin 1977,p.9:

> I suppose that the most central usage of the term'real' is its
> use to characterize material things of ordinary size--things
> which a baby can handle and (preferably) put into his (sic) mouth.
> From this,the usage of the term 'real' is extended,first,to big-
> ger things--things which are too big for us to handle,like rail-
> way trains,houses,mountains,the earth and the stars,and also
> to smaller things--things like dust particles or mites.It is
> further extended,of course,to liquids and then also to air,to
> gases and to molecules and atoms.
> What is the principle behind the extension?It is,I suggest,that
> the entities which we conjecture to be real should be able to
> exert a causal effect upon the prima facie real things;that is,
> upon material things of an ordinary size:that we can explain
> changes in the ordinary material world of things by causal effect
> of entities conjectured to be real.

18. A.Einstein in Louis de Broglie,Physicien et Penseur,Albin MIchel, 1952,p.7 and Einstein,Podolsky,Rosen,Physical Review 47 (1935),p.777 where the offer the following criterion of physical reality:

> If,without in any way disturbing a system,we can predict with
> certainty (i.e. with probability equal to unity) the value of a
> physical quantity,then there exists an element of reality corre-
> sponding to this physical quantity.

Also cf. the related discussions by F.Selleri,'Quantum Reality as an empirical Problem' in The Concept of Physical Reality,E.Bitsakis (ed),Zacharopoulos,Athens 1983 and E.Bitsakis,'Is it possible to save causality and locality in Quantum Mechanics?',Open Questions in Quantum Mechanics,Reidel 1986.

19. Hacking in his RI states the criterion thus:

> We are completely convinced of the reality of electrons when
> we are regularly set out to build--and often enough we succeed
> in building--new kinds of devices that use well-understood causal
> properties of electrons to interfere in more hypothetical parts
> of nature,(p.265)

20. N.Cartwright HLPL,in particular Essay 9 'How the Measurement Problem ia an Artefact of the Mathematics,pp.163-216.

UNIFICATION OF THE CONCEPTS OF QUANTUM ENSEMBLES AND

POTENTIAL POSSIBILITIES

> A.A. Tyapkin
> Joint Institute for Nuclear Research
> Head Post Office
> P.O. Box 79
> Moscow
> USSR

ABSTRACT. This report mainly deals with a possibility of establishing the objective status of quantum mechanics and with closer definition of some principles in the quantum measurements. The report also touches on some aspects in the development of the interpretation of quantum mechanics related to the explanation of interference effects in experiments with single quantum objects.

1. INTRODUCTION

The orthodox interpretation of quantum mechanics adopted by the majority of physicists is far from completely explaining the main peculiarities of its laws. If one attempts to deepen the understanding of quantum mechanics, one must pay much attention to analysis of all earlier concepts in order to clarify the principles which must certainly be laid in the basis of a fuller interpretation of the theory. Philosophical orientation towards acknowledging absolute objectivity of quantum mechanics laws shall contribute to the fruitfullness of this analysis.

Now two practically opposite opinions about gnosiological essence of quantum mechanics laws have formed.

According to the first one, quantum mechanics is a theory where the properties of microobjects do not show themselves independently, but only when investigated by macroscopic means of observation. The development of this point of view has led to the formation of the principle that the properties of microscopic objects are relative to the means of observation. Despite its obviously positivistic character, this principle is also defended by those who consider themselves adherents of materialistic philosophy.

The opposite gnosiological view acknowledges the objective status of quantum theory with no concessions to positivism, i.e. without necessarily relating its laws to the measuring activities of an individual. According to this point of view, quantum mechanics does not describe properties of microscopic objects themselves, but those of quantum systems which also include macroconditions of motion of microparticles. The latter should not at all be identified with macroscopic observational instruments. Only in a particular case of considering a laboratory test experiment the macroscopically given conditions of motion of microparticles are realized in physical instruments which are not measuring instruments, but those producing processes of quantum transitions from the initially given state of motion to other possible final states.

Any laboratory experiment both in quantum and classical physics always employs two types of physical instruments with radically different gnosiological functions. Some instruments produce naturally existing physical phenomena under the laboratory conditions. (These are the Pisa Tower in Galilei's experiment, particle accelerators and nuclear targets in modern high-energy physics experiments). The function of instruments of the second type is quite different. These are the instrument for observation of physical phenomena occuring under natural or laboratory conditions (e.g. optical and radio - telescopes - in the former case, and microscopes, bubble chambers, other particles detectors - in the latter case). They include, of course, a wide group of measuring equipment of definite physical quantities.

When discussing the problem of quantum mechanics, the greatest experts in this field allowed a gross inaccuracy in terminology: instruments of both types were called measuring instruments or observation instruments. This inaccuracy of outstanding theoreticians in discussion of quantum physics experiments gave rise to a serious confusion and complicated the gnosiologically correct analysis of the contents of quantum mechanics. As a result, the term "observation instrument" and the notion "measurement" itself has been wrongly used even by those who are developing a new interpretation of quantum mechanics which essentially differs from the orthodox one.

Elimination of the above confusion is necessary for clarifying the objective contents of quantum mechanics and for the fuller and deeper understanding of quantum laws.

2. COMPATIBILITY OF TWO WELL-KNOWN QUANTUM CONCEPTS

All attempts to develop new approaches to the understanding of quantum mechanics were aimed at avoiding the subjectivity and limitation of the Copenhagen interpretation of the essence of quantum phenomena. Further move in this direction is possible on the basis of unification of positive features of two well-known concepts in the interpretation of quantum mechanics. One is the concept of quantum ensembles. It was put forward by von Neumann and developed by the Soviet scientists K.V.Nikolsky and D.I. Blokhintsev. The other concept, called the concept of potential possibilities, was developed by the Soviet scientist V.A.Fock. Some underlying ideas of this concept were earlier expressed by A.Einstein and V.Heisenberg.

Reflecting different sides of the objective contents of quantum theory, these concepts have been opposed to each other up to now. The present report shows a possibility of eliminating obstacles hindering unification of these approaches into one concept of quantum ensembles of possibilities. To do this, one must accept V.A.Fock's objection to regarding the original quantum ensemble as a statistical collective. Actually, such collectives as associations for which a certain distribution of probabilities is given appear in quantum mechanics only in the final state, the type of the probability distribution obtained depending on the character of physical effect (of the analyzer) on the initial quantum systems. Therefore the original association of identical quantum systems with a certain wave function corresponding to it must be compared not with the concept of probabilities but with the concept of potential possibilities of manifestation of probability properties determined by the type of effect on original systems.

In classical statistical mechanics separate measurements give always a single random sample from the initial statistical ensemble.

In quantum mechanics we have in principle quite another situation, when every element of the statistical ensemble itself is formed only in the final state as a result of the action of a certain analyzer (or selector) on a single initial quantum system.

3. EXAMPLE OF THE ACTION ON A SINGLE QUANTUM SYSTEM

Now we discuss the next example of the two-digit quantum value-projections of half-integral spin (1/2 h) for different fixed directions.

Let us have a beam of particles with half-integral spin in some initial state described by a certain wave function Ψ_o. This beam passes throuch a magnet Z-analyzer of spin projections along the Z-axis which is perpendicular to the beam direction. Then the beam splits into beams of particles with the Z_+ and Z_- projections in the states ψ_+^Z and ψ_-^Z, respectively. These beams have intensities $J_+^Z = |a|^2 J_o$ and $J_-^Z = |b|^2 J_o$, respectively, where J_o is the intensity of the primary beam, "a" and "b" are the amplitudes of the eigenstates ψ_+^Z and ψ_-^Z in expansion of the initial wave function $\Psi_o = a \psi_+^Z + b \psi_-^Z$.

For the fixed running time D_+ and D_- detectors count $N_+ \sim |a|^2 J_o$ and $N_- \sim |b|^2 J_o$, respectively, when they are behind the analyzer. In this case every separate particle is registered only by one detector D_+ or D_-.

The concept of potential possibilities reflects the illegitimacy of the supposition about a realization of the Z_+ or Z_- - state for the single particle before action of the Z-analyzer.

The considered device can be transformed into a Z_+-selector if behind the Z-analyzer there is an absorber in the beam of particles in the Z_--state. Such a selective instrument prepares a new state ψ_+^Z which may be an initial state for further action on single particles.

The evident results are $N_+ \sim |a|^2 J_o$ and $N_- = 0$ for the Z-analyzer with D_+ and D_- detectors in the beam behind the Z_+-selector. This result ($N_- = 0$) proves that the new starting quantum ensemble does not contain particles in the Z_--state.

However we will obtain a nonzero result for particles in the Z_--state ($N_- \sim 1/4|a|^2 J_o$), when a Y_--selector (which gives only particles in the ψ_-^Y-state) is between the Z_+-selector and the Z-analyzer with D_+ and D_- detectors.

Really, particles are in the Z_+-state in front of the Y_--separator. They are described by the wave function ψ_+^Z. To predict an action of the Y_--separator, it is necessary to present with wave function as

$\psi_+^Z = \frac{1}{\sqrt{2}} \psi_+^Y + \frac{1}{\sqrt{2}} \psi_-^Y$. The intensity of this beam equals $J_+^Z = |a|^2 J_o$.

Behind the Y_--separator we have a beam of particles in the Y_--state with an intensity $J_-^Y = \frac{1}{2}|a|^2 J_o$. The wave function of this state may be presented in the form of expansion
$$\psi_-^Y = \frac{1}{\sqrt{2}} \psi_+^Z + \frac{1}{\sqrt{2}} \psi_-^Z.$$

Consequently, after the action of the Z-analyzer we shall have two beams of particles in the $Z_\pm$-states with intensities $J_\pm^Z = \frac{1}{2} J_-^Y = \frac{1}{4}|a|^2 J_o$ each, i.e. $N_- \neq 0$.

Thus an action on the identical initial systems introduces fundamental changes in them, if the action causes a transition to a Y-state which is an additional one for the initial Z-state. In the general case, the action on the initial quantum system (the former is usualle called a measurement though there are no much grounds for it) may produce significant changes in the initial system, cause a new quality to appear, of which the system was evidently free from the beginning. This is the main difference between the quantum ensembles of potential possibilities and the statistical ensembles in classical statistical mechanics.

4. CONCLUSION

Thus, to understand quantum phenomena as objectively occurring in nature irrespective of their being comprehended by a person, one must also take into consideration the following. A physical process treated as a measurement (a checking experiment) is first of all to produce quantum transitions fron initial states into final ones under controllable conditions; in nature these transitions occur by themselves without special preparation and analyzing devices. It is only detection of the transitions and the detecting devices producing macroscopic signals that have a direct relation to observation of the process by a person.

With this specification, the concept of quantum ensembles of potential possibilities allows completely avoiding subjectivity in the interpretation of quantum laws. This concept is not however free from limitation and any difficulties of the Copenhagen interpretation applied to explanation of paradoxes in correlation experiments and of interference effects in experiments with single quantum objects. These difficulties can be only avoided by interpreting deep physical sense of particle-wave dualism as applied to a separate test of each individual objects of the quantum ensemble. As 20 years ago I still see the only possible way of solving this problem: one must admit the existence of a real wave having no relation to the particle energy and affecting only a probability of its macroscopic manifestations /1,2/. This idea was also expressed in Ref./3/. Recently F.Selleri has proposed an experiment to find this wave which he fitly called an "empty wave"/4/. Carrying out this experiment will be of fundamental significance for physics and the theory of knowledge.

REFERENCES

1. Tyapkin A.A., 1966: Report at the Conference devoted to the 40th Anniversary of Quantum Mechanics, Dubna. Published in:

   a) 1968: Preprint E4-3687, JINR, Dubna, (in English). 'Development of Statistical Interpretation of Quantum Mechanics by Means of the Joint Coordinate-Momentum Representation'.

   b) 1970: In the Proceedings of the Dubna Conference 1966 Philosophical Aspects of Quantum Mechanics, Moscow, "Nauka", p. 139-180 (in Russian).

   c) 1972: L'interpretazione Materialistica Della Meccanica Quantistica. Fisica e Filosofia in URSS, Feltrinelli Editore Milano, p. 413-448 (in Italian).

2. This idea was formulated in the work 1 a) as follows:... The hidden be Broglie waves can correspond to the real physical process in vacuum which is related to the motion of the mechanical microobject and manifesting its existence only by affecting the virtual process of vacuum effects on the same mechanical system. Attention should be paid to the principal difference of this assumption from the earlier discussed ideas of the wavepilot or an idea of a double solution according to which the wave controls the microparticle motion in space and time but not the stochastic process of virtual interaction with vacuum. ... However, the very idea of separating quantum-mechanical object both into the micro-particle itself carrying real energy and capable of performing real (not virtual) effects used in measurements and into the hidden directly not observed wave process in vacuum producing stochastic effects on the micro-particle, the very idea is still the only possibility to uniquely explain without any logical and physical contradiction the interference effects of a single particle. ... Unfortunately, this possibility of explaining has not been discussed seriously either by N. Bohr or any other physicists.

3. Selleri F., 1969: Lett.Nuovo Cim., Ser.1, $\underline{1}$, p. 908.

4. Selleri F., 1982: Found. of Phys. , $\underline{12}$, No.11, p. 1087.

SUBJECT INDEX

action at a distance   71
Aharonov-Bohm type interferometer   190
amplitudes   44, 45, 47
anticorrelation experiments   272, 273, 274
Aspect's experiment   72, 88, 113, 133, 136, 179
Aspect, Dalibare, Roger experiment   112
Atomicity   55, 340
Atomism   21
    Democretian   15-19, 25
    Epicurian   25, 26
    Greek   21-26
    Platonic   19

Becoming   224
Bell's inequality   8, 9, 105, 111, 112, 117, 122, 278, 284, 286
    theorem   71-74, 81, 88, 131
    Bell-type inequalities   105, 115
    additional assumptions   121, 123, 124, 125, 128, 129, 131
    strong inequalities   121, 124, 125, 128, 129, 131
    weak inequalities   121, 122, 125, 128, 131
Boolean algebra   116, 118, 120
Boolean field   375, 376, 377
Borel probability measures   268
Born and Jordan algebras   322
Bose-Einstein Statistics   278
Bosom-Fermion unification   250
Brown-Twiss experiment   275
Brownian motion   244, 245, 246, 248, 249, 301

causal interpretation of Q.M.   93, 94, 95, 96, 134, 144, 146, 148, 151, 153
causality   137, 277, 326, 339, 347, 349, 391
Clauser and Horne hypothesis   85, 121
Clauser and Horne inequality   86, 107
Clauser and Horne postulates   71, 74, 77
collision operator   223, 230, 231, 232
Colmogorov instability   224
Colmogorov systems   225
complexity theory   396
comprehensibility of nature   15
consciousness   167, 177

conservation laws   237
Copenhagen Interpretation of Q.M.   71, 135, 136, 138, 143, 145, 148, 151, 152, 153, 154, 156, 178, 273, 301, 389, 437, 438
CP-invariance   159, 162
CP-violation   163, 165

De Broglie interpretation of Q.M.   143, 144, 145, 146
delayed-choice experiments   184, 185
determinism   126, 129, 131, 200, 202, 203, 225, 240, 332, 337, 345, 348, 362, 402
deterministic models   126, 127, 129
Dirac's algebra   315
double slit experiment   38, 144, 145, 146, 272
double solution theory   143
Drummond-Gardiner generalized P-representations   293
dynamical equations   200, 238
dynamical systems   199, 201, 203, 223, 230, 232
dynamics   224, 227

electromagnetic vacuum   281
empiricism   427, 428, 432
empty waves   8, 148, 149, 150, 154
empty waves generator   151, 152
enhancement   105
    non-enhancement   105-111, 285, 286
entangled systems   9, 10, 11
entropy   223, 224, 225, 236, 257, 265
EPR correlations   315, 323
EPR paradox   5, 7, 8, 111, 121, 122, 133, 136, 162
EPR type experiments   131, 159, 161, 389
EPR-B argument   71-74, 81, 84, 364
ergodicity   202

factorizability condition   75, 106
factorization   10, 37, 134
Faddeev-Popov ghost field   308
field theory   308
first representation theorem   164
Fock space   243, 244, 246, 250, 252
force   24, 25
free will   135
Friedrichs model   232
fundamental interactions   405, 407

Glauber-Sudarshan representation   290, 291, 292, 295

SUBJECT INDEX

Gleason's theorem 66

Hamilton function 98
Hamilton-Jacobi equation 91, 93, 97-101
Hardy class space 257
Hegelian logic 373, 374
Heisenberg operator 237
hidden variable theories 133, 134, 138, 302, 303, 390
Hilbert space 49-50
H-Theorem 134

imaginary exponentials 384, 385
incompleteness of Q.M. 165
indeterminism 232, 233
indistinguishability 6, 18, 137
indivisibility 19
information 236, 237
instability 230, 239, 241
interference experiments 273
interference in phase space 294
invariance 192, 193, 194, 315
invariants 200, 202, 203, 338
irregular dynamical motion 238
irreversibility 223, 224 225, 257, 318, 320, 329, 330
irreversible evolution 224

Jánossy-Máray experiment 8, 149
Jauch-Piron states 117, 118, 119, 120

knowledge 5, 6, 11, 13
Kolmogorov-Sinai entropy 225
K-property 228, 232

Lamb shift 232
language 395
Laplace's algebra 315, 317, 322
law of large numbers 34, 35, 36, 339
Liapunov observable 225
Lie algebra 229
Lionville generator 227
local action 82, 83
local explanatory theory 81
local models 75, 125
locality 71-88, 105, 106, 279
Lorentz invariance 192

Mach-Zehnder interferometer 155
Machida-Namiki theory 189, 190, 197
macroscopic quantum effects 180
Markov processes 135
measure theory 61, 63
measurement 43-50, 257, 260, 261, 262
mechanistic conception 337
mechanistic epistemology 337
meta physics 21, 328
mind 178
Möllenstedt-Bayh experiment 149
momentum space 301

Naimark theorem 263
Navier-Stokes equations 238, 239, 240
necessity 339, 343
negentropy 351
NP problem 397, 398
neutron interferometer 191, 193
neutron interferometry 207, 209, 213
neutron interferometry type experiment 190, 192, 193
non-equilibrium theory 237, 238
non Hilbert formalism 49
non-local correlations 160, 161
non-locality 71, 112, 133, 134, 135, 137, 279, 316, 392
non-separability 160, 316

objective reality 43
objectivity 337, 418
observable 122, 238
orthogonality 53, 57, 58

permanent properties 200
phase space 92, 118, 226, 265, 289, 294
photon bunching 275
photon counting statistics 275
photon trajectory 186
physical constants 409, 410, 412, 414
Planck's elementary quantum 405, 406, 408
Poincaré catastrophe 230
Poincaré theorem 231, 267
Positivism 4, 336, 338, 432, 437
Possible 344
potentiality 8, 335, 339, 342, 344, 437, 439, 440
potentiality waves 138
probabilistic character of Q.M. 189
probabilistic local models 131

SUBJECT INDEX

probability 3, 16, 31, 32, 38, 44, 121, 123, 235, 327, 336, 353, 405, 415
    addition of probabilities 32
    axiomatic theory 31
    classical 37, 41, 61, 65, 243, 272, 275, 335, 338, 340-343, 419
    conditional 317, 320, 326, 362
    ensemble interpretation 8, 353, 359, 360, 361, 364, 365, 366
    epistemological interpretation 417
    in ergodic systems 202, 206
    joint probability 37, 127, 160, 290, 317, 318, 319
    Kolmogorov's probability theory 62
    Kolmogorov's theory 62
    limit frequency interpretation 33, 34, 338, 418, 433
    objective interpretation 417
    objectivistic conception 338, 354
    Popper's interpretation 417, 418
    probabilistic experiments 214
    probability as modality 430, 431
    propensity interpretation 38, 419, 422, 423, 426
    pseudo expectation 46, 47
    pseudo-probability 46
    quasi-probability 289, 290, 293
    quantum probabilities 35-37, 61, 63, 65, 102, 178, 204, 243, 244, 289, 335, 343, 347, 363, 407, 419
    relativistic probability calculus 328
    single system interpretation 346, 364
    subjective interpretation 338, 354, 355, 356, 363, 417
    transition probability 51, 52, 54, 56, 319, 320
    Van Fraassen's conception 427, 430, 431
psychokinesis 331
Pythia 395, 397

quantum jumps, 182, 183
quantum logic 51, 52, 53, 56, 115, 116, 118
    Piron version 115
quantum Mach principle 12
quantum noise 243
quantum optics 271, 289
quantum potential 91, 93-96, 136, 138
quantum Schrödinger field ψ, 93
quantum states 7
quantum statistical determinism 347, 348, 349
quantum stochastic calculus 243, 245, 248, 251
quantum theory - realist interpretation 243

random events generators (REG) 168, 169, 170, 173
rationality 22, 23

REG cumulative deviations   171, 172, 173, 174, 175
REG series   176, 177
randomness   398, 399, 400, 401, 402
random processes   17, 167
Rayleigh-Bernard instability   239
realism   336, 427, 428, 429, 431, 432, 433
    local   121, 122, 162
reductionism   337
relativistic hydrodynamics   134
reversibility   30, 318

satisfiability problem   396, 397
Schrödinger conception   184
Schrödinger's cat   181
Schrödinger's interferometer   182
Schrödinger's philosophy   15-19, 21, 23, 24
separability   6, 7, 10, 13
separated systems   7, 10, 11
Sikorsky's theorem   66
single-atom experiments   179
solipsism   367
spin superposition   211
spectral theorem   66
square integrability conditions   249
SQUID potentials   181
stability of conditions   339
state   43, 45, 119
    mixture   10, 34, 345, 366
    pure   118, 344, 346
    superposition   342, 345
    statistical   340
statistical independence   277
statistical interpretation of Q.M.   346
statistical law   204
statistical principles   17
Statistical Mechanics   235, 240, 257, 265
stochasticity   16
stochastic calculus   246, 250, 251
    dynamics   301
    electrodynamics   280, 287
    integral   247
    interpretation of Q.M.   134, 136, 138
    modal   63, 287
    optics   271, 279, 283, 285, 287
    processes   225, 361, 362
    quantization   301, 302, 303, 304, 305, 307, 309, 310
subquantal aether   134
subquantal level   390
Summhammer's experiment   179

superoperator 231
superposition 8, 185, 211
    classical 342
    minimal 57
superposition principle, 39, 40, 342, 345, 346
system
    chaotic 361, 367
    dynamical 199, 223, 224, 230, 238, 266
    physical 356, 357, 359
    stochastic 361

time 111, 112, 224, 225, 232, 255
    algebra 229
    operator 227, 228, 232, 255, 256, 257, 263, 265
totality 337, 392
trajectory 94, 199, 200, 240
transfinite ordinals 381-384
transition from classical to quantum mechanics 96
Tyche 359, 399

uncertainty relations 18, 61, 255, 262, 402
unitary processes 251
universality claim 81

vacuum 22, 280, 281
virtual gravitons 413
virtual photons 411
Von Neumann's theorem 302

wave function 46
wave packet 94
wave packet's reduction 135, 151, 164, 189, 194, 346
wave particle duality 5, 251, 271, 273
wave with singularity 146, 147, 148, 155
Weiner-Segal isomorphism 244, 245
Wheeler's dragon 323
Wigner function 239, 296, 303
World 3, 4